CLIMATE CHANGE MITIGATION

Greenhouse Gas Reduction and Biochemicals

CLIMATE CHANGE MITIGATION

Greenhouse Gas Reduction and Biochemicals

Edited by

Jimmy Alexander Faria Albanese, PhD, and M. Pilar Ruiz, PhD

Apple Academic Press Inc. | Apple Academic Press Inc.
3333 Mistwell Crescent | 9 Spinnaker Way
Oakville, ON L6L 0A2 | Waretown, NJ 08758
Canada | USA

©2016 by Apple Academic Press, Inc.

First issued in paperback 2021

Exclusive worldwide distribution by CRC Press, a member of Taylor & Francis Group
No claim to original U.S. Government works

ISBN 13: 978-1-77463-571-1 (pbk)
ISBN 13: 978-1-77188-242-2 (hbk)

Library and Archives Canada Cataloguing in Publication

Climate change mitigation: greenhouse gas reduction and biochemicals / edited by Jimmy Alexander Faria Albanese, PhD, and M. Pilar Ruiz, PhD.

Includes bibliographical references and index.
ISBN 978-1-77188-242-2 (bound)
1. Biomass energy. 2. Biochemistry. 3. Biotechnology. 4. Greenhouse gas mitigation.
5. Climate change mitigation. I. Faria Albanese, Jimmy Alexander, author, editor II. Ruiz, M. Pilar, author, editor

TP339.C55 2015   662'.88   C2015-901789-0

Library of Congress Cataloging-in-Publication Data

Climate change mitigation: greenhouse gas reduction and biochemicals / editors, Jimmy Alexander Faria Albanese, PhD, and M. Pilar Ruiz, PhD.

pages cm
Includes bibliographical references and index.
ISBN 978-1-77188-242-2 (alk. paper)
1. Greenhouse gas mitigation. 2. Climate change mitigation. 3. Organic compounds--Environmental aspects. I. Faria Albanese, Jimmy Alexander. II. Ruiz, M. Pilar (Maria Pilar)

TD885.5.G73C5728 20115   363.738'747--dc23   2015008791

Apple Academic Press also publishes its books in a variety of electronic formats. Some content that appears in print may not be available in electronic format. For information about Apple Academic Press products, visit our website at **www.appleacademicpress.com** and the CRC Press website at **www.crc-press.com**

CLIMATE CHANGE MITIGATION

Greenhouse Gas Reduction and Biochemicals

Edited by

Jimmy Alexander Faria Albanese, PhD, and M. Pilar Ruiz, PhD

Apple Academic Press Inc. | Apple Academic Press Inc.
3333 Mistwell Crescent | 9 Spinnaker Way
Oakville, ON L6L 0A2 | Waretown, NJ 08758
Canada | USA

©2016 by Apple Academic Press, Inc.

First issued in paperback 2021

Exclusive worldwide distribution by CRC Press, a member of Taylor & Francis Group
No claim to original U.S. Government works

ISBN 13: 978-1-77463-571-1 (pbk)
ISBN 13: 978-1-77188-242-2 (hbk)

This book contains information obtained from authentic and highly regarded sources. Reprinted material is quoted with permission and sources are indicated. Copyright for individual articles remains with the authors as indicated. A wide variety of references are listed. Reasonable efforts have been made to publish reliable data and information, but the authors, editors, and the publisher cannot assume responsibility for the validity of all materials or the consequences of their use. The authors, editors, and the publisher have attempted to trace the copyright holders of all material reproduced in this publication and apologize to copyright holders if permission to publish in this form has not been obtained. If any copyright material has not been acknowledged, please write and let us know so we may rectify in any future reprint.

Trademark Notice: Registered trademark of products or corporate names are used only for explanation and identification without intent to infringe.

Library and Archives Canada Cataloguing in Publication

Climate change mitigation: greenhouse gas reduction and biochemicals / edited by Jimmy Alexander Faria Albanese, PhD, and M. Pilar Ruiz, PhD.

Includes bibliographical references and index.
ISBN 978-1-77188-242-2 (bound)
1. Biomass energy. 2. Biochemistry. 3. Biotechnology. 4. Greenhouse gas mitigation.
5. Climate change mitigation. I. Faria Albanese, Jimmy Alexander, author, editor II. Ruiz, M. Pilar, author, editor

TP339.C55 2015   662'.88   C2015-901789-0

Library of Congress Cataloging-in-Publication Data

Climate change mitigation: greenhouse gas reduction and biochemicals / editors, Jimmy Alexander Faria Albanese, PhD, and M. Pilar Ruiz, PhD.

pages cm
Includes bibliographical references and index.
ISBN 978-1-77188-242-2 (alk. paper)
1. Greehnouse gas mitigation. 2. Climate change mitigation. 3. Organic compounds--Environmental aspects. I. Faria Albanese, Jimmy Alexander. II. Ruiz, M. Pilar (Maria Pilar)

TD885.5.G73C5728 20115   363.738'747--dc23   2015008791

Apple Academic Press also publishes its books in a variety of electronic formats. Some content that appears in print may not be available in electronic format. For information about Apple Academic Press products, visit our website at **www.appleacademicpress.com** and the CRC Press website at **www.crc-press.com**

About the Editors

JIMMY ALEXANDER FARIA ALBANESE, PhD

Jimmy Faria is Senior Scientist at Abengoa Research, a R&D division of Abengoa. He is a chemical engineer and obtained a PhD from the University of Oklahoma (USA) in 2012. His research at the School of Chemical, Biological and Material Science at the University of Oklahoma (USA) is focused on the catalytic conversion of biomass-derived compounds in a novel nanoparticlestabilized emulsion system developed in this group, as well as on the synthesis, characterization and applications of amphiphilic nanohybrids (e.g., enhanced oil recovery).

M. PILAR RUIZ, PhD

Maria Pilar Ruiz-Ramiro is Senior Scientist at Abengoa Research, a R&D division of Abengoa. She is a chemical engineer and obtained a PhD from the University of Zaragoza (Spain) in 2008. She later worked as Research Associate with Daniel E Resasco at the School of Chemical, Biological and Material Science at the University of Oklahoma (USA). Her research is focused on the thermochemical conversion of biomass, synthesis and characterization of carbon solids (carbon nanotubes, biomass char and soot), and the development of nanostructured catalysts for biofuels upgrading reactions.

Contents

.

Acknowledgment and How to Cite

The editor and publisher thank each of the authors who contributed to this book. The chapters in this book were previously elsewhere. To cite the work contained in this book and to view the individual permissions, please refer to the citation at the beginning of each chapter. Each chapter was read individually and carefully selected by the editor; the result is a book that provides a nuanced look preparing for future climate change. The chapters included are broken into five sections, which describe the following topics:

- Chapter 1, by Kirschbaum, was selected to provide the initial foundations for why the topic of this book is vital to our world.
- In Chapter 2, Burger and his colleagues' article builds the platform wider with considerations of what true "sustainability" means in terms of future generations' access to energy.
- In chapter 3, we provide the basic information for this second part of the book: how biological feedstock relates to future fuel production. Climate change is inevitably coming, and it behooves us to create alternatives now that will not only offer us possibilities that may allow us to maintain our current lifestyles without using fossil fuels, but which will also contribute fewer greenhouse gases themselves.
- Gouveia's article in chapter 4 narrows the focus to a specific biofuel feedstock: microalgae.
- Lanzafame and her colleagues, in chapter 5, give us another aspect of biofuel production's potential to build a sustainable future, focusing on catalysis and solar energy.
- In chapter 6, Caiazzo and his colleagues assess the ways in which the cultivation of biomass feedstocks may impact the solar energy reflected off the earth in comparison to traditional fuel production.
- Chapter 7 acknowledges that bioenergy is not a simplistic, automatic solution for climate change. The authors investigate the changes in energy consumption and emissions that biocrop cultivation may cause.
- Along the same lines, Schmer and his colleague in chapter 8 research the emissions, petroleum offsets, energy efficiency, and fertilizer use involved with growing corn as a feedstock.

- Khanal and his colleagues in chapter 9 examine lignocellulosic feedstocks as a potentially better option for climate change offset—but point out that fuel crops may impact the hydrologic cycle and water resources.
- This leads us to the third section, where we begin in chapter 10 by looking at the potential to use biofuel production in a credit system to replace carbon-fuel emissions. The authors underline again that there is no simplistic answer: biofuel emissions and economic factors must be considered to reach a more accurate assessment of what are the best courses of action to prepare for climate change.
- In chapter 11, Ahlgren and Di Lucia follow up chapter 10 with a review of modeling results to begin finding the answers to questions presented in earlier chapters. They look at this within the context of EU policy.
- Ultimately, in the final section, chapter 12, we want to underline that as complicated and difficult as the challenge may be that we face, the answers are accessible, affordable, and urgent.

List of Contributors

Serina Ahlgren
Department of Energy and Technology, Swedish University of Agricultural Sciences, Uppsala, Sweden

Craig D. Allen
United States Geological Survey, Fort Collins Science Center, Jemez Mountains Field Station, Los Alamos, New Mexico, United States of America

Christopher J. Anderson
Dept. of Agronomy, Iowa State University, Ames, Iowa, United States of America

Robert P. Anex
Dept. of Biological Systems Engineering, University of Wisconsin-Madison, Madison, Wisconsin, United States of America

Steven R. H. Barrett
Laboratory for Aviation and the Environment, Department of Aeronautics and Astronautics, Massachusetts Institute of Technology, Cambridge, MA, USA

Joseph R. Burger
Department of Biology, University of New Mexico, Albuquerque, New Mexico, United States of America

James H. Brown
Department of Biology, University of New Mexico, Albuquerque, New Mexico, United States of America and Santa Fe Institute, Santa Fe, New Mexico, United States of America

William R. Burnside
Department of Biology, University of New Mexico, Albuquerque, New Mexico, United States of America

Fabio Caiazzo
Laboratory for Aviation and the Environment, Department of Aeronautics and Astronautics, Massachusetts Institute of Technology, Cambridge, MA, USA

Gabriele Centi
Dipartimento di Ingegneria Elettronica, Chimica ed Ingegneria Industriale, University of Messina, IN-STM/CASPE (Laboratory of Catalysis for Sustainable Production and Energy) and ERIC (European Research Institute of Catalysis), V.le F. Stagno D'Alcontres 31, 98166 Messina, Italy

Shih-Chih Chen
Department of Accounting Information at Southern Taiwan University of Science and Technology, Tainan 710, Taiwan

Ana D. Davidson
Department of Biology, University of New Mexico, Albuquerque, New Mexico, United States of America and Instituto de Ecología, Universidad Nacional Autónoma de México, México Distrito Federal, México

Lorenzo Di Lucia
Department of Technology and Society, Lund University, PO Box 118, SE 221 00 Lund, Sweden

André P. C. Faaij
Department of Science, Technology and Society, Copernicus Institute, University of Utrecht, Heidelberglaan 2, 3584 CS Utrecht, The Netherlands

Jimmy Faria
Abengoa Research, Seville, Spain

Ronald F. Follett
Soil-Plant Nutrient Research Unit, United States Department of Agriculture-Agricultural Research Service (USDA-ARS), Ft. Collins, Colorado, United States of America

Trevor S. Fristoe
Department of Biology, University of New Mexico, Albuquerque, New Mexico, United States of America

Luisa Gouveia
LNEG - National Laboratory of Energy and Geology. Bioenergy Unit, Estrada do Paço do Lumiar 22 - 1649-038, Lisbon, Portugal

Marcus J. Hamilton
Department of Biology, University of New Mexico, Albuquerque, New Mexico, United States of America, Santa Fe Institute, Santa Fe, New Mexico, United States of America, and Department of Anthropology, University of New Mexico, Albuquerque, New Mexico, United States of America

Daryl E. Herzmann
Dept. of Agronomy, Iowa State University, Ames, Iowa, United States of America

Virginia L. Jin
Agroecosystem Management Research Unit, United States Department of Agriculture-Agricultural Research Service (USDA-ARS), Lincoln, Nebraska, United States of America

Sami Khanal
School of Environment and Natural Resources, Ohio State University, Wooster, OH, United States of America

Miko U. F. Kirschbaum
Landcare Research, Private Bag 11052, Palmerston North 4442, New Zealand

Chun Kung
Institute of Poyang Lake Eco-Economics at Jiangxi University of Finance and Economics, Nanchang 330013, China

Paola Lanzafame
Dipartimento di Ingegneria Elettronica, Chimica ed Ingegneria Industriale, University of Messina, IN-STM/CASPE (Laboratory of Catalysis for Sustainable Production and Energy) and ERIC (European Research Institute of Catalysis), V.le F. Stagno D'Alcontres 31, 98166 Messina, Italy

Jobien Laurijssen
Department of Science, Technology and Society, Copernicus Institute, University of Utrecht, Heidelberglaan 2, 3584 CS Utrecht, The Netherlands

Robert Malina
Laboratory for Aviation and the Environment, Department of Aeronautics and Astronautics, Massachusetts Institute of Technology, Cambridge, MA, USA

Norman Mercado-Silva
Department of Biology, University of New Mexico, Albuquerque, New Mexico, United States of America and School of Natural Resources and the Environment, University of Arizona, Tucson, Arizona, United States of America

Robert B. Mitchell
Grain, Forage and Bioenergy Research Unit, United States Department of Agriculture-Agricultural Research Service (USDA-ARS), Lincoln, Nebraska, United States of America

Jeffrey C. Nekola
Department of Biology, University of New Mexico, Albuquerque, New Mexico, United States of America

Jordan G. Okie
Department of Biology, University of New Mexico, Albuquerque, New Mexico, United States of America

Siglinda Perathoner
Dipartimento di Ingegneria Elettronica, Chimica ed Ingegneria Industriale, University of Messina, IN-STM/CASPE (Laboratory of Catalysis for Sustainable Production and Energy) and ERIC (European Research Institute of Catalysis), V.le F. Stagno D'Alcontres 31, 98166 Messina, Italy

María Pilar Ruiz-Ramiro
Abengoa Research, Seville, Spain

Marty R. Schmer
Agroecosystem Management Research Unit, United States Department of Agriculture-Agricultural Research Service (USDA-ARS), Lincoln, Nebraska, United States of America

Juan Carlos Serrano-Ruiz
Abengoa Research, Seville, Spain

Mark D. Staples
Laboratory for Aviation and the Environment, Department of Aeronautics and Astronautics, Massachusetts Institute of Technology, Cambridge, MA, USA

Jeroen C. J. M. van den Bergh
Institute for Environmental Science and Technology & Department of Economics and Economic History, Universitat Autònoma de Barcelona, Edifici Cn–Campus, UAB 08193, Bellaterra, Spain and Faculty of Economics and Business Administration & Institute for Environmental Studies, Vrije Universiteit, Amsterdam, The Netherlands

Gary E. Varvel
Agroecosystem Management Research Unit, United States Department of Agriculture-Agricultural Research Service (USDA-ARS), Lincoln, Nebraska, United States of America

Kenneth P. Vogel
Grain, Forage and Bioenergy Research Unit, United States Department of Agriculture-Agricultural Research Service (USDA-ARS), Lincoln, Nebraska, United States of America

Philip J. Wolfe
Laboratory for Aviation and the Environment, Department of Aeronautics and Astronautics, Massachusetts Institute of Technology, Cambridge, MA, USA

Tao Wu
School of Economics at Jiangxi University of Finance and Economics, Nanchang 330013, China

Hualin Xie
Institute of Poyang Lake Eco-Economics at Jiangxi University of Finance and Economics, Nanchang 330013, China

Steve H. L. Yim
Laboratory for Aviation and the Environment, Department of Aeronautics and Astronautics, Massachusetts Institute of Technology, Cambridge, MA, USA

Wenyun Zuo
Department of Biology, University of New Mexico, Albuquerque, New Mexico, United States of America

Introduction

Climate change is a significant threat to humanity's future. Culturally, politically, economically, and personally, however, we are all deeply embedded in a system that continues to send us on a collision course that leads directly toward this threat. At this point, climate change is inevitable. What we must do now is to find ways to prepare—and to do all we can to slow our race to disaster.

Meanwhile, around the globe, human life, to greater and lesser extents, depends on materials and energy sources extracted from the planet, which are then transformed by technology into the goods and services so necessary to our lifestyles. This unsustainable consumption jeopardizes the biosphere, while climate change will jeopardize consumption in a variety of interlocking ways.

This means that a transition to a lower-carbon economy is unavoidable. Postponing this transition will only make it more expensive, while levels of greenhouse gas concentrations in the atmosphere will continue to climb. The optimal time for a safer climate policy is right now.

Biochemical research is vitally necessary for the transition we must make, and it will be an essential component of any climate policy. To that end, this compendium has collected the most recent and relevant research in this field—not merely to summarize answers that have already been found, but also to point the way toward future investigations that are still urgently needed.

Jimmy Alexander Faria Albanese, PhD, and M. Pilar Ruiz, PhD

For policy applications, such as for the Kyoto Protocol, the climate-change contributions of different greenhouse gases are usually quantified through their global warming potentials. They are calculated based on the cumulative radiative forcing resulting from a pulse emission of a gas over a specified time period. However, these calculations are not explicitly linked to an assessment of ultimate climate-change impacts. A new metric, the climate-change impact potential (CCIP), is presented in Chapter 1, by Kirschbaum, that is based on explicitly defining the climate-change perturbations that lead to three different kinds of climate-change impacts. These kinds of impacts are: (1) those related directly to temperature increases; (2) those related to the rate of warming; and (3) those related to cumulative warming. From those definitions, a quantitative assessment of the importance of pulse emissions of each gas is developed, with each kind of impact assigned equal weight for an overall impact assessment. Total impacts are calculated under the RCP6 concentration pathway as a base case. The relevant climate-change impact potentials are then calculated as the marginal increase of those impacts over 100 years through the emission of an additional unit of each gas in 2010. These calculations are demonstrated for CO_2, methane and nitrous oxide. Compared with global warming potentials, climate-change impact potentials would increase the importance of pulse emissions of long-lived nitrous oxide and reduce the importance of short-lived methane.

The discipline of sustainability science has emerged in response to concerns of natural and social scientists, policymakers, and lay people about whether the Earth can continue to support human population growth and economic prosperity. Yet, sustainability science has developed largely independently from and with little reference to key ecological principles that govern life on Earth. Chapter 2, by Burger and colleagues, uses a macroecological perspective to highlight three principles that should be integral to sustainability science: 1) physical conservation laws govern the flows of energy and materials between human systems and the environment, 2) smaller systems are connected by these flows to larger systems in which they are embedded, and 3) global constraints ultimately limit flows at smaller scales. Over the past few decades, decreasing per capita rates of consumption of petroleum, phosphate, agricultural land, fresh water, fish, and wood indicate that the growing human population has surpassed the capacity of the Earth to supply enough of these essential resources to sustain even the current population and level of socioeconomic development.

The production of fuels from renewable sources, such as biomass, is an interesting alternative to mitigate climate change and to progressively displace petroleum as a raw material for the transportation sector. A number of technologies have been developed in recent years to achieve conversion of diverse biomass feedstocks (e.g., sugars, vegetable oils and lignocellulose) into a variety of biofuels (e.g., bioethanol, biodiesel, higher alcohols and green hydrocarbons). Chapter 3, by Serrano-Ruiz and colleagues, provides a brief overview of these technologies, taking into consideration aspects such as the complexity of the process and the type of fuel produced (e.g., conventional and advanced). Particular emphasis is given to aqueous-phase catalytic routes to process biomass derivatives such as glycerol, hydroxymethyl furfural, levulinic acid and γ-valerolactone into green hydrocarbons.

Microalgae are an emerging research field due to their high potential as a source of several biofuels in addition to the fact that they have a high-nutritional value and contain compounds that have health benefits. They are also highly used for water stream bioremediation and carbon dioxide mitigation. Therefore, the tiny microalgae could lead to a huge source of compounds and products, giving a good example of a real biorefinery approach. Chapter 4, by Gouveia, shows and presents examples of experimental microalgae-based biorefineries grown in an autotrophic mode at a laboratory scale.

The use of biomass, bio-waste and CO_2 derived raw materials, the latter synthesized using H_2 produced using renewable energy sources, opens new scenarios to develop a sustainable and low carbon chemical production, particularly in regions such as Europe lacking in other resources. In Chapter 5, Lanzafame and colleagues provide a tutorial review that discusses first this new scenario with the aim to point out, between the different possible options, those more relevant to enable this new future scenario for the chemical production, commenting in particular the different drivers (economic, technological and strategic, environmental and sustainability and socio-political) which guide the selection. The case of the use of non-fossil fuel based raw materials for the sustainable production of light olefins is discussed in more detail, but the production of other olefins and polyolefins, of drop-in intermediates and other platform molecules are also analysed. The final part discusses the role of catalysis in establishing this new scenario, summarizing the development of catalysts with respect to industrial targets, for (i) the production of light olefins by catalytic dehydration of ethanol and by CO_2 conversion via FTO process, (ii) the catalytic synthesis of butadiene from ethanol, butanol and butanediols, and (iii) the catalytic synthesis of HMF and its conversion to 2,5-FDCA, adipic acid, caprolactam and 1,6-hexanediol.

Lifecycle analysis is a tool widely used to evaluate the climate impact of greenhouse gas emissions attributable to the production and use of biofuels. In Chapter 6, Caiazzo and colleagues employ an augmented lifecycle framework that includes climate impacts from changes in surface albedo due to land use change. The authors consider eleven land-use change scenarios for the cultivation of biomass for middle distillate fuel production, and compare their results to previous estimates of lifecycle greenhouse gas emissions for the same set of land-use change scenarios in terms of CO_{2e} per unit of fuel energy. They find that two of the land-use change scenarios considered demonstrate a warming effect due to changes in surface albedo, compared to conventional fuel, the largest of which is for replacement of desert land with salicornia cultivation. This corresponds to 222 gCO_{2e}/MJ, equivalent to 3890% and 247% of the lifecycle GHG emissions of fuels derived from salicornia and crude oil, respectively. Nine of the land-use change scenarios considered demonstrate a cooling effect, the largest of which is for the replacement of tropical rainforests with soybean cultivation. This corresponds to −161 gCO_{2e}/MJ, or −28% and −178% of the lifecycle greenhouse gas emissions of fuels derived from soybean and crude oil, respectively. These results indicate that changes in surface albedo have the potential to dominate the climate impact of biofuels, and the article concludes that accounting for changes in surface albedo is necessary for a complete assessment of the aggregate climate impacts of biofuel production and use.

One of the most important concerns facing Taiwan is lack of energy security. Chapter 7, by Kung and colleagues, examines to what extent the Taiwan energy security can be enhanced through bioenergy production and how bioenergy affects net greenhouse gases emissions. Ethanol, conventional bioelectricity and pyrolysis based electricity are analyzed and emissions from fertilizer use and land use change are also incorporated. The study employs the Modified Taiwan Agricultural Sector Model (MTASM) for economic and environmental analysis. The results indicate that Taiwan indeed increases its energy security from bioenergy production but net greenhouse gases emissions are also increased. Emissions from fertilizer use and land use change have significant impacts on emissions reduction and pyrolysis does not always provide net greenhouse emissions offset. Some policy implications including goal determination, land availability and emissions trading systems are also provided for potential policy decision making.

Low-carbon biofuel sources are being developed and evaluated in the United States and Europe to partially offset petroleum transport fuels. Chapter 8, by Schmer and colleagues, evaluates current and potential biofuel production systems from a long-term continuous no-tillage corn (*Zea mays* L.) and switchgrass (*Panicum virgatum* L.) field trial under differing harvest strategies and nitrogen (N) fertilizer intensities to determine overall environmental sustainability. Corn and switchgrass grown for bioenergy resulted in near-term net greenhouse gas (GHG) reductions of −29 to −396 grams of CO_2 equivalent emissions per megajoule of ethanol per year as a result of direct soil carbon sequestration and from the adoption of integrated biofuel conversion pathways. Management practices in switchgrass and corn resulted in large variation in petroleum offset potential. Switchgrass, using best management practices produced 3919±117 liters of ethanol per hectare and had 74±2.2 gigajoules of petroleum offsets per hectare which was similar to intensified corn systems (grain and 50% residue harvest under optimal N rates). Co-locating and integrating cellulosic biorefineries with existing dry mill corn grain ethanol facilities improved net energy yields (GJ ha^{-1}) of corn grain ethanol by >70%. A multi-feedstock, landscape approach coupled with an integrated biorefinery would be a viable option to meet growing renewable transportation fuel demands while improving the energy efficiency of first generation biofuels.

Chapter 9, by Khanal and colleagues, shows likely changes in precipitation (P) and potential evapotranspiration (PET) resulting from policy-driven expansion of bioenergy crops in the United States to create significant changes in streamflow volumes and increase water stress in the High Plains. Regional climate simulations for current and biofuel cropping system scenarios are evaluated using the same atmospheric forcing data over the period 1979–2004 using the Weather Research Forecast (WRF) model coupled to the NOAH land surface model. PET is projected to increase under the biofuel crop production scenario. The magnitude of the mean annual increase in PET is larger than the inter-annual variability of change in PET, indicating that PET increase is a forced response to the biofuel cropping system land use. Across the conterminous U.S., the change in mean streamflow volume under the biofuel scenario is estimated to range from negative 56% to positive 20% relative to a business-as-usual baseline scenario. In Kansas and Oklahoma, annual streamflow volume is reduced by an average of 20%, and this reduction in streamflow volume is due primarily to increased PET. Predicted increase in mean annual P under the biofuel crop production scenario is lower than its inter-annual variability, indicating that additional simulations would be necessary to determine conclusively whether predicted change in P is a response to biofuel crop production. Although estimated changes in streamflow volume include the influence of P change, sensitivity results show that PET change is the significantly dominant factor causing streamflow change. Higher PET and lower streamflow due to biofuel feedstock production are likely to increase water stress in the High Plains. When pursuing sustainable biofuels policy, decision-makers should consider the impacts of feedstock production on water scarcity.

Global biomass potentials are considerable but unequally distributed over the world. Countries with Kyoto targets could import biomass to substitute for fossil fuels or invest in bio-energy projects in the country of biomass origin and buy the credits (Clean Development Mechanism (CDM) and Joint Implementation (JI)). Chapter 10, by Laurijssen and Faaij, analyzes which of those options is optimal for transportation fuels and looks for the key variables that influence the result. In two case studies (Mozambique and Brazil), the two trading systems are compared for the amount of credits generated, land-use and associated costs. The authors found costs of 17–30 euro per ton of carbon for the Brazilian case and economic benefits of 11 to 60 euros per ton of carbon avoided in the Mozambique case. The impact of carbon changes related to direct land-use changes was found to be very significant (both positive and negative) and can currently only be included in emission credit trading, which can largely influence the results. In order to avoid indirect land-use changes (leakage) and consequent GHG emissions, it is crucial that bioenergy crop production is done in balance with improvements of management of agriculture and livestock management. Whatever trading option is economically most attractive depends mainly on the emission baseline in the exporting (emission credit trading) or importing (physical trading) country since both bio- and fossil fuel prices are world market prices in large scale trading systems where transportation costs are low. Physical trading could be preferential since besides the GHG reduction one could also benefit from the energy. It could also generate considerable income sources for exporting countries. This study could contribute to the development of a methodology to deal with bio fuels for transport, in Emission Trading (ET), CDM and the certification of traded bio fuels.

The issue of indirect land use changes (ILUC) caused by the promotion of transport biofuels has attracted considerable attention in recent years. In Chapter 11, Ahlgren and Di Lucia reviewed the current literature on modelling work to estimate emissions of greenhouse gases (GHG) caused by ILUC of biofuels. The authors also reviewed the development of ILUC policies in the EU. Their review of past modelling work revealed that most studies employ economic equilibrium modelling and focus on ethanol fuels, especially with maize as feedstock. It also revealed major variation in the results from the models, especially for biodiesel fuels. However, there has been some convergence of results over time, particularly for ethanol from maize, wheat and sugar cane. The review of EU policy developments showed that the introduction of fuel-specific ILUC factors has been officially suggested by policymakers to deal with the ILUC of biofuels. The values proposed as ILUC factors in the policymaking process in the case of ethanol fuels are generally in line with the results of the latest modelling exercises, in particular for first-generation ethanol fuels from maize and sugar cane, while those for biodiesel fuels are somewhat higher. If the proposed values were introduced into EU policy, no (first-generation) biodiesel fuel would be able to comply with the EU GHG saving requirements. The authors identified a conflict between the demand from EU policymakers for exact, highly specific values and the capacity of the current models to supply results with that level of precision. They concluded that alternative policy approaches to ILUC factors should be further explored.

There is a widespread sense that a sufficiently stringent climate mitigation policy, that is, a considerable reduction of greenhouse gas emissions to avoid extreme climate change, will come with very high economic costs for society. This is supported by many cost–benefit analyses (CBA) and policy cost assessments of climate policy. All of these, nevertheless, are based on debatable assumptions. In Chapter 12, van den Bergh argues instead that safe climate policy is not excessively expensive and is indeed cheaper than suggested by most current studies. To this end, climate CBA and policy cost assessments are critically evaluated, and as a replacement twelve complementary perspectives on the cost of climate policy are offered.

PART I

FOUNDATIONS

CHAPTER 1

Climate-Change Impact Potentials as an Alternative to Global Warming Potentials

MIKO U. F. KIRSCHBAUM

1.1 INTRODUCTION

Climate-change policies aim to prevent ultimate adverse climate-change impacts, stated explicitly by the UNFCCC as "preventing dangerous anthropogenic interference with the climate system." This has led to the adoption of specific climate-change targets to avoid exceeding certain temperature thresholds, such as the "2° target" agreed to in Copenhagen in 2009. The UNFCCC also stated that this aim should be achieved through measures that are "comprehensive and cost-effective." To achieve comprehensive and cost-effective climate-change mitigation requires an assessment of the relative marginal contribution of different greenhouse gases (GHGs) to ultimate climate-change impacts.

Currently, the importance of the emission of different GHGs is usually quantified through their global warming potentials (GWPs), which are calculated as their cumulative radiative forcing over a specified time horizons under constant GHG concentrations (e.g. Lashof and Ahuja 1990,

Climate-Change Impact Potentials as an Alternative to Global Warming Potentials. © *Kirschbaum MUF.* Environmental Research Letters **9,**3 *(2014), doi:10.1088/1748-9326/9/3/034014. Licensed under Creative Commons Attribution 3.0 Unported License, http://creativecommons.org/licenses/by/3.0.*

Fuglestvedt et al 2003). Typical time horizons are 20, 100 and 500 years, with 100 years used most commonly, such as for the Kyoto Protocol. Setting targets in terms of avoiding specified peak temperatures is, however, conceptually inconsistent with a metric that is based on cumulative radiative forcing (e.g. Smith et al 2012). Climate-change metrics were also discussed at a 2009 IPCC expert workshop that noted shortcomings of GWPs and laid out requirements for appropriate metrics, but proposed no alternatives (Plattner et al 2009). Other important issues related to GHG accounting were discussed by Manne and Richels (2001), Fuglestvedt et al (2003, 2010), Johansson et al (2006), Tanaka et al (2010), Peters et al (2011a, 2011b), Manning and Reisinger (2011), Johansson (2012), Kendall (2012), Ekholm et al (2013) and Brandão et al (2013).

Out of these and earlier discussions emerged proposals for alternative metrics. Most prominent among these is the global temperature change potential (GTP), proposed by Shine et al (2005, 2007), which is based on assessing the temperature that might be reached in future years and can be linked directly to adopted temperature targets. A key difference between GWPs and GTPs is that GWPs are measures of the cumulative GHG impact, whereas GTPs are measures of the direct or instantaneous GHG impact. Some impacts, most notably sea-level rise, are not functions of the temperature in future years, but of the cumulative warming leading up to those years (Vermeer and Rahmstorf 2009). Even if the global temperature were to reach and then stabilized at 2 °C above pre-industrial levels, sea levels would continue to rise for centuries (Vermeer and Rahmstorf 2009, Meehl et al 2012). Mitigation efforts that focus solely on maximum temperature increases thus provide no limit on future sea levels rise and only partly address the totality of climate-change impacts.

To be consistent with the policy aim of preventing adverse climate-change impacts, GHG metrics must include all relevant impacts. It is therefore necessary to explicitly define the climate-change perturbations that lead to specific kinds of impacts. The present paper proposes a new metric for comparing GHGs as an alternative to GWPs, termed climate-change impact potential (CCIP). It is based on an explicit definition and quantification of the climate perturbations that lead to different kinds of climatic impacts.

1.2 REQUIREMENTS OF AN IMPROVED METRIC

1.2.1 KINDS OF CLIMATE-CHANGE IMPACTS

There are at least three different kinds of climate-change impacts (Kirsch-baum 2003a, 2003b, 2006, Fuglestvedt et al 2003, Tanaka et al 2010) that can be categorized based on their functional relationship to increasing temperature as:

1. the impact related directly to elevated temperature;
2. the impact related to the rate of warming; and
3. the impact related to cumulative warming.

1.2.1.1 DIRECT-TEMPERATURE IMPACTS

Impacts related directly to temperature increases are easiest to focus on, and are the basis of the notion of keeping warming to 2 °C above pre-industrial temperatures. It is also the explicit metric for calculating GTPs (Shine et al 2005). It is the relevant measure for impacts such as heat waves (e.g. Huang et al 2011) and other extreme weather events (e.g. Webster et al 2005). Coral bleaching, for example, has occurred in nearly all tropical coral-growing regions and is unambiguously related to increased temperatures (e.g. Baker et al 2008).

1.2.1.2 RATE-OF-WARMING IMPACTS

The rate of warming is a concern because higher temperatures may not be inherently worse than cooler conditions, but change itself will cause problems for both natural and socio-economic systems. A slow rate of change will allow time for migration or other adjustments, but faster rates of change may give insufficient time for such adjustments (e.g. Peck and Teisberg 1994).

For example, the natural distribution of most species is restricted to narrow temperature ranges (e.g. Hughes et al 1996). As climate change makes their current habitats climatically unsuitable for many species (Parmesan and Yohe 2003), it poses serious and massive extinction risks (e.g. Thomas et al 2004). The rate of warming will strongly influence whether species can migrate to newly suitable habitats, or whether they will be driven to extinction in their old habitats.

1.2.1.3 CUMULATIVE-WARMING IMPACTS

The third kind of impact includes impacts such as sea-level rise (Vermeer and Rahmstorf 2009) which is quantified by cumulative warming, as sea-level rise is related to both the magnitude of warming and the length of time over which oceans and glaciers are exposed to increased temperatures. Lenton et al (2008) listed some possible tipping points in the global climate system, including shut-off of the Atlantic thermohaline circulation and Arctic sea-ice melting. If the world passes these thresholds, the global climate could shift into a different mode, with possibly serious and irreversible consequences. Their likely occurrence is often linked to cumulative warming. Cumulative warming is similar to the calculation of GWPs except that GWPs integrate only radiative forcing without considering the time lag between radiative forcing and resultant effects on global temperatures. The difference between GWPs and integrated warming are, however, only small over a 100-year time horizon and diminish even further over longer time horizons (Peters et al 2011a).

1.2.2 THE RELATIVE IMPORTANCE OF DIFFERENT KINDS OF IMPACTS

For devising optimal climate-change mitigation strategies, it is also necessary to quantify the importance of different kinds of impacts relative to each other. Without any formal assessment of their relative importance being available in the literature, they were therefore assigned here the same relative weighting. However, the different kinds of impacts change differently over time so that the importance of one kind of impact also changes over time relative to the importance of the others.

The notion of assigning them equal importance can therefore be implemented mathematically only under a specified emission pathway and at a defined point in time. This was done by expressing each impact relative to the most severe impact over the next 100 years under the "representative concentration pathway" (RCP) with radiative forcing of 6 W m^{-2} (RCP6; van Vuuren et al 2011).

1.2.3 CUMULATIVE DAMAGES OR MOST SEVERE DAMAGES?

Any focus on maximum temperature increases, such as the "2° target," explicitly targets the most extreme impacts. However, that ignores the lesser, but still important, impacts that occur before and after the most extreme impacts are experienced. Hence, the damage function used here sums all impacts over the next 100 years. Summing impacts is different from summing temperatures to derive initial impacts. For example, the damage from tropical cyclones is linked to sea-surface temperatures in a given year (Webster et al 2005). Total damages to society, however, are the sum of cyclone damages in all years over the defined assessment horizon.

1.2.4 IMPACT SEVERITY

Climate-change impacts clearly increase with increases in the underlying climate perturbation, but how strongly? By 2012, global temperatures had increased by nearly 1 °C above pre-industrial temperatures (Jones et al 2012), equivalent to about 0.01 °C yr^{-1}, with about 20 cm sea-level rise (Church and White 2011), and there are increasing numbers of unusual weather events that have been attributed to climate change (e.g. Schneider et al 2007, Trenberth and Fasullo 2012). By the time temperature increases reach 2°, or sea-level rise reaches 40 cm, would impacts be twice as bad or increase more sharply? If impacts increase sharply with increasing perturbations, then overall damages would be largely determined by impacts at the times of highest perturbations, whereas with a less steep impact

response function, impacts at times with lesser perturbations would contribute more to overall damages.

Schneider et al (2007) comprehensively reviewed and discussed the
quantification of climate-change impacts and their relationship to underlying climate perturbations but concluded that a formal quantification of impacts was not yet possible. This was due to remaining scientific uncertainty, and the intertwining of scientific assessments of the likelihood of the
occurrence of certain events and value judgements as to their significance.

For example, Thomas et al (2004) quantified the likelihood of species
extinction under climate change and concluded that by 2050, 18% of species would be "committed to extinction" under a low-emission scenario,
which approximately doubled to 35% under a high-emission scenario.
Given the functional redundancy of species in natural ecosystems, their
impact on ecosystem function, and their perceived value for society, doubling the loss of species would presumably more than double the perceived
impact of the loss of those species. The scientifically derived estimate of
species loss therefore does not automatically translate into a usable damage response function. It requires additional value judgements, such as an
assessment of the importance of the survival of species, including those
without economic value.

It is also difficult to quantify the impact related to the low probability
of crossing key thresholds (Lenton et al 2008). It may be possible to agree
on the importance of crossing some irreversible thresholds, but it is difficult to confidently derive probabilities of crossing them. But despite these
uncertainties, some kind of damage response function must be used to
quantify the marginal impact of extra emission units.

As it is difficult, if not impossible, to employ purely objective means
of generating impact response functions, we have to resort to what Stern
called a "subjective probability approach. It is a pragmatic response to the
fact that many of the true uncertainties around climate-change policy cannot themselves be observed and quantified precisely" (Stern 2006). Different workers have used some semi-quantitative approaches, such as polling
of expert opinion (e.g. Nordhaus 1994), or the generation of complex uncertainty distributions from a limited range of existing studies (Tol 2012),
but none of these overcomes the essentially subjective nature of devising
impact response functions.

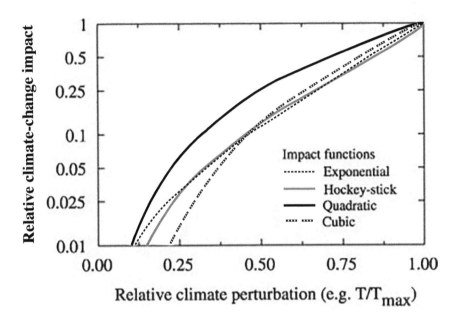

FIGURE 1: Quantification of climate-change impacts as a function of relative climate perturbations. This is illustrated for the exponential relationship used here, the "hockey-stick" function of Hammitt et al (1996) and quadratic and cubic impact functions. It is shown for different relative climate perturbations, such as temperature changes, relative to the maximum perturbations anticipated over the next 100 years.

Figure 1 shows some possible response functions that relate an under-lying climate perturbation to its resultant impact. This is quantified relative to maximum impacts anticipated over the next 100 years for perturbations such as temperature. The current temperature increase of about 1 °C is approximately 1/3 of the temperature increase expected under RCP6 over the next 100 years, giving a relative perturbation of 0.33. For the quantification of CCIPs, impacts had to be expressed as functions of relative climate perturbations to enable equal quantitative treatment of all three kinds of climatic impacts.

Economic analyses tend to employ quadratic or cubic responses function (e.g. Nordhaus 1994, Hammitt et al 1996, Roughgarden and Schneider 1999, Tol 2012), but there is concern that these functions that are based only on readily quantifiable impacts may give insufficient weight to the small probability of extremely severe impacts (e.g. Weitzman 2012, 2013, Lemoine and McJeon 2013). A response function that includes these extreme impacts would increase much more sharply than quadratic or cubic response functions (e.g. Weitzman 2012).

The relationship used here uses an exponential increase in impacts with increasing perturbations to capture the sharply increasing damages with larger temperature increases (as shown by Hammitt et al 1996 and Weitzman 2012). Warming by 3/4 of the expected maximum warming, for example, would have about 10 times the impact as warming by only 1/4 of maximum warming. The graph also shows the often-used power relationships (e.g. Hammitt et al 1996, Boucher 2012), shown here with powers of 2 (quadratic) and 3 (cubic), and a more extreme impacts function (hockey-stick function) presented by Hammitt et al (1996). Compared to the power functions, the exponential relationship calculates relatively modest impacts for moderate climate perturbations that increase more sharply for more extreme climate perturbations. It is thus very similar to the "hockey-stick" relationship of Hammitt et al (1996).

1.2.5 DISCOUNT FACTORS

Should near-term impacts be treated as more important than more distant impacts? If one applies discount rates of 4%, for example, it would ren-der impacts occurring in just 17 years as being only half as important as impacts occurring immediately. The choice of discount rates is hence one of the most critical components of any impact analysis, and the influential Stern report (Stern 2006) derived a fairly bleak outlook on the seriousness of climate change, largely due to using an unusually low discount rate of 1.4%.

While the use of large discount rates is warranted in purely economic analyses, it is questionable in environmental assessments as it essentially treats the lives and livelihood of our children and grandchildren as less important than our own, which is hard to justify ethically (e.g. Schelling 1995, Sterner and Persson 2008). On the other hand, using a 0 discount rate would treat impacts in perpetuity as equally important as short-term impacts. This raises at least the practical problem that it becomes increasingly difficult to predict events and their significance into the more distant future.

The calculation of GWPs essentially uses 0 discount rates, but ignores impacts beyond the end of the assessment period (Tanaka et al 2010). This avoids a preferential emphasis on the impacts on one generation over another, yet avoids the unmanageable requirement of having to assess impacts in perpetuity. This approach is also used here for calculating CCIPs.

1.3 CALCULATION METHODS

1.3.1 QUANTIFYING CLIMATE-CHANGE IMPACT POTENTIALS

To quantify the three different kinds of impacts, it is necessary to first calculate the climate perturbations underlying them. The perturbation $P_{y,T}$ in year y, related to direct-temperature impacts, is simply calculated as:

$$P_{y,T} = T_y - T_p \tag{1}$$

where T_y and T_p are the temperatures in year y and pre-industrially. The temperature in 1900 is taken as the pre-industrial temperature.

The rate of temperature change, $P_{y,\Delta}$, is calculated as the temperature increase over a specified time frame:

$$P_{y,\Delta} = (T_y - T_{y-d}) / d \tag{2}$$

where d is the length of the calculation interval, set here to 100 years. Shorter calculation intervals could be used, in principle, but extra emission units would then affect both the starting and end points for calculating rates of change, leading to complex and sometimes counter-intuitive consequences. The choice of 100 years is further discussed below.

The cumulative temperature perturbation, $P_{y,\Sigma}$, is calculated as the sum of temperatures above pre-industrial temperatures:

$$P_{y,\Sigma} = \sum_{i=p}^{y} (T_i - T_p) \tag{3}$$

where T_i is the temperature in every year i from pre-industrial times to the year y. All three perturbations are then normalized to calculate relative perturbations, Q, as:

$$Q_{y,T} = P_{y,T} / \max(P_{T,RCP6}) \tag{4a}$$

$$Q_{y,\Delta} = P_{y,\Delta} / \max(P_{\Delta,RCP6}) \tag{4b}$$

$$Q_{y,\Sigma} = P_{y,\Sigma} / \max(P_{\Sigma,RCP6}) \tag{4c}$$

where the P-terms are the calculated perturbations under a chosen emissions pathway, and the max-terms are the maximum perturbations calculated under RCP6 over the next 100 years. With this normalization, each kind of climate impact can be treated mathematically the same.

Impacts, I, are then derived from relative perturbations as:

$$I_{y,T} = [(e^{Qy, T})^s] - 1 \tag{5a}$$

$$I_{y,\Delta} = [(e^{Qy, \Delta})^s] - 1 \tag{5b}$$

$$I_{y,\Sigma} = [(e^{Qy, \Sigma})^s] - 1 \tag{5c}$$

where s is a severity term that describes the relationship between perturbations and impacts (figure 1). The work presented here uses s = 4 (as discussed in section 2.4 above).

Temperatures from 1900 to 2010 were based on the HadCRUT4 data set of Jones et al (2012). They were used to set initial temperatures for calculating rates of warming and cumulative warming up to 2010. Temperatures beyond 2010 were added to base temperatures and together determined respective perturbations over the next 100 years.

The relevant impacts were then calculated using equation (4), and summed over 100 years. To calculate CCIPs, these calculation steps were followed four times. The first set of calculations was based on RCP6 and was only used to derive $\max(P_{RCP6})$ which was needed for subsequent normalizations. This normalization made it possible to assign each kind of impact equal importance at their highest perturbations over the next 100 years under RCP6.

The second set of calculations used a chosen emission pathway, RCP6, or a different one as specified below, to calculate background gas concentrations and perturbations. The final two sets of calculations used the same chosen emission pathway and added either 1 tonne of CO_2 or of a different gas. The calculations then derived marginal extra impacts of extra emission units under the three different kinds of impacts. CCIPs of each gas were then calculated as the ratios of marginal impacts of different gases relative to those of CO_2.

These calculations aim to estimate impacts over the coming 100 years, and how those impacts might be modified through pulse emissions of different GHGs. They use the best estimates of relevant background conditions based on emerging science and updated emission scenarios. These calculations would need to be repeated every few years with new scientific understanding and newer emission projections to provide updated guidance of the importance of different GHG over the next 100-year period.

1.3.2 CALCULATING RADIATIVE FORCING AND TEMPERATURE CHANGES

The calculations of radiative forcing and temperature followed the approach of Kirschbaum et al (2013), including the carbon cycle based on the Bern model and radiative forcing calculations provided by the IPCC. Calculations also included the replacement of a molecule of CO_2 by CH_4 in the biogenic production of CH_4, and its partial conversion back to CO_2 when CH_4 was oxidized (Boucher et al 2009). Global temperature calculations included a term for the thermal inertia of the climate system. Full calculation details are given in the supplemental information (available at stacks.iop.org/ERL/9/034014/mmedia).

1.4 RESULTS

1.4.1 IMPACTS UNDER BUSINESS-AS-USUAL CONCENTRATIONS

A quantification of the marginal impact of additional units of each gas must be based on background conditions that include quantification of the impacts that are expected to occur without those additional emission units. Figure 2 shows the relative perturbations underlying the three kinds of impacts and resultant calculated climate-change impacts.

Under RCP6, direct and cumulative-warming impacts continue to increase throughout the 21st century, with greatest impacts reached by 2109. Rate-of-warming impacts reach their maximum by about 2080 and then start to fall slightly (figure 2(a)). While the underlying climate perturbations increase fairly linearly over the next 100 years, this leads to sharply increasing impacts towards the end of the assessment period (figure 2(b)). This pattern is most pronounced for cumulative-warming impacts. The irregular pattern in calculated rates of warming is related to the unevenness in the observed temperature records up to 2010 as rates of warming are calculated from the temperature difference over the preceding 100 years.

1.4.2 PHYSICO-CHEMICAL EFFECTS OF EXTRA GHG EMISSIONS

To calculate the marginal impact of pulse emissions of extra GHG units, it is necessary to first establish their physico-chemical consequences. Concentration increases are greatest immediately after the emission of extra units. They then decrease exponentially for CH_4 (figure 3(b)) and N_2O (figure 3(c)). CO_2 concentrations also decrease but follow a more complex pattern (figure 3(a)). For CH_4, the decrease is quite rapid, with a time constant of 12 years, but is more prolonged for N_2O, with a time constant of 120 years.

These concentration changes exert radiative forcing. It is also highest immediately after the emission of each gas and decreases thereafter. It decreases proportionately faster than the concentration decrease because of increasing saturation of the relevant infrared absorption bands. This is most pronounced for CO_2 (figure 3(a)), for which RCP6 projects large concentration increases (figure 3(d)), which makes the remaining CO_2 molecules from 2010 pulse emissions progressively less effective (e.g. Reisinger et al 2011). For N_2O, RCP6 projects only moderate concentration increases. The infrared absorption bands of N_2O are also less saturated than for CO_2 so that the effectiveness of any remaining molecules remains high. RCP6 projects little change in the CH_4 concentration. Radiative forcing then drives temperature changes (figure 3) that lag radiative forcing by 15–20 years due to the thermal inertia of the climate system.

1.4.3 MARGINAL IMPACTS OF EXTRA EMISSION UNITS

From the information in figures 2 and 3, one can calculate marginal increases in impacts due to a 2010 pulse emission of each gas (figure 4).

Extra units of CO_2 emitted in 2010 cause the largest temperature increase in about 2025 (figure 3(a)). Base temperatures, however, are still fairly mild in 2025 (figure 2(a)) so that the extra warming at that time increases direct-temperature impacts only moderately (figure 4(a)). Even though the extra warming from CO_2 added in 2010 diminishes over time (figure 3(a)), it adds to increasing base temperatures (figure 2(a)) to cause increasing ultimate impacts (figure 4(a)). This pattern is strongest for cumulative-warming impacts. The patterns for N_2O (figure 4(c)) are similar to those for CO_2 because the longevity of N_2O in the atmosphere is similar to that of CO_2.

CH_4 emitted in 2010, however, modifies direct-temperature impacts only over the first few decades after its emission (figure 3(b)). While later warming could potentially have greater impacts, the residual warming several decades after its emission becomes so small to have very little effect. For cumulative-warming impacts, however, the greatest marginal impact of CH_4 additions also occurs at the end of the assessment period. Even though CH_4 emissions exert their warming early in the 21st century, that warming is effectively remembered in the cumulative temperature record and leads to the largest ultimate impact when it combines with large cumulative-warming base impacts (figure 4(b)).

For rate-of-warming impacts and direct-temperature impacts, there are distinctly different patterns for the different gases that are principally related to the longevity of the gases in the atmosphere. For cumulative-warming impacts, however, the patterns are similar for all gases, with the marginal impact from a 2010 pulse emission being muted for the first 50–80 years and then increasing sharply over the remainder of the 100-year assessment period. This is because cumulative warming can be increased in much the same way for contributions made earlier as from on-going warming. Even though different gases make their additions to cumulative warming at different times, that increased perturbation has the largest impact when it adds to large base values (see figure 2(a)) so that for all gases, the largest marginal impacts occur at the end of the 100-year assessment period (figure 4).

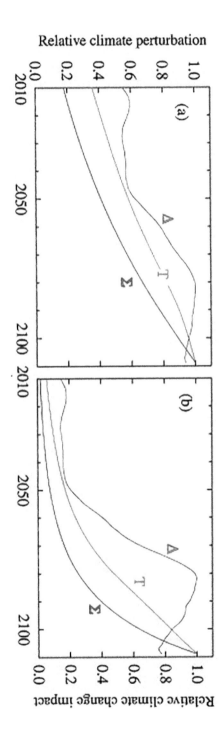

FIGURE 2: Relative climate perturbations (a) and resultant impacts (b) for the three kinds of impacts calculated under RCP6. T refers to direct-temperature impacts, Δ to rate-of-warming impacts, and Σ to cumulative-warming impacts. All values are expressed relative to their calculated maxima over the next 100 years. Maximum perturbations to 2109 under RCP6 were 2.6 °C, 0.016 °C yr^{-1} and 206 °C yr, respectively.

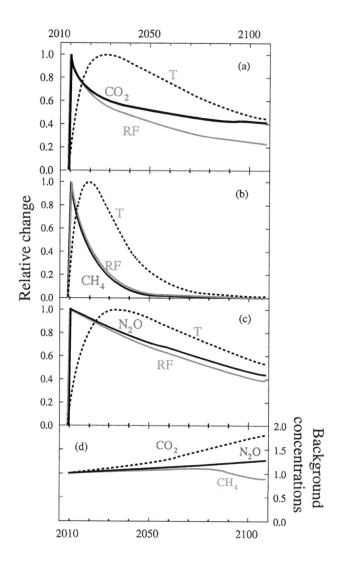

FIGURE 3: Calculated increases in the concentrations of CO_2 (a), CH_4 (b) and N_2O (c) due to pulse emissions of each gas in 2010 and the resultant radiative forcing and temperature increases. Also shown are relative changes in background gas concentrations according to RCP6 (d). All values in (a)–(c) are expressed relative to their highest values over the next 100 years, and concentrations in (d) are relative to 2010 concentrations.

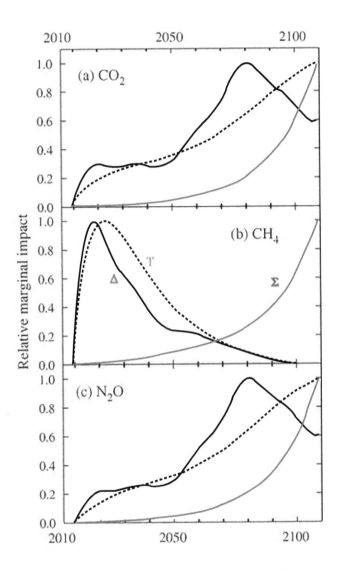

FIGURE 4: Change in the three kinds of impacts due to the addition of one unit of CO_2 (a), biogenic CH_4 (b) and N_2O (c) in 2010. Symbols as for figure 2. All numbers are normalized to the highest marginal impacts calculated over the next 100 years.

1.4.4 CLIMATE-CHANGE IMPACT POTENTIALS

Marginal impacts can then be summed over the 100 years after the pulse emission of each gas and expressed relative to CO_2 (table 1). Under RCP6, CCIPs for biogenic and fossil CH_4 are 20 and 23, respectively, compared to a 100-year GWP of 25. These lower values are due to the much lower direct-temperature and rate-of-warming impacts. Peak warming from CH_4 emissions in 2010 occurs at a time when background temperature increases are still fairly mild so that the extra warming from CH_4 (figure 4(b)) causes less severe impacts than the warming from CO_2 (figure 4(a)) that is still strong many decades later when it combines with higher background temperatures to cause more severe additional impacts.

In contrast, cumulative-warming impacts under RCP6 are 34 and 37 (for biogenic and fossil CH_4), which are greater than the corresponding values for cumulative radiative forcing. The earlier radiative forcing from CH_4 ensures that all radiative forcing leads to warming within the assessment period. For CO_2 and N_2O, on the other hand, radiative forcing overestimates their warming impact because of the thermal inertia of the climate system. Some of the radiative forcing exerted towards the end of the 100-year assessment period only leads to warming after the end of the assessment period providing relatively more cumulative radiative forcing than cumulative warming.

Fossil-fuel-derived CH_4 has higher CCIPs than biogenic CH_4 by about three units. Biogenic CH_4 production means that a molecule of carbon is converted to CH_4, which lowers the atmospheric CO_2 concentration and thereby reduces its overall climatic impact. After it has been oxidized, any CH_4, however, continues its radiative forcing as CO_2, which increases its overall impact (Boucher et al 2009), with the same effect for both fossil and biogenic CH_4.

CCIPs of CH_4 become progressively smaller when they are calculated under higher concentration pathways (table 1). This is caused by much higher impact damages being reached under higher concentration pathways so that the earlier warming contribution of CH_4 relative to CO_2 becomes increasingly less important. This affects direct-temperature impacts and rate-of-warming impacts, whereas cumulative temperature impacts remain similar under the different RCPs.

For N_2O, the CCIP is greater than the 100-year GWP (348 versus 298 under RCP6). This is mainly due to the reducing effectiveness of infra-red absorption of extra CO_2 under increasing background concentrations, which increases the relative importance of the emission of other gases. This interaction with base-level gas concentrations is not included in GWPs as they are calculated under constant background gas concentrations.

TABLE 1: Cumulative radiative forcing, the three kinds of impacts calculated separately and combined to calculate CCIPs over 100 years. Calculations are done under constant 2010 GHG concentrations, and under four different RCPs. All numbers are expressed relative to CO_2. Calculations are done separately for biogenic (B) and fossil-derived (F) CH_4. CCIPs are calculated as the average of the three individually calculated kinds of impacts. Calculations under RCP6 are shown in bold as the reference condition used here. Constant 2010 concentrations were taken to be 387, 1.767 and 0.322 ppmv for CO_2, CH_4 and N_2O, respectively. Numbers for cumulative radiative forcing are given only for comparison. Currently used 100-year GWPs are 25 for CH4 and 298 for N2O.

		Cumulative radiative forcing	Direct-temperature impacts	Rate-of-warming impacts	Cumulative-warming impacts	CCIPs
CH_4 (B)	Const	22	23	34	32	29
CH_4 (F)		24	25	36	34	32
N_2O		282	285	285	288	286
CH_4 (B)	RCP3	24	24	32	35	30
CH_4 (F)		27	27	35	37	33
N_2O		306	313	313	313	313
CH_4 (B)	RCP4.5	26	16	19	34	23
CH_4 (F)		29	19	22	37	26
N_2O		331	341	342	328	337
CH_4 (B)	**RCP6**	**27**	**12**	**13**	**34**	**20**
CH_4 (F)		**30**	**15**	**16**	**37**	**23**
N_2O		**338**	**359**	**356**	**329**	**348**
CH_4 (B)	RCP8.5	29	5.0	3.9	34	14
CH_4 (F)		32	7.8	6.7	37	17
N_2O		365	437	438	351	408

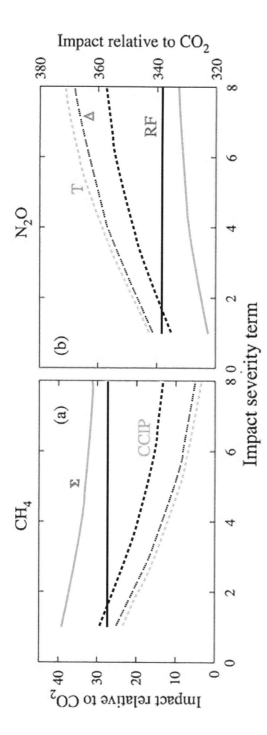

FIGURE 5: Dependence of cumulative radiative forcing (RF) and CCIPs and its components on the climate-impact severity terms for biogenic CH_4 (a) and N_2O (b). T refers to direct-temperature impacts, Δ to rate-of-warming impacts, Σ to cumulative-warming impacts, RF to radiative forcing and CCIP to the derived combined index.

1.4.5 THE IMPORTANCE OF CLIMATE-CHANGE SEVERITY

The relative importance of different gases also depends strongly on the underlying climate-change severity term (figure 5). With increases in the severity term, the importance of short-lived CH_4 decreases considerably (figure 5(b)), whereas the importance of N_2O increases slightly (figure 5(a)). This is because the greatest temperature and rate of change perturbations are projected to occur at the end of the assessment period when CH_4 adds little to those perturbations, while N_2O adds even more than CO_2. As the climate-change severity term increases, it progressively increases the importance of impacts at these later periods and thereby greatly reduces the importance of CH_4.

1.5 DISCUSSION

In this work, climate-change impact potentials are presented as an alternative metric for comparing different GHGs. Why use a new metric? The ultimate aim of climate-change mitigation is to avert adverse climate-change impacts. Hence, there is an obvious logic for policy setting to start with a clear definition of the different kinds of climatic impacts that are to be avoided. Climate-change metrics are needed to support that climate-change policy with the same definition and quantification of climate-change impacts so that the effects of different GHGs can be compared. Mitigation efforts can then be targeted at the gases through which mitigation efforts can be achieved most cost-effectively.

The key aim of metrics should be the quantification of the marginal impact of pulse emissions of extra GHG units. CCIPs aim to provide that measure. They aim to achieve that by combining simple calculations of the relevant physics and atmospheric chemistry with an assessment of the key impacts on nature and society. This full assessment is needed to underpin the development of the most cost-effective mitigation strategies.

The calculation of CCIPs begins with setting the most likely background conditions with respect to gas concentrations and background temperatures in order to quantify the marginal impact of an extra emission unit of a GHG. The use of CCIPs thus requires a periodic re-evaluation of background conditions to devise new optimal mitigation strategies. It is necessary for mitigation efforts to be continuously refocused to achieve the most cost-effective climate-change impact amelioration (Johansson et al 2006). This is first because the relative importance of extra GHG units diminishes with an increase in their own background concentrations because of increasing saturation of the relevant infrared absorption bands. Conversely, the extra warming caused by additional emission units has a greater impact when background temperatures are already higher as it can contribute towards raising temperatures into an increasingly dangerous range. The marginal impact of extra emission units can, therefore, only be quantified under a specified emissions pathway and time horizon.

Along the chain of causality from greenhouse gas emissions to ultimate climate-change impacts, the relevance of respective metrics increases but the uncertainty associated with their calculation increases as well (e.g., Fuglestvedt et al 2003). This relates to scientific uncertainty, the value judgements needed about the relative importance of different impacts, and the ethical considerations of accounting for impacts occurring at different times. GWPs are at one extreme of this continuum, requiring a minimum of assumptions in their calculation, but they only quantify a precursor of ultimate impacts. CCIPs try to go several steps further by quantifying specific climate perturbations that are more directly related to different kinds of climate-change impacts. The functions used to calculate CCIPs still retain simplicity and transparency.

The use of CCIPs instead of 100-year GWPs would reduce the short-term emphasis on CH_4 as CH_4 emitted in 2010 will have disappeared from the atmosphere by the time the most damaging temperatures or rates of warming will be reached. This conclusion is similar to that reached by studies based on GTP calculations (e.g. Shine et al 2007). However, even CH_4 contributes to cumulative-warming impacts. Using CCIPs would thus make CH_4 less important without rendering it irrelevant. Over time, and if future GHG emissions remain high, CH_4 is likely to become more important as the time of emission gets closer to the times when the most severe impacts may be anticipated (Shine et al 2007, Smith et al 2012). CH_4 would then increasingly contribute not only to cumulative-warming impacts but also to direct-temperature and rate-of-warming impacts. CCIPs would need to be recalculated periodically in line with continuously changing expectations of the future.

CCIPs also change with background conditions, and it is considered likely that CCIPs calculated under RCP6 are the most relevant. Recent concentration trends, even during times of global economic crisis (e.g. Peters et al 2013), point towards higher concentration pathways. The limited willingness of the international community to seriously address climate change also suggests that higher concentration pathways will be more likely. RCP6 was therefore used here as the most likely background condition from which to assess the marginal impacts of the emission of extra GHG units.

The derivation of CCIPs explicitly defined and quantified three distinct kinds of impacts. They were all related to temperature as even climate impacts such as flooding that may be more directly related to rainfall intensity can be related to temperature-based perturbations as the underlying climatic driver. One impact that cannot be related to temperature is the direct impact of elevated CO_2 itself. Increasing CO_2 leads to ocean acidification (e.g. Kiessling and Simpson 2011) and shifts the ecological balance between plant species, especially benefiting C_3 plants at the expense of C_4 plants (e.g. Galy et al 2008). On the other hand, increasing CO_2 is beneficial through increasing biological productivity and may partly negate the pressures on food production from increasing temperatures and precipitation changes (e.g. Jaggard et al 2010). With these divergent impacts its overall net impact remains uncertain, and may even be regarded as either positive or negative, and it was, therefore, not included in the CCIP calculations.

Another critically important factor is the steepness of the relationship between underlying climate perturbations and their resultant impact (figures 1, 5). The more steeply impacts increase with increasing perturbation, the more it shifts the importance of extra warming to the times when impacts are already high. With a less steep impact curve, warming at times with lower background temperatures also makes significant contributions towards the overall impact load. The present work used a response function similar to the "hockey-stick" function first presented by Hammitt et al (1996). This steep response curve emphasizes the contribution of different gases at the times of highest impacts while reducing the importance of their contributions at times of lower background impacts. It thus reduces the importance of short-lived gases such as CH_4.

It is also important over what time interval relevant rate-of-warming impacts are calculated. The present work used an assessment period of 100 years and calculated the rate of warming from the temperature increase over the preceding 100 years. With these assumptions, the starting temperatures are always part of the immutable past and extra emissions affect only the end-point temperatures. However, a calculation interval of 100 years may be regarded as too long (e.g. Peck and Teisberg 1994) and a shorter interval might be seen as more appropriate.

Shortening the calculation interval to less than 100 years, however, creates complex interactions because extra emission units then affect both the starting and end point for calculating the rate of warming, and results can become complex and counter-intuitive. For instance, if rates of warming were calculated over 50 years, extra methane emissions would paradoxically reduce the sum of calculated rate-of-warming impacts. How does that occur? While extra methane would increase temperatures over a few decades after its emission, it would increase the short-term rate of warming (calculated from, say, 1970–2020), but it would reduce the rate of warming calculated from 2020 to 2070. Since higher rates of warming are anticipated later during the 21st century (see figure 2), the gains from decreasing the more damaging rates of warming at the later time would be greater than the harm from marginally worsening the milder rates-of-warming impacts in the shorter term.

Whether extra emissions would be considered to do ultimate harm or good would thus depend on the timing of those respective increased and decreased perturbations relative to the base perturbations and the length of period that is assessed as most appropriate for assessing rate-of-warming impacts. Exploring these complex interactions is beyond the scope of the present paper, and the present work had to restrict itself to the simpler case of calculating rates-of-warming impacts by the temperature change over 100 years.

The calculation of CCIPs cannot be based purely on objective science, but has to combine scientific insights with value judgements and assumptions about future background conditions. They relate to the steepness of the impacts function, the choice of background scenario, the inclusion or exclusion of time discounting, the length of assessment horizon, the relative weighting assigned to the different kinds of impacts, the length of the time period for quantifying the rate of change and others. These choices all have a bearing on calculated CCIPs. It may be seen as unfortunate that CCIPs cannot be developed without recourse to a number of key assumptions. However, society makes these assumptions implicitly whenever it decides on adopting any policies related to climate change. The process that is followed formally and explicitly in this paper is similar to the process followed implicitly in all discussions of the importance of climate change, and that has led to the current level of concern and partial willingness to pursue mitigative measures.

Various possible metrics to account for different GHG emissions have been proposed in the past (Ekholm et al 2013). They fall under three broad categories: (1) using measures of cumulative radiative forcing, such as for GWPs; (2) rate of warming, like that explored by Peck and Teisberg (1994); and (3) a number of proposals that are predicated on impacts related directly to elevated temperature, such as the Global Damage Potential (Kandlikar 1996, Hammitt et al 1996, see also Boucher 2012), and the Global Temperature Change Potential, GTP, (Shine et al 2005, 2007). The present work is the first to derive a metric explicitly based on all three kinds of impacts.

Metrics may also restrict themselves to the use of physico-chemical quantities, such as the GWP or the GTP, or employ detailed economic analyses to derive ultimate cost or damage functions (e.g. Kandlikar 1996, Manne and Richels 2001). Including explicit models to calculate damages aims to get closer to an explicit calculation of the ultimate impacts that matter, but it greatly reduces the transparency of resultant metrics (Johansson 2012). It also tends to bias analyses towards those aspects that can be quantified more readily, such as economic impacts, while other impacts, such as the perceived loss from the extinction of species, or the damage from low-probability, but high impact, events tend to be ignored (e.g. Weitzman 2012). The present work restricts itself to using simple models for calculating physico-chemical processes that allowed a number of critical assumptions to remain explicit and transparent. It thereby aims to retain the transparency needed for adoption in international policy or research applications.

1.6 CONCLUSIONS

Global Warming Potentials calculated over 100 years are the current default metric to compare different GHGs. They have become the default metric despite the recognition that they are not directly related to the ultimate climate-change impacts that society is trying to avert. To achieve mitigation objectives most cost-effectively, and to be able to target an optimal mix of GHGs, requires a clearer definition of what is to be avoided. This, in turn, necessitates a more complex analysis than provided by the use of GWPs.

Over the years, there have been several proposals of alternative accounting metrics. A key difference between these different metrics lies in their damage functions that may be related directly to elevated temperature (e.g. Kandlikar 1996, Shine et al 2005, 2007), or to the rate of warming (Peck and Teisberg 1994), or to a measure of cumulative radiative forcing (as for GWPs). However, no previously proposed metric explicitly included all three different kinds of climate perturbations that all contribute towards overall impacts (e.g. Fuglestvedt et al 2003, Brandão et al 2013). Instead, previous work derived respective metrics based on only one of these kinds of impacts and thus implicitly negated the importance of the other kinds of impacts. CCIPs are the first attempt to explicitly develop a metric that is based on all three kinds of impacts.

Climate change continues to be a significant threat for the future of humanity, and mitigation is needed to avert those threats as much as possible. The global community, however, is showing only a limited willingness to allocate sufficient resources to avert serious long-term impacts. The development of CCIPs aims to assist in using those limited resources as cost-effectively as possible.

REFERENCES

1. Baker A C, Glynn P W and Riegl B 2008 Climate change and coral reef bleaching: an ecological assessment of long-term impacts, recovery trends and future outlook Estuar. Coast. Mar. Sci. 80 435–71
2. Boucher O 2012 Comparison of physically- and economically-based CO2-equivalences for methane Earth Syst. Dyn. 3 49–61

3. Boucher O, Friedlingstein P, Collins B and Shine K P 2009 The indirect global warming potential and global temperature change potential due to methane oxidation Environ. Res. Lett. 4 044007

4. Brandão M et al 2013 Key issues and options in accounting for carbon sequestration and temporary storage in life cycle assessment and carbon footprinting Int. J. Life Cycle Assess. 18 230–40

5. Church J A and White N J 2011 Sea-level rise from the late 19th to the early 21st century Surv. Geophys. 32 585–602

6. Ekholm T, Lindroos T J and Savolainen I 2013 Robustness of climate metrics under climate policy ambiguity Environ. Sci. Policy 31 44–52

7. Fuglestvedt J S, Berntsen T K, Godal O, Sausen R, Shine K P and Skodvin T 2003 Metrics of climate change: assessing radiative forcing and emission indices Clim. Change 58 267–331

8. Fuglestvedt J S et al 2010 Transport impacts on atmosphere and climate: metrics Atmos. Environ. 44 4648–77

9. Galy V, Francois L, France-Lanord C, Faure P, Kudrass H, Palhol F and Singh S K 2008 C4 plants decline in the Himalayan basin since the Last Glacial Maximum Quat. Sci. Rev. 27 1396–409

10. Hammitt J K, Jain A K, Adams J L and Wuebbles D J 1996 A welfare based index for assessing environmental effects of greenhouse-gas emissions Nature 381 301–3

11. Huang C R, Barnett A G, Wang X M, Vaneckova P, FitzGerald G and Tong S L 2011 Projecting future heat-related mortality under climate change scenarios: a systematic review Environ. Health Perspect. 119 1681–90

12. Hughes L, Cawsey E M and Westoby M 1996 Climatic range sizes of Eucalyptus species in relation to future climate change Glob. Ecol. Biog. Lett. 5 23–9

13. Jaggard K W, Qi A M and Ober E S 2010 Possible changes to arable crop yields by 2050 Phil. Trans. R. Soc. B 365 2835–51

14. Johansson D J A 2012 Economics- and physical-based metrics for comparing greenhouse gases Clim. Change 110 123–41

15. Johansson D J A, Persson U M and Azar C 2006 The cost of using global warming potentials: analysing the trade off between CO2, CH4 and N2O Clim. Change 77 291–309

16. Jones P D, Lister D H, Osborn T J, Harpham C, Salmon M and Morice C P 2012 Hemispheric and large-scale land-surface air temperature variations: an extensive revision and an update to 2010 J. Geophys. Res. 117 D05127

17. Kandlikar M 1996 Indices for comparing greenhouse gas emissions: integrating science and economics Energy Econ. 18 265–81

18. Kendall A 2012 Time-adjusted global warming potentials for LCA and carbon footprints Int. J. Life Cycle Assess. 17 1042–9

19. Kiessling W and Simpson C 2011 On the potential for ocean acidification to be a general cause of ancient reef crises Glob. Change Biol. 17 56–67

20. Kirschbaum M U F 2003a Can trees buy time? An assessment of the role of vegetation sinks as part of the global carbon cycle Clim. Change 58 47–71

21. Kirschbaum M U F 2003b To sink or burn? A discussion of the potential contributions of forests to greenhouse gas balances through storing carbon or providing biofuels Biomass Bioenergy 24 297–310

22. Kirschbaum M U F 2006 Temporary carbon sequestration cannot prevent climate change Mitig. Adapt. Strateg. Glob. Change 11 1151–64

23. Kirschbaum M U F, Saggar S, Tate K R, Thakur K and Giltrap D 2013 Quantifying the climate-change consequences of shifting land use between forest and agriculture Sci. Tot. Environ. 465 314–24

24. Lashof D A and Ahuja D R 1990 Relative contributions of greenhouse gas emissions to global warming Nature 344 529–31

25. Lemoine D and McJeon H C 2013 Trapped between two tails: trading off scientific uncertainties via climate targets Environ. Res. Lett. 8 034019

26. Lenton T M, Held H, Kriegler E, Hall J W, Lucht W, Rahmstorf S and Schellnhuber H J 2008 Tipping elements in the Earth's climate system Proc. Natl Acad. Sci. USA 105 1786–93

27. Manne A S and Richels R G 2001 An alternative approach to establishing trade-offs among greenhouse gases Nature 410 675–7

28. Manning M and Reisinger A 2011 Broader perspectives for comparing different greenhouse gases Phil. Trans. R. Soc. A 369 1891–905

29. Meehl G A et al 2012 Relative outcomes of climate change mitigation related to global temperature versus sea-level rise Nature Clim. Change 2 576–80

30. Nordhaus W D 1994 Expert opinion on climatic-change Am. Sci. 82 45–51

31. Parmesan C and Yohe G 2003 A globally coherent fingerprint of climate change impacts across natural systems Nature 421 37–42

32. Peck S C and Teisberg T J 1994 Optimal carbon emissions trajectories when damages depend on the rate or level of global warming Clim. Change 28 289–314

33. Peters G P, Aamaas B, Berntsen T and Fuglestvedt J S 2011a The integrated global temperature change potential (iGTP) and relationships between emission metrics Environ. Res. Lett. 6 044021

34. Peters G P, Aamaas B, Lund M T, Solli C and Fuglestvedt J S 2011b Alternative 'Global Warming' metrics in life cycle assessment: a case study with existing transportation data Environ. Sci. Technol. 45 8633–41

35. Peters G P et al 2013 The challenge to keep global warming below 2 °C Nature Clim. Change 3 4–6

36. Plattner G K, Stocker P, Midgley P and Tignor M 2009 IPCC Expert Meeting On the Science of Alternative Metrics (Available at: www.ipcc.ch/pdf/supporting-material/expert-meeting-metrics-oslo.pdf)

37. Reisinger A, Meinshausen M and Manning M 2011 Future changes in global warming potentials under representative concentration pathways Environ. Res. Lett. 6 024020

38. Roughgarden T and Schneider S H 1999 Climate change policy: quantifying uncertainties for damages and optimal carbon taxes Energy Policy 27 415–29

39. Schelling T C 1995 Intergenerational discounting Energy Policy 23 395–401

40. Schneider S H et al 2007 Assessing key vulnerabilities and the risk from climate change Climate Change 2007: Impacts, Adaptation and Vulnerability. Contribution of WGII to the Fourth Assessment Report of the IPCC ed M L Parry, O F Canziani, J P Palutikof, P J van der Linden and C E Hanson (Cambridge: Cambridge University Press) 779–810

41. Shine K P, Berntsen T K, Fuglestvedt J S, Skeie R B and Stuber N 2007 Comparing the climate effect of emissions of short- and long-lived climate agents Phil. Trans. R. Soc. A 365 1903–14

42. Shine K P, Fuglestvedt J S, Hailemariam K and Stuber N 2005 Alternatives to the global warming potential for comparing climate impacts of emissions of greenhouse gases Clim. Change 68 281–302

43. Smith S M, Lowe J A, Bowerman N H A, Gohar L K, Huntingford C and Allen M R 2012 Equivalence of greenhouse-gas emissions for peak temperature limits Nature Clim. Change 2 535–8

44. Stern N H 2006 The Economics of Climate Change (Available at www.hmtreasury. gov.uk/independent_reviews/stern_review_economics_climate_change/stern_re-view_report.cfm)

45. Sterner T and Persson U M 2008 An even sterner review: introducing relative prices into the discounting debate Rev. Environ. Econ. Policy 2 61–76

46. Tanaka K, Peters G P and Fuglestvedt J S 2010 Policy update: multicomponent climate policy: why do emission metrics matter? Carbon Manag. 1 191–7

47. Thomas C D et al 2004 Extinction risk from climate change Nature 427 145–8

48. Tol R S J 2012 On the uncertainty about the total economic impact of climate change Environ. Resour. Econ. 53 97–116

49. Trenberth K E and Fasullo J T 2012 Climate extremes and climate change: the Russian heat wave and other climate extremes of 2010 J. Geophys. Res.-Atmos. 117 D17103

50. van Vuuren D P et al 2011 The representative concentration pathways: an overview Clim. Change 109 5–31

51. Vermeer M and Rahmstorf S 2009 Global sea level linked to global temperature Proc. Natl Acad. Sci. USA 106 21527–32

52. Webster P J, Holland G J, Curry J A and Chang H R 2005 Changes in tropical cyclone number, duration, and intensity in a warming environment Science 309 1844–6

53. Weitzman M L 2012 GHG targets as insurance against catastrophic climate damages J. Public Econ. Theory 14 221–44

54. Weitzman M L 2013 A precautionary tale of uncertain tail fattening Environ. Resour. Econ. 55 159–73

CHAPTER 2

The Macroecology of Sustainability

JOSEPH R. BURGER, CRAIG D. ALLEN, JAMES H. BROWN,
WILLIAM R. BURNSIDE, ANA D. DAVIDSON, TREVOR S.
FRISTOE, MARCUS J. HAMILTON, NORMAN MERCADO-SILVA,
JEFFREY C. NEKOLA, JORDAN G. OKIE, AND WENYUN ZUO

"Sustainability" has become a key concern of scientists, politicians, and lay people—and for good reason. There is increasing evidence that we have approached, or perhaps even surpassed, the capacity of the planet to support continued human population growth and socioeconomic development [1]–[3]. Currently, humans are appropriating 20%–40% of the Earth's terrestrial primary production [4]–[6], depleting finite supplies of fossil fuels and minerals, and overharvesting "renewable" natural resources such as fresh water and marine fisheries [7]–[10]. In the process, we are producing greenhouse gases and other wastes faster than the environment can assimilate them, altering global climate and landscapes, and drastically reducing biodiversity [2]. Concern about whether current trajectories of human demography and socioeconomic activity can continue in the face of such environmental impacts has led to calls for "sustainability." A seminal event was the Brundtland commission report [11], which defined "sustainable development (as) development that meets the needs of the present without compromising the ability of future generations to meet their own needs."

One result has been the emergence of the discipline of sustainability science. "Sustainability science (is) an emerging field of research dealing with the interactions between natural and social systems, and with how those interactions affect the challenge of sustainability: meeting the needs of present and future generations while substantially reducing poverty and conserving the planet's life support systems" (*Proceedings of the National Academy of Sciences of the USA* [PNAS], http://www.pnas. org/site/misc/sustainability.shtml). It is the subject of numerous books, at least three journals (*Sustainability Science* [Springer]; *Sustainability: Science, Practice, & Policy* [ProQuest-CSA]; *International Journal of Sustainability Science and Studies* [Polo Publishing]), and a special section of the PNAS. In "A Survey of University-Based Sustainability Science Programs", conducted in 2007, (http://sustainabilityscience.org/content .html?contentid=1484), the American Association for the Advancement of Science listed 103 academic programs, including 64 in the United States and Canada, and many more have been established subsequently.

Interestingly, despite the above definition, the majority of sustainability science appears to emphasize social science while largely neglecting natural science. A survey of the published literature from 1980 through November 2010 using the Web of Science reveals striking results. Of the 23,535 published papers that include "sustainability" in the title, abstract, or key words, 48% include "development" or "economics". In contrast, only 17% include any mention of "ecology" or "ecological", 12% "energy", 2% "limits", and fewer than 1% "thermodynamic" or "steady state". Any assessment of sustainability is necessarily incomplete without incorporating these concepts from the natural sciences.

2.1 HUMAN MACROECOLOGY

A macroecological approach to sustainability aims to understand how humans are integrated into and constrained by the Earth's systems [12]. In just the last 50,000 years, *Homo sapiens* has expanded out of Africa to become the most dominant species the Earth has ever experienced. Near-exponential population growth, global colonization, and socioeconomic development have been fueled by extracting resources from the environment and transforming them into people, goods, and services. Hunter-gatherers had subsistence economies based on harvesting local biological resources for food and fiber and on burning wood and dung to supplement energy from human metabolism. With the transition to agricultural societies after the last ice age [13] and then to industrial societies within the last two centuries, per capita energy use has increased from approximately 120 watts of human biological metabolism to over 10,000 watts, mostly from fossil fuels [3],[14]. Modern economies rely on global networks of extraction, trade, and communication to rapidly distribute vast quantities of energy, materials, and information.

The capacity of the environment to support the requirements of contemporary human societies is not just a matter of political and economic concern. It is also a central aspect of ecology—the study of the interactions between organisms, including humans, and their environments. These relationships always involve exchanges of energy, matter, or information. The scientific principles that govern the flows and transformations of these commodities are fundamental to ecology and directly relevant to sustainability and to the maintenance of ecosystem services, especially in times of energy scarcity [15]. A macroecological perspective highlights three principles that should be combined with perspectives from the social sciences to achieve an integrated science of sustainability.

2.2 PRINCIPLE 1: THERMODYNAMICS AND THE ZERO-SUM GAME

The laws of thermodynamics and conservation of energy, mass, and chemical stoichiometry are universal and without exception. These principles are fundamental to biology and ecology [16]–[18]. They also apply equally to humans and their activities at all spatial and temporal scales. The laws of thermodynamics mean that continual flows and transformations of energy are required to maintain highly organized, far-from-equilibrium states of complex systems, including human societies. For example, increased rates of energy use are required to fuel economic growth and development, raising formidable challenges in a time of growing energy scarcity and insecurity [3],[15],[19]. Conservation of mass and stoichiometry means that the planetary quantities of chemical elements are effectively finite [15],[18].

Human use of material resources, such as nitrogen and phosphorus, alters flows and affects the distribution and local concentrations in the environment [18]. This is illustrated by the Bristol Bay salmon fishery, which is frequently cited as a success story in sustainable fisheries management [20],[21]. In three years for which good data are available (2007–2009), about 70% of the annual wild salmon run was harvested commercially, with one species, sockeye, accounting for about 95% of the catch [22]. From a management perspective, the Bristol Bay sockeye fishery has been sustainable, because annual runs have not declined. Additional implications for sustainability, however, come from considering the effect of human harvest on the flows of energy and materials in the upstream ecosystem (Figure 1). When humans take about 70% of Bristol Bay sockeye runs as commercial catch, this means a 70% reduction in the number of mature salmon returning to their native waters to spawn and complete their life cycles. It also means a concomitant reduction in the supply of salmon to support populations of predators, such as grizzly bears, bald eagles, and indigenous people, all of which historically relied on salmon for a large proportion of their diet [23],[24]. Additionally, a 70% harvest means annual removal of more than 83,000 metric tonnes of salmon biomass, consisting of approximately 12,000, 2,500, and 330 tonnes of carbon, nitrogen, and phosphorus, respectively (see Text S1 for sources and calculations). These marine-derived materials are no longer deposited inland in the Bristol Bay watershed, where they once provided important nutrient subsidies to stream, lake, riparian, and terrestrial ecosystems [24]–[27]. So, for example, one apparent consequence is that net primary production in one oligotrophic lake in the Bristol Bay watershed has decreased "to about 1/3 of its level before commercial fishing" [28]. Seventy percent of Bristol Bay salmon biomass and nutrients are now exported to eastern Asia, western Europe, and the continental US, which are the primary markets for commercially harvested wild Alaskan salmon. Our macroecological assessment of the Bristol Bay fishery suggests that "sustainable harvest" of the focal salmon species does not consider the indirect impacts of human take on critical resource flows in the ecosystem (Figure 1). So the Bristol Bay salmon fishery is probably not entirely sustainable even at the "local" scale.

2.3 PRINCIPLE 2: SCALE AND EMBEDDEDNESS

Most published examples of sustainability focus on maintaining or improving environmental conditions or quality of life in a localized human system, such as a farm, village, city, industry, or country ([29],[30] and articles following [31]). These socioeconomic systems are not closed or isolated, but instead are open, interconnected, and embedded in larger environmental systems. Human economies extract energy and material resources from the environment and transform them into goods and services. In the process, they create waste products that are released back into the environment. The laws of conservation and thermodynamics mean that the embedded human systems are absolutely dependent on these flows: population growth and economic development require increased rates of consumption of energy and materials and increased production of wastes. The degree of dependence is a function of the size of the economy and its level of socioeconomic development [3]. Most organic farms import fuel, tools, machinery, social services, and even fertilizer, and export their products to markets. A small village in a developing country harvests food, water, and fuel from the surrounding landscape.

Large, complex human systems, such as corporations, cities, and countries, are even more dependent on exchanges with the broader environment and consequently pose formidable challenges for sustainability. Modern cities and nation states are embedded in the global economy, and supported by trade and communication networks that transport people, other organisms, energy, materials, and information. High densities of people and concentrations of socioeconomic activities require massive inputs of energy and materials and produce proportionately large amounts of wastes. Claims that such systems are "sustainable" usually only mean that they are comparatively "green"—that they aim to minimize environmental impacts while offering their inhabitants happy, healthy lifestyles.

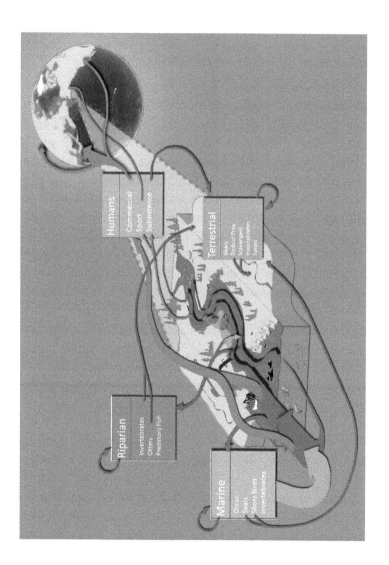

FIGURE 1: Pictorial illustration of important flows of salmon and contained biomass, energy, and nutrients within and out of the Bristol Bay ecosystem. Darker arrows depict the flows within the ecosystem, lighter arrows depict inputs due to growth in fresh water or the sea, and small arrows represent human harvest. Seventy percent of salmon are extracted by humans and are no longer available to the Bristol Bay ecosystem.

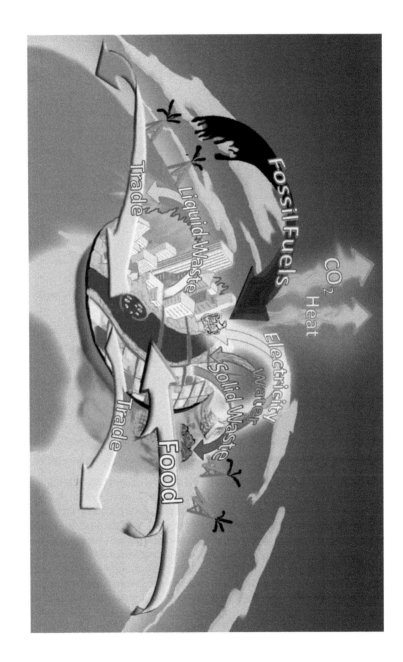

FIGURE 2: Pictorial illustration of important flows of resources into and wastes out of Portland, Oregon. This "most sustainable city in America" depends on exchanges with the local, regional, and global environments and economies in which it is embedded.

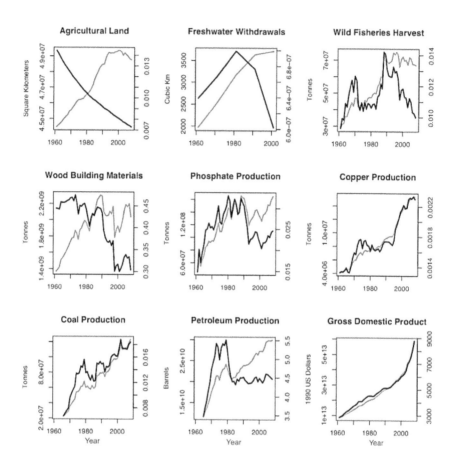

FIGURE 3: Global trends in total and per capita consumption of resources and GDP from 1961 to 2008. Total global use/production is represented by the grey line using the axis scale on the left side of each diagram. Per capita use/production is represented by the black line using the axis scale on the right side of each diagram. Per capita values represent the total values divided by global population size as reported by the World Resources Institute (http://earthtrends.wri.org/). The y-axes are untransformed and scaled to allow for maximum dispersion of variance. Individual sources for global use/production values are as follows: Agricultural land in square-km is from the World Development Indicators Database of the World Bank (http://data.worldbank.org/data-catalog/world-development-indicators) and represents the sum of arable, permanent crop, and permanent pasture lands (see also [46]). Freshwater withdrawal in cubic-km from 1960, 1970, 1980, and 1990 is from UNESCO (http://webworld.unesco.org/water/ihp/db/shiklomanov/part%273/HTML/Tb_14.html) and for 2000 from The Pacific Institute (http://www.worldwater.org/data.html). Wild

fisheries harvest in tonnes is from the FAO Fishery Statistical Collection Global Capture Production Database (http://www.fao.org/fishery/statistics/global-capture-production/en) and is limited to diadromous and marine species. Wood building material production in tonnes is based on the FAO ForeSTAT database (http://faostat.fao.org/site/626/default. aspx), and represents the sum of compressed fiberboard, pulpwood+particles (conifer and non-conifer [C & NC]), chips and particles, hardboard, insulating board, medium density fiberboard, other industrial roundwood (C & NC), particle board, plywood, sawlogs+veneer logs (C & NC), sawnwood (C & NC), veneer sheets, and wood residues. Phosphate, copper, and combustible coal production in tonnes is based on World Production values reported in the USGS Historical Statistics for Mineral and Material Commodities (http:// minerals.usgs.gov/ds/2005/140/). Global coal production data is limited to 1966–2008. Petroleum production in barrels from 1965 to 2008 is based on The Statistical Review of World Energy (http://www.bp.com/sectiongenericarticle800.do?categoryId=9037130& contentId=7068669) and represents all crude oil, shale oil, and oil sands plus the liquid content of natural gas where this is separately recovered. These data are reported in 1,000 barrels/day units, and were transformed to total barrels produced per year. GDP in 1990 US dollars are from the World Resources Institute (http://earthtrends.wri.org/). All data were accessed May 15–June 15, 2011.

A macroecological perspective on the sustainability of local systems emphasizes their interrelations with the larger systems in which they are embedded, rather than viewing these systems in isolation. Portland, Oregon offers an illuminating example. The city of Portland and surrounding Multnomah County, with a population of 715,000 and a median per capita income of US$51,000, bills itself and is often hailed by the media as "the most sustainable city in America" (e.g., SustainLane.com, 2008). On the one hand, there can be little question that Portland is relatively green and offers its citizens a pleasant, healthy lifestyle, with exemplary bike paths, parks, gardens, farmers' markets, and recycling programs. About 8% of its electricity comes from renewable non-hydroelectric sources (http://apps3. eere.energy.gov/greenpower/resources/tables/topten.shtml). On the other hand, there also can be no question that Portland is embedded in and completely dependent on environments and economies at regional, national, and global scales (Figure 2). A compilation and quantitative analysis of the flows into and out of the city are informative (see Text S1 for sources and calculations). Each year the Portland metropolitan area consumes at least 1.25 billion liters of gasoline, 28.8 billion megajoules of natural gas, 31.1 billion megajoules of electricity, 136 billion liters of water, and 0.5 million tonnes of food, and the city releases 8.5 million tonnes of carbon as CO_2, 99 billion liters of liquid sewage, and 1 million tonnes of solid waste into the environment. Total domestic and international trade amounts to 24 million tonnes of materials annually. With respect to these flows, Portland is not conspicuously "green"; the above figures are about average for a US city of comparable size (e.g., [32]).

A good way to see the embedding problem is to imagine the consequences of cutting off all flows in and out, as military sieges of European castles and cities attempted to do in the Middle Ages. From this point of view and in the short term of days to months, some farms and ranches would be reasonably sustainable, but the residents of a large city or an apartment building would rapidly succumb to thirst, starvation, or disease. Viewed from this perspective, even though Portland may be the greenest and by some definitions "the most sustainable city in America", it is definitely not self-sustaining. Massive flows of energy and materials across the city's boundaries are required just to keep its residents alive, let alone provide them with the lifestyles to which they have become accustomed. Any complete ecological assessment of the sustainability of a local system should consider its connectedness with and dependence on the larger systems in which it is embedded.

2.4 PRINCIPLE 3: GLOBAL CONSTRAINTS

For thousands of years, humans have harvested fish, other animals, and plants with varying degrees of "sustainability" and lived in settlements that depend on imports and exports of energy and materials. Throughout history, humans have relied on the environment for goods and services and used trade to compensate for imbalances between extraction, production, and consumption at local to regional scales. What is different now are the enormous magnitudes and global scales of the fluxes of energy and materials into and out of human systems. Every year fisheries export thousands of tonnes of salmon biomass and the contained energy and nutrients from the Bristol Bay ecosystem to consumers in Asia, Europe, and the US. Every year Portland imports ever larger quantities of energy and materials to support its lifestyle and economy. Collectively, such activities, replicated thousands of times across the globe, are transforming the biosphere.

Can the Earth support even current levels of human resource use and waste production, let alone provide for projected population growth and economic development? From our perspective, this should be the critical issue for sustainability science. The emphasis on local and regional scales—as seen in the majority of the sustainability literature and the above two examples—is largely irrelevant if the human demand for essential energy and materials exceeds the capacity of the Earth to supply these resources and if the release of wastes exceeds the capacity of the biosphere to absorb or detoxify these substances.

Human-caused climate change is an obvious and timely case in point. Carbon dioxide has always been a waste product of human metabolism— not only the biological metabolism that consumes oxygen and produces carbon dioxide as it converts food into usable energy for biological activities, but also the extra-biological metabolism that also produces CO_2 as it burns biofuels and fossil fuels to power the maintenance and development of hunter-gatherer, agricultural, and industrial-technological societies. Only in the last century or so, however, has the increasing production of CO_2 by humans overwhelmed the Earth's capacity to absorb it, increasing atmospheric concentrations and warming the planet more each decade. So, for example, efforts to achieve a "sustainable" local economy for a coastal fishing village in a developing country will be overwhelmed if, in only a few decades, a rising sea level caused by global climate change inundates the community. This shows the importance of analyzing sustainability on a global as well as a local and regional scale.

A macroecological approach to sustainability science emphasizes how human socioeconomic systems at any scale depend on the flows of essential energy and material resources at the scale of the biosphere as a whole. The finite Earth system imposes absolute limits on the ecological processes and human activities embedded within it. The impossibility of continued exponential growth of population and resource use in a finite world has long been recognized [33]–[35]. But repeated failures to reach the limits in the predicted time frames have caused much of the economic establishment and general public to discredit or at least discount Malthusian dynamics. Now, however, there is increasing evidence that humans are pushing if not exceeding global limits [2],[3],[36],[37]. For example, the Global Footprint Network estimates that the ecological footprint, the amount of land required to maintain the human population at a steady state [9], had exceeded the available land area by more than 50% by 2007, and the imbalance is increasing (http://www.footprintnetwork.org/en/index .php/GFN).

Here we present additional evidence that humans have approached or surpassed the capacity of the biosphere to provide essential and often non-substitutable natural resources. Figure 3 plots trends in the total and per capita use of agricultural land, fresh water, fisheries, wood, phosphate, petroleum, copper, and coal, as well as gross domestic product (GDP), from 1961 to 2008. Note that only oil, copper, coal, and perhaps fresh water show consistent increases in total consumption. Consumption of the other resources peaked in the 1980s or 1990s and has since declined. Dividing the total use of each resource by the human population gives the per capita rate of resource use, which has decreased conspicuously for all commodities except copper and coal. This means that production of these commodities has not kept pace with population growth. Consumption by the present generation is already "compromising the ability of future generations to meet their own needs." And this does not account for continued population growth, which is projected to increase the global population to 9–10 billion by 2050 and would result in substantial further decreases in per capita consumption.

Figure 3 shows results consistent with other analyses reporting "peak" oil, fresh water, and phosphate, meaning that global stocks of these important resources have been depleted to the point that global consumption will soon decrease if it has not already done so [10],[37]. Decreased per capita consumption of essential resources might be taken as an encouraging sign of increased efficiency. But the increase in efficiency is also a response to higher prices as a result of decreasing supply and increasing demand. We have included plots for copper and coal to show that overall production of some more abundant commodities has kept pace with population growth, even though the richest stocks have already been exploited. This is typical in ecology: not all essential resources are equally limiting at any given time. Diminishing supplies of some critical resources, such as oil, phosphorus, arable land, and fresh water, jeopardize the capacity to maintain even the current human population and standard of living.

What are the consequences of these trends? Many economists and sustainability scientists suggest that there is little cause for concern, at least in the short term of years to decades. They give several reasons: i) the finite stocks have not been totally exhausted, just depleted; there are still fish in the sea, and oil, water, phosphate, copper, and coal in the ground; they are just getting harder to find and extract; ii) conservation and substitution can compensate for depletion, allowing economies to grow and provide for increases in population and standard of living; iii) production depends more on the relationship between supply and demand as reflected in price than on absolute availability; and iv) the socioeconomic status of contemporary humans depends not so much on raw materials and conventional goods as on electronic information, service industries, and the traditional economic variables of money, capital, labor, wages, prices, and debt.

There are several reasons to question this optimistic scenario. First, the fact that GDP has so far kept pace with population does not imply that resource production will do likewise. Indeed, we have shown that production of some critical resources is not keeping pace. Second, there is limited or zero scope to substitute for some resources. For most of them, all known substitutes are inferior, scarcer, and more costly. For example, there is no substitute for phosphate, which is an essential requirement of all living things and a major constituent of fertilizer. No other element has the special properties of copper, which is used extensively in electronics. Despite extensive recycling of copper, iron, aluminum, and other metals, there is increasing concern about maintaining supplies as the rich natural ores have been depleted (e.g., [38], but see [39]). Third, several of the critical resources have interacting limiting effects. For example, the roughly constant area of land in cultivation since 1990 indicates that modern agriculture has fed the increasing human population by achieving higher yields per unit area. But such increased yields have required increased inputs of oil for powering machinery, fresh water for irrigation, and phosphate for fertilizer. Similarly, increased use of finite fossil fuels has been required to synthesize nitrogen fertilizers and to maintain supplies of mineral resources, such as copper, nickel, and iron, as the richest ores have been depleted and increased energy is required to extract the remaining stocks. An optimistic scenario would suggest that increased use of coal and renewable energy sources such as solar and wind can substitute for depleted reserves of petroleum, but Figure 3 shows a similar pattern of per capita consumption for coal as for other limiting resources, and the capacity of renewables to substitute for fossil fuels is limited by thermodynamic constraints due to low energy density and economic constraints of low energy and monetary return on investment [40]–[43]. Fourth, these and similar results (e.g., [3]) are starting to illuminate the necessary interdependencies between the energetic and material currencies of ecology and the monetary currencies of economics. The relationship between decreasing supply and increasing demand is causing prices of natural resources to increase as they are depleted, and also causing prices of food to increase as fisheries are overharvested and agriculture requires increasing energy and material subsidies [2],[8],[43]. The bottom line is that the growing human population and economy are being fed by unsustainable use of finite resources of fossil fuel energy, fertilizers, and arable land and by unsustainable harvests of "renewable resources" such as fish, wood, and fresh water. Furthermore, attaining sustainability is additionally complicated by inevitable yet unpredictable changes in both human socioeconomic conditions and the extrinsic global environment [44]. Sustainability will always be a moving target and there cannot be a single long-term stable solution.

Most sustainability science focuses on efforts to improve standards of living and reduce environmental impacts at local to regional scales. These efforts will ultimately and inevitably fail unless the global system is sustainable. There is increasing evidence that modern humans have already exceeded global limits on population and socioeconomic development, because essential resources are being consumed at unsustainable rates. Attaining sustainability at the global scale will require some combination of two things: a decrease in population and/or a decrease in per capita resource consumption (see also [45]). Neither will be easy to achieve. Whether population and resource use can be reduced sufficiently and in time to avoid socioeconomic collapse and attendant human suffering is an open question.

Critics will point out that our examination of sustainability from a macroecological and natural science perspective conveys a message of "doom and gloom" and does not offer "a way forward". It is true that humanity is faced with difficult choices, and there are no easy solutions. But the role of science is to understand how the world works, not to tell us what we want to hear. The advances of modern medicine have cured some diseases and improved health, but they have not given us immortality, because fundamental limits on human biology constrain us to a finite lifespan. Similarly, fundamental limits on the flows of energy and materials must ultimately limit the human population and level of socioeconomic development. If civilization in anything like its present form is to persist, it must take account of the finite nature of the biosphere.

2.5 CONCLUSION

If sustainability science is to achieve its stated goals of "dealing with the interactions between natural and social systems" so as to "[meet] the needs of present and future generations while substantially reducing poverty and conserving the planet's life support systems", it must take account of the ecological limits on human systems and the inherently ecological nature of the human enterprise. The human economy depends on flows of energy and materials extracted from the environment and transformed by technology to create goods and services. These flows are governed by physical conservation laws. These flows rarely balance at local or regional scales. More importantly, however, because these systems are all embedded in the global system, the flows of critical resources that currently sustain socioeconomic systems at these scales are jeopardized by unsustainable consumption at the scale of the biosphere. These ecological relationships will determine whether "sustainability" means anything more than "green", and whether "future generations [will be able] to meet their own needs".

REFERENCES

1. Goodland R (1995) The concept of environmental sustainability. Ann Rev Ecol System 26: 1–24.
2. Rockström J,Steffen W,Noone K,Persson A,Chapin F. S 3rd,et al. (2009) A safe operating space for humanity. Nature 461: 472–475.

3. Brown J. H,Burnside W. R,Davidson A. D,DeLong J. P,Dunn W. C,et al. (2011) Energetic limits to economic growth. BioScience 61: 19–26.

4. Vitousek P. M,Mooney H. A,Lubchenco J,Melillo J. M (1997) Human domination of Earth's ecosystems. Science 277: 494–499.

5. Imhoff M. L,Bounoua L,Ricketts T,Loucks C,Harriss R,et al. (2004) Global patterns in human consumption of net primary production. Nature 429: 870–873.

6. Haberl H,Erb K. H,Krausmann F,Gaube V,Bondeau A,et al. (2007) Quantifying and mapping the human appropriation of net primary production in earth's terrestrial ecosystems. Proc Natl Acad Sci U S A 104: 12942–12947.

7. Pauly D,Watson R,Alder J (2005) Global trends in world fisheries: impacts on marine ecosystems and food security. Phil Trans Roy Soc B 360: 5–12.

8. Worm B,Barbier E. B,Beaumont N,Duffy J. E,Folke C,et al. (2006) Impacts of biodiversity loss on ocean ecosystem services. Science 314: 787–790.

9. Wackernagel M,Rees W. E (1996) Our ecological footprint: reducing human impact on the earth. New Society Publications.

10. Gleick P. H,Palaniappan M (2010) Peak water limits to freshwater withdrawal and use. Proc Natl Acad Sci U S A 107: 11155–11162.

11. Brundtland G. H (1987) Our common future. World Commission on Environment and Development.

12. Burnside W. R,Brown J. H,Burger O,Hamilton M. J,Moses M,et al. (2012) Human macroecology: linking pattern and process in big picture human ecology. Biol Rev 87: 194–208.

13. Day J. W Jr,Boesch D. F,Clairain E. J,Kemp G. P,Laska S. B,et al. (2007) Restoration of the Mississippi Delta: lessons from hurricanes Katrina and Rita. Science 315: 1679–1684.

14. Moses M. E,Brown J. H (2003) Allometry of human fertility and energy use. Ecol Let 6: 295–300.

15. Day J. W Jr,et al. (2009) Ecology in times of scarcity. BioScience 59: 321–331.

16. Odum H. T (1971) Environment, power and society. New York: Wiley-Interscience.

17. Odum H. T (2007) Environment, power, and society for the twenty-first century: the hierarchy of energy. Columbia University Press.

18. Sterner R. W,Elser J. J (2002) Ecological stoichiometry: the biology of elements from molecules to the biosphere. Princeton: Princeton University Press.

19. Czúcz B,Gathman J. P,Mcpherson G. U. Y (2010) The impending peak and decline of petroleum production: an underestimated challenge for conservation of ecological integrity. Con Biol 24: 948–956.

20. Hilborn R,Quinn T. P,Schindler D. E,Rogers D. E (2003) Biocomplexity and fisheries sustainability. Proc Natl Acad Sci U S A 100: 6564–6568.

21. Hilborn R (2006) Salmon-farming impacts on wild salmon. Proc Natl Acad Sci U S A 103: 15277.

22. ADG&F (2010) Alaska historical commercial salmon catches, 1878–2010. Alaska Department of Game and Fish, Division of Commercial Fisheries.

23. Coupland G,Stewart K,Patton K (2010) Do you never get tired of salmon? Evidence for extreme salmon specialization at Prince Rupert harbour, British Columbia. J Anthro Arch 29: 189–207.

24. Cederholm C. J,Kunze M. D,Murota T,Sibatani A (1999) A Pacific salmon carcasses: essential contributions of nutrients and energy for aquatic and terrestrial ecosystems. Fisheries 24: 6–15.

25. Gende S. M,Edwards R. T,Willson M. F,Wipfli M. S (2002) Pacific salmon in aquatic and terrestrial ecosystems. BioScience 52: 917–928.

26. Naiman R. J,Bilby R. E,Schindler D. E,Helfield J. M (2002) Pacific salmon, nutrients, and the dynamics of freshwater and riparian ecosystems. Ecosystems 5: 399–417.

27. Schindler D. E,Scheuerell M. D,Moore J. W,Gende S. M,Francis T. B,et al. (2003) Pacific salmon and the ecology of coastal ecosystems. Front Ecol Enviro 1: 31–37.

28. Schindler D. E,Leavitt P. R,Brock C. S,Johnson S. P,Quay P. D (2005) Marine-derived nutrients, commercial fisheries, and production of salmon and lake algae in Alaska. Ecology 86: 3225–3231.

29. International Council for Science (2002) Science and technology for sustainable development. Paris: International Council for Science.

30. Millenium Ecosystem Assessment (2005) Ecosystems and human well-being: synthesis. Island Press.

31. Turner B. L,Lambin E. F,Reenberg A (2007) The emergence of land change science for global environmental change and sustainability. Proc Natl Acad Sci U S A 104: 20666–20671.

32. Hillman T,Ramaswami A (2010) Greenhouse gas emission footprints and energy use benchmarks for eight US cities. Environ Science & Tech 44: 1902–1910.

33. Malthus T. R (1798) An essay on the principle of population. Prometheus.

34. Ehrlich P. R (1968) The population bomb. New York.

35. Meadows D. H (1972) The limits of growth. A report for The Club of Rome.

36. Holdren J. P (2008) Science and technology for sustainable well-being. Science 319: 424–434.

37. Nel W. P,Van Zyl G (2010) Defining limits: energy constrained economic growth. Applied Energy 87: 168–177.

38. Gordon R. B,Bertram M,Graedel T. E (2006) Metal stocks and sustainability. Proc Natl Acad Sci U S A 103: 1209–1214.

39. Kesler S. E,Wilkinson B. H (2008) Earth's copper resources estimated from tectonic diffusion of porphyry copper deposits. Geology 36: 255.

40. Fargione J,Hill J,Tilman D,Polasky S,Hawthorne P (2008) Land clearing and the biofuel carbon debt. Science 319: 1235–1238.

41. Hall C. A. S,Day J. W (2009) Revisiting the limits to growth after peak oil. American Scientist 97: 230–237.

42. Smil V (2008) Energy in nature and society: general energetics of complex systems. MIT Press.

43. The Royal Society (2009) Reaping the benefits: science and the sustainable intensification of global agriculture. London: The Royal Society.

44. Milly P. C,Betancourt J,Falkenmark M,Hirsch R. M,Kundzewicz Z. W,et al. (2008) Stationarity is dead: whither water management? Science 319: 573–574.

45. Cohen J. E (1996) How many people can the earth support? WW Norton & Company.

46. Foley J. A,Ramankutty N,Brauman K. A,Cassidy E. S,Gerber J. S (2011) Solutions for a cultivated planet. Nature 478: 337–342.

There are several supplemental files that are not available in this version of the article. To view this additional information, please use the citation on the first page of this chapter.

PART II

BIOMASS IN ENERGY
AND CHEMICAL INDUSTRIES

CHAPTER 3

Biological Feedstocks for Biofuels

JUAN CARLOS SERRANO-RUIZ, MARÍA PILAR RUIZ-RAMIRO,
AND JIMMY FARIA

The production of fuels from renewable sources, such as biomass, is an interesting alternative to mitigate climate change and to progressively displace petroleum as a raw material for the transportation sector. A number of technologies have been developed in recent years to achieve conversion of diverse biomass feedstocks (e.g., sugars, vegetable oils and lignocellulose) into a variety of biofuels (e.g., bioethanol, biodiesel, higher alcohols and green hydrocarbons). This chapter provides a brief overview of these technologies, taking into consideration aspects such as the complexity of the process and the type of fuel produced (e.g., conventional and advanced). Particular emphasis is given to aqueous-phase catalytic routes to process biomass derivatives such as glycerol, hydroxymethyl furfural, levulinic acid and g-valerolactone into green hydrocarbons.

Finite fossil fuels such as petroleum, natural gas and coal currently supply most of the energy consumed in the world, and their consumption is associated with the release of greenhouse gases such as CO_2. Renewable sources such as solar, wind, hydroelectric and geothermal activity

Originally published in An Introduction to Green Chemistry Methods *by Rafael Luque & Juan Carlos Colmenares, (London: Future Science, 2013), pp. 117–131. Used with the permission of the authors and the publisher.*

have been proposed as alternatives to fossil fuels, with biomass being the only sustainable source of organic carbon available on Earth and, consequently, the perfect replacement for petroleum in the production of fuels and chemicals [1].

Most of the petroleum consumed today is used to produce fuels for the transportation sector, responsible for almost a third of the energy consumed in the world. Consequently, large-scale processes for biofuel production will be necessary to effectively allow displacement of petroleum from our energy system and to promote a low-carbon transport system (as biofuels are considered carbon-neutral fuels). Besides, the implementation of a strong biofuel industry generates important economic benefits for the society as new rural economies can be boosted while energy dependency on abroad is reduced.

Two biofuels are currently well implemented in the transportation system: biodiesel, produced by transesterification of vegetable oils with methanol; and bioethanol, produced by bacterial fermentation of biomass sugars. These biofuels are known as first-generation biofuels, as they are derived from edible biomass. Biodiesel and bioethanol are also known as conventional biofuels as they are commercially produced by simple and well-known technologies. Apart from conventional bioethanol and biodiesel, other more advanced fuels (with better characteristics as fuels and produced by more refined technologies) have recently appeared. Examples of these advanced biofuels include higher alcohols (e.g., butanol) and liquid hydrocarbon fuels (green hydrocarbons).

A summary of the technologies available today for the production of biofuels is represented in Figure 8.1. As indicated above, edible sugars (derived from corn or sugar cane, or from nonedible lignocellulose) can be converted into bioethanol by biological fermentation, in a well-known process that resembles that of wine making. Biodiesel is derived from vegetable oils by simple low-temperature transesterification. The simplicity of fermentation and transesterification technologies, along with their partial compatibility with the gasoline and diesel infrastructures, respectively, has allowed conventional biofuels to dominate the biofuel market. However, a strong demand for better fuels in recent years, boosted by government programs [2], has facilitated the onset of new technologies such as aqueous-phase processing, biomass-to-liquids, pyrolysis coupled with catalytic upgrading, hydrotreating of vegetable oils and advanced microbial routes [3,4]. A short summary with the most relevant aspects of these technologies will be provided in the following sections.

3.1 TECHNOLOGIES FOR THE PRODUCTION OF BIOFUELS

3.1.1 FERMENTATION OF SUGARS

The production of bioethanol by fermentation of biomass sugars is the most relevant technology today in terms of production volume. Therefore, bioethanol is the predominant biomass-derived fuel at present, representing almost 90% of the total biofuel production in the world [201]. Bioethanol is mixed with gasoline in limited amounts serving as a high-octane additive to improve combustion. Simultaneously, it allows the reduction of pollutant emissions such as CO, NOx and SOx. On the other hand, bioethanol has important compatibility, energy-density and water-absorption issues that limit its penetration in the fuel market. Therefore, ethanol can not be used as a pure fuel in current spark ignition engines, which only tolerate low concentration blends (v/v: 5–15%; E5–E15) with conventional gasoline. This is indeed an important issue, as countries such as the USA (the first bioethanol producer) have experienced the so-called 'blend wall' (the point at which the production of bioethanol is larger than the amount needed to blend gasoline at 10%) [4–6]. In addition, ethanol contains less energy that gasoline (thereby ultimately penalizing the gas mileage of vehicles running on mixtures of bioethanol and gasoline) and promotes the absorption of water in the fuel, potentially leading to phase separation issues [4,202].

Figure 8.2 shows a summarized scheme of the fermentation technology used for the production of bioethanol. The process begins with the extraction and subsequent deconstruction of the carbohydrate polymers forming part of the structure of biomass. At this point, edible biomass is readily depolymerized into sugar monomers by simple extraction with water, whereas recalcitrant lignocellulose requires extra pretreatment (to break/weaken the lignin protection) and hydrolysis (to break unreactive cellulose into monomers) steps, thereby increasing the complexity of the plant and finally leading to extra costs for the process. The released sugar monomers are subsequently fermented to ethanol using a variety of microorganisms (e.g., yeast, bacteria and mold), in a well-known process similar to that used in beer and wine making. Common microorganisms used for bioethanol production include yeasts such as *Saccharomyces cerevisiae*. This yeast possesses a mixture of invertase (to catalyze the conversion of sucrose into glucose and fructose) and zymases (to ferment glucose and fructose into ethanol. Alternatively, bacteria such as *Zymmomonas mobilis* can be used in the fermentation process in virtue of the higher specific productivity, ethanol yield and alcohol tolerance of this microorganism [7].

The stringent conditions required for microbes to survive obligates the carrying out of fermentation at near ambient temperatures, while keeping low concentrations of ethanol. These requirements have important implications for the process. First, low reaction rates are achieved, thereby making it necessary to increase the size of the bioreactor to make the process economically feasible. Second, dilute aqueous solutions of ethanol (maximum 15 wt%) are obtained after fermentation, thereby requiring further concentration (by distillation and dehydration steps) to fuel grade ethanol ($\geq$99 wt%). This water removal step is costly and typically accounts for 30–40% of the energy consumed in the overall process. Butanol, a compound with energy density and polarity similar to gasoline, can be also produced by fermentation of sugars [8,9]. Unlike ethanol, butanol can be readily implemented in the hydrocarbon-based transportation infrastructure (e.g., tanks, pipes, filling stations and engines), while its hydrophobic character prevents excessive water absorption. Several companies are biologically producing biobutanol at present [203–205].

3.2 ADVANCED MICROBIAL TECHNOLOGIES

Microorganisms utilized for fermentation can be genetically engineered to convert sugars to advanced biofuels (e.g., higher alcohols and hydrocarbons) instead of bioethanol. Compared with classical fermentation, these advanced microbial technologies achieve important improvements as hydrocarbons, and some higher alcohols, separate spontaneously from the aqueous broth, thereby avoiding poisoning of microbes by the accumulated products and facilitating separation/collection of products from the reaction medium. Advanced microbial routes have experienced tremendous progress in the last few years, mainly due to metabolic engineering (i.e., the genetic manipulation of a microbe to change its inherent metabolism favoring pathways to desired products). As indicated in Figure 8.3, a variety of advanced biofuels can be biologically produced in modified organisms. Therefore, hydrophobic C_4-C_7 alcohols can be produced from sugars by a nonfermentative pathway [10,11] reaching concentrations of isobutanol of 22 g/l (86% of the theoretical maximum yield for the glucose to isobutanol process). Companies such as Gevo Inc. (CO, USA) are currently exploiting this process to generate isobutanol and downstream related green hydrocarbon products.

Green hydrocarbon fuels can be produced following the isoprenoid route [12]. Therefore, C_{10}, C_{15} and C_{20} isoprene-derived activated units can be derived from this route. For example, farnesene, a C_{15} branched olefin, is produced in genetically modified *Escherichia coli* and *S. cerevisiae* at yields of 14 g/l [101]. Sugars can be also converted to fatty acids following the natural pathway to accumulate energy in organisms. By modifying this natural route, straight-chain hydrocarbons can be biologically produced in modified microorganisms [13].

3.3 TRANSTERIFICATION OF VEGETABLE OILS

Vegetable oils derived from soybeans, palm or sunflower can be used as feedstocks for the production of first-generation biodiesel. Biodiesel is the second most abundant biofuel globally with an annual production of 1.6 × 10^{10} l [206]. Biodiesel is sulfur free and, as in the case of ethanol, its use allows reduction of particulate matter and hydrocarbons. Figure 8.4 summarizes the methodology employed to produce biodiesel. Edible (e.g., palm, sunflower, canola, rapeseed and soybean) and nonedible (e.g., low-quality waste oils and nonedible plants such as *Jathropha curcas*, or *Camelina* and algae) feedstocks can be used in the process, although nonedible biomass is preferred as they do not compete with food supply. The process starts with the oil extraction, which is rich in triglycerides (TGs). The oil is subsequently reacted with methanol at mild temperatures (50–80°C) in the presence of a base, allowing conversion of TGs into a mixture of fatty acid methyl esters (the components of biodiesel) and glycerol (1,2,3-pro-panetriol). Glycerol, which is produced in large amounts (100 kg per ton of biodiesel) is subsequently separated from the fatty acid methyl ester mixture by decantation and further washing/drying steps.

In the same way as bioethanol, biodiesel possesses important compatibility and energy-density drawbacks. Biodiesel usage in current diesel vehicles is limited to low-concentration mixtures (B blends) because of its corrosiveness. Biodiesel also contains less energy per volume than regular diesel, and its higher cloud point compared with regular diesel increases the risk of plugging filters or small orifices when operating at cold temperatures.

3.4 HYDROTREATING OF VEGETABLE OILS

Green diesel and jet fuels can be produced from plant oils and animal fats by means of deoxygenation reactions under hydrogen pressure in a catalytic process denoted as hydrotreating. A variety of feedstocks, reaction conditions and catalysts are employed for hydrotreating of oils. Noble metal-based catalysts (Pt and Pd) and, especially, metal sulphides (Co, Mo, Ni and W) supported on alumina (similar to those used in the petrochemical industry) are typical materials used in this process. Some advantages of metal sulphides include low cost, high resistance to impurities and the possibility of using them in the current petroleum facilities. The acidity and porosity of the support are important aspects to control during hydrotreating as acidity promotes undesirable cracking and microporosity avoids proper diffusion of bulky TG molecules.

A large number of feedstocks, including edible and nonedible oils, waste cooking oils, animal fats and mixtures of vegetable/petroleum oils, are currently utilized in hydrotreating. The utilization of waste and residues derived from existing industrial processes is important in that it can help to improve the economics of the hydrotreating. However, the hydrotreating of waste feedstocks is troublesome due to the presence of impurities, thereby requiring catalysts resistant to S and N, alkalis and phospholipids.

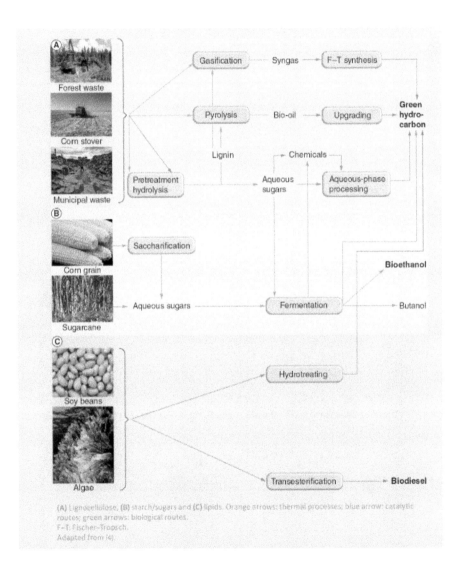

(A) Lignocellolose, (B) starch/sugars and (C) lipids. Orange arrows: thermal processes; blue arrow: catalytic routes; green arrows: biological routes.
F–T: Fischer–Tropsch.
Adapted from (4).

FIGURE 1: Most relevant technologies for the production of biofuels.

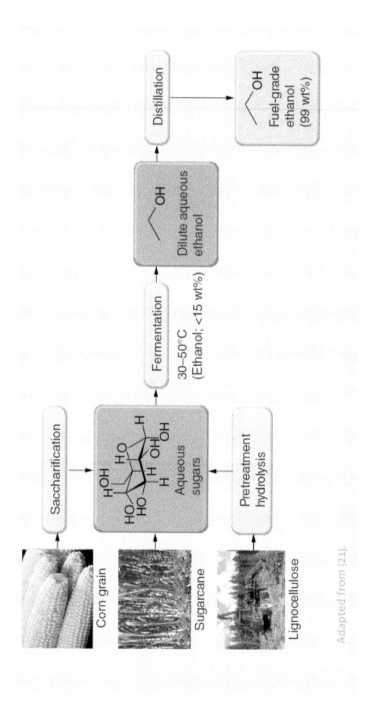

FIGURE 2: Fermentation process utilized for the production of bioethanol.

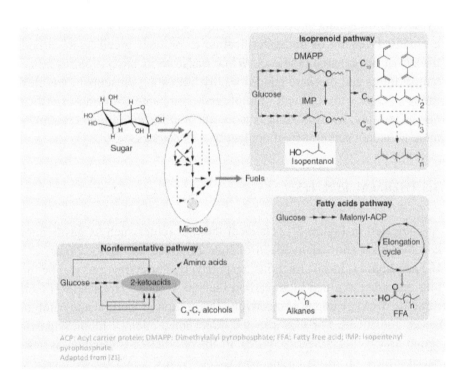

Figure 3: Main advanced microbial routes for the production of advanced biofuels.

Hydrogen consumption is elevated in hydrotreating. Since hydrogen is expensive and typically derived from fossil sources, strategies to reduce hydrogen consumption are highly necessary. Reaction conditions, the degree of unsaturation of the TGs and the reactor type (continuous fixed bed vs batch) are important parameters affecting hydrogen consumption.

Several companies such as Neste Oil (Espoo, Finland) and Honeywell UOP (IL, USA) are currently producing hydrotreated diesel and jet fuels by several trademark processes [207,208]. Hydrotreating can be seen as a modified version of transesterification to produce green hydrocarbons instead of oxygenated biodiesel. While transesterification requires milder temperatures and pressure conditions, hydrotreating presents higher flexibility to cope with different kind of feeds compared with transesterification. Therefore, hydrotreating is more adapted for the implementation of more abundant and non-food-competitive feedstocks.

3.5 THERMAL ROUTES

As indicated in Figure 8.1, classical thermal routes such as gasification and pyrolysis allow conversion of lignocellulose into gas (syngas, CO/H_2) and liquid (bio-oil) fractions, respectively. These fractions can be further upgraded to green hydrocarbon fuels by Fischer–Tropsch (F–T) and catalytic deoxygenation, respectively. Integration of thermal and catalytic routes is difficult. For example, the direct integration of biomass gasification and F–T synthesis requires an intermediate gas-cleaning system, because the F–T unit is highly sensitive to typical impurities delivered in gasifiers, such as tars, NH_3, HCl and sulfur compounds. The cleaning unit is expensive as the number of impurities is high and the cleaning requirements are very stringent. In addition, the syngas delivered from lignocellulosic sources is typically enriched in CO (H_2/CO: 0.5), and F–T synthesis requires syngas with a H_2/CO ratio closer to two.

In the case of pyrolysis, the bio-oil generated is a complicated mixture of more than 400 highly oxygenated compounds, including acids, alcohols, aldehydes, esters, ketones and aromatic species. Bio-oils are highly unstable, reactive and corrosive liquids that can not be used as transportation fuels. In addition, the high oxygen content of bio-oils decreases the energy density (16–19 vs 46 MJ/kg of regular gasoline). Therefore, extensive oxygen removal is required for bio-oils to be used as fuels. Hydrodeoxygenation (i.e., treatment of the bio-oil at moderate temperatures and high hydrogen pressures) over metal catalysts [14] and zeolite upgrading [15] (similar to the catalytic cracking approach used in petroleum refining) are, at the present time, the most used strategies to achieve bio-oil deoxygenation.

3.6 AQUEOUS-PHASE PROCESSING OF SUGARS & PLATFORM MOLECULES

Aqueous solutions of sugars and important derivatives obtained from them (i.e., platform molecules) can be catalytically processed in the aqueous phase to generate green hydrocarbons for the transportation sector. Unlike thermal routes, this approach uses milder reaction conditions allowing for better control of the chemistry and, consequently, the generation of liquid hydrocarbon fuels in a selective fashion. However, the production of liquid hydrocarbon transportation fuels (e.g., C_5-C_{12} for gasoline, C_9-C_{16} for jet fuel and C_{10}-C_{20} for diesel applications) from biomass derivatives involves deep deoxygenation (e.g., dehydration, hydrogenation and hydrogenolysis), combined with molecular weight enlargement by C-C coupling processes (e.g., aldol condensation, ketonization and oligomerization) [3]. For these routes to be economically feasible it is crucial to reduce the number of processing steps and to achieve deoxygenation of biomass feedstocks with minimal consumption of hydrogen from an external source.

As shown in Figure 8.5, several platform molecules have shown great potential for their conversion into liquid hydrocarbon fuels by following the deoxygenation plus C-C coupling approach. As described in the section entitled 'transterification of vegetable oils', glycerol is produced in large amounts as a byproduct of the trasesterification process, and it can also be produced by fermentation of biomass sugars. Consequently, glycerol represents an interesting feedstock for the large volume production of fuels. Dumesic and coworkers have proposed a two-step route for the conversion of concentrated aqueous solutions of glycerol into liquid hydrocarbon fuels [16]. This route involves the integration of aqueous-phase reforming of glycerol to syngas and subsequent F–T synthesis. During the first step, concentrated aqueous solutions of glycerol are gasified to a mixture of H_2, CO and CO_2 over supported platinum catalysts at moderate temperatures (e.g., 250°C).

Unlike gasification, glycerol reforming over platinum takes place at temperatures within the range employed for F–T, which allows for the effective integration of both processes.

Furfural and hydroxymethylfurfural, important compounds obtained by chemical dehydration of biomass-derived sugars, can serve as a platform for green diesel and jet fuels by means of cascade processes involving dehydration, hydrogenation and aldol-condensation reactions [17]. The carbonyl group in the furan compounds is the reactive center for C-C coupling by typically base-catalyzed aldol condensation with acetone. The aldol adduct formed can subsequently undergo a second condensation with another furanic molecule, thereby increasing the molecular weight of the final hydrocarbon product obtained after complete oxygen removal from the hydrogenated aldol adducts.

Sugars can be catalytically converted into green hydrocarbons in a process that combines deep oxygen removal and adjustment of the molecular weight using a minimum number of reactors and with minimal utilization of external hydrogen [18]. The process involved a two-step cascade catalytic approach (involving sugar reforming/reduction and C-C coupling processes) that allows conversion of aqueous solutions of sugars and polyols into the full range of liquid hydrocarbon fuels. First, aqueous sugars and polyols are deoxygenated/reformed over a Pt-Re catalyst into an organic phase mixture of monofunctional compounds (e.g., acids, alcohols, ketones and heterocycles) in the C_4-C_6 range. Interestingly, the hydrogen required for deoxygenation is internally supplied in the process by mean of reforming of a fraction of the sugar feedstock. In a second step, organic monofunctionals undergo C-C coupling reactions according to their functional group (e.g., acids: ketonization; ketones and aldehydes: aldol condensation; and alcohols: dehydration and alkylation).

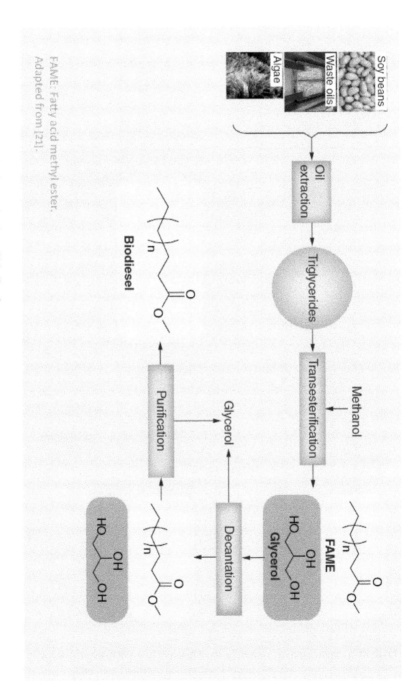

FIGURE 4: Process utilized for the production of biodiesel.

FAME: Fatty acid methyl ester.
Adapted from [21].

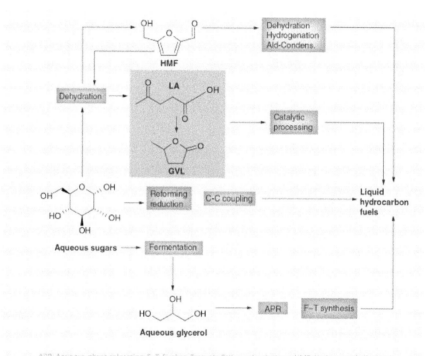

APR, Aqueous-phase reforming; F–T, Fischer-Tropsch; GVL, γ-valerolactone; HMF, Hydroxymethylfurfural;
LA, Levulinic acid.
Adapted from [4].

FIGURE 5: Main catalytic routes for the aqueous-phase catalytic processing of sugars and
some platform chemicals into liquid hydrocarbon transportation fuels.

Two biomass derivatives, levulinic acid (LA; obtained from sugars or hydroxymethylfurfural through dehydration processes) and g-valerolactone (obtained by hydrogenation of LA), are important feedstocks for green hydrocarbon production. Oxygen removal is achieved by dehydration/ hydrogenation and decarboxylation, while ketonization and oligomerization allow the increasing of the molecular weight and the adjusting of the structure of the final hydrocarbon product [19,20]. First, aqueous LA is hydrogenated to g-valerolactone, which serves as a starting point for liquid hydrocarbon fuel production with the intermediate production of pentanoic acid or C_4 alkenes. In a subsequent step, pentanoic acid undergoes ketonization to nonanone, while butenes are oligomerized to a distribution of alkenes centered at C_{12}. Both components are subsequently converted to green hydrocarbons by well-known chemistries.

3.7 CONCLUSION

Biofuels represents an interesting alternative to petroleum-based fuels to decrease the carbon footprint of the transportation sector. Conventional biofuels such as bioethanol and biodiesel are currently being commercialized, although their penetration into the transport infrastructure has been limited so far. Alternatively, advanced biofuels such as higher alcohols and green hydrocarbons offer excellent properties as fuels, although the production technologies are less mature and require intense process improvements. Therefore, several promising routes such as gasification of biomass to syngas coupled with F–T, pyrolysis integrated with catalytic upgrading processes, aqueous-phase catalytic processing of biomass-derived sugars and derivatives, hydrotreating of vegetable oils and related feedstocks, and advanced microbial routes have recently appeared and developed fast. This chapter has intended to summarize these routes offering the reader the most relevant aspects of these technologies.

REFERENCES

1.   Ragauskas AJ, Williams CK, Davison BH et al. The path forward for biofuels and biomaterials. Science 311, 484–489 (2006).

2. Regalbuto JR. The sea change in US biofuels' funding: from cellulosic ethanol to green gasoline. Biofuels Bioprod. Bioref. 5(5), 495–504 (2011).

3. Martin-Alonso D, Bond JQ, Dumesic JA. Catalytic conversion of biomass to biofuels. Green Chem. 12, 1493–1513 (2010).

4. Serrano-Ruiz JC, Dumesic JA. Catalytic routes for the conversion of biomass into liquid hydrocarbon transportation fuels. Energy Environ. Sci. 4, 83–99 (2011).

5. Service RF. Is there a road ahead for cellulosic ethanol? Science 329, 784–785 (2010).

6. Tyner WE, Dooley FJ, Viteri DA. Alternative pathways for fulfilling the RFS mandate. J. Agr. Econ. 93, 465–472 (2010).

7. The Royal Society. Sustainable Biofuels: Prospects and Challenges. The Royal Society, London, UK (2008).

8. Dürre P. Biobutanol: an attractive biofuel. Biotechnol. J. 2, 1525–1534 (2007).

9. Ezeji TC, Qureshi N, Blaschek HP. Bioproduction of butanol from biomass: from genes to bioreactors. Curr. Opin. Biotechnol. 18, 220–227 (2007).

10. Atsumi S, Hanai T, Liao JC. Non-fermentative pathways for synthesis of branched-chain higher alcohols as biofuels. Nature 451, 86–89 (2008).

11. Zhang KS, Sawaya MR, Eisemberg DS, Liao JC. Expanding metabolism for biosynthesis of nonnatural alcohols. Proc. Natl Acad. Sci. USA 105, 20653–20658 (2009).

12. Peralta-Yahya PP, Keasling JD. Advanced biofuel production in microbes. Biotechnol. J. 5, 147–162 (2010).

13. Schirmer A, Rude MA, Li X, Popova E, Cardayre SB. Microbial biosynthesis of alkanes. Science 329, 559–562 (2010).

14. Elliott DC. Historical developments in hydroprocessing bio-oils. Energy Fuels 21, 1792–1815 (2007).

15. Carlson TR, Vispute TP, Huber GW. Green gasoline by catalytic fast pyrolysis of solid biomass derived compounds. ChemSusChem. 1, 397–400 (2008).

16. Soares RR, Simonetti DA, Dumesic JA. Glycerol as a source of fuels and chemicals by low temperature catalytic processing. Angew. Chem. 118, 4086–4089 (2006).

17. Huber GW, Chheda J, Barret C, Dumesic JA. Production of liquid alkanes by aqueous-phase processing of biomassderived carbohydrates. Science 308, 1447–1450 (2005).

18. Kunkes EL, Simonetti DA, West RM, Serrano-Ruiz JC, Gärtner CA, Dumesic JA. Catalytic conversion of biomass to monofunctional hydrocarbons and targeted liquid-fuel classes. Science 322, 417–421 (2008).

19. Serrano-Ruiz JC, Wang D, Dumesic JA. Catalytic upgrading of levulinic acid to 5-nonanone. Green Chem. 12, 574–577 (2010).

20. Bond JQ, Martin-Alonso D, Wang D, West RM, Dumesic JA. Integrated catalytic conversion of g-valerolactone to liquid alkenes for transportation fuels. Science 327, 1110–1114 (2010).

21. Serrano-Ruiz JC, Ramos- Fernandez EV, Sepulveda- Escribano A. From biodiesel and bioethanol to liquid hydrocarbon fuels: new hydrotreating and advanced micr bial technologies, Energy Environ. Sci. 5, 5638–5652 (2012).

PATENT

22. Reninger NS, McPhee DJ: US0098645 (2008).

WEBSITES

23. IEA Energy Technology Essentials. Biofuel production. www.iea.org/techno/essentials2.pdf
24. US Environmental Protection Agency. Water phase separation in oxygenated gasoline. www.epa.gov/oms/regs/fuels/rfg/waterphs.pdf
25. The Energy Blog. Dupont, BP to produce biobutanol. http://thefraserdomain.type-pad.com/energy/2006/06/dupont_bp_to_pr.html#more
26. Butamax™ Advanced Biofuels LLC. The biobutanol advantage. http://www.butamax.com/the-biobutanol-advantage.ashx
27. Green Biologics. www.butanol.com/index.html
28. Biodiesel Magazine. Report: 12 billion gallons of biodiesel by 2020. www.biodieselmagazine.com/article.jsp?article_id=4080
29. Neste Oil. Production technology. www.nesteoil.com/default.asp?th=1,41,11991,12243,12335,12337
30. Honeywell UOP. Honeywell green diesel. www.uop.com/processingsolutions/biofuels/greendiesel

CHAPTER 4

From Tiny Microalgae
to Huge Biorefineries

LUISA GOUVEIA

4.1 INTRODUCTION

Biofuel and bioproduct production from microalgae have several advantages when compared to the 1st and 2nd biofuel generation having: high areal productivity, minimal competition with conventional agriculture, environmental benefits by recycling nutrients (N and P) from waste waters and mitigating carbon dioxide from air emissions. In addition, all components of microalgae can be separated and transformed into different valuable products. The high metabolic versatility of microalgae and cyanobacteria metabolisms, offer interesting applications in several fields, such as nutrition (human and animal), nutraceuticals, therapeutic products, fertil-

This article has been updated from the original. The original article can be found in: Oceanography *2,1 (2014). Reprinted with permission from the author.*

izers, plastics, isoprene, biofuels and environment (such as water stream bioremediation and carbon dioxide mitigation).

The high content of antioxidants and pigments (carotenoids such as fucoxanthin, lutein, beta-carotene and/or astaxanthin and phycobilliproteins) and the presence of long-chain polyunsaturated fatty acids (PUFAs) and proteins (essential amino acids methionine, threonine and tryptophan), makes microalgae an excellent source of nutritional compounds. Co-extraction of other high-value products (PUFAs, such as eicosapentaenoic acid (EPA), docosahexaenoic acid (DHA), and arachidonic acid (AA)) will also be evaluated since these compounds may enhance the nutritional or nutraceutical value of the microalgal oil.

Microalgae have also been screened for new pharmaceutical compounds with biological activity, such as antibiotics, antiviral, anticancer, enzyme inhibitory agents and other therapeutic applications. They have been reported to potentially prevent or reduce the impact of several lifestyle-related diseases (Shibata et al., 2003, 2006, 2007) with antimicrobial (antibacterial, antifungal, antiprotozoal) and antiviral (including anti-HIV) functions and they also have cytotoxic, antibiotic, and anti-tumour properties as well as having biomodulatory effects such as immunosuppressive and anti-inflammatory roles (Burja et al., 2001; Singh et al., 2005). *Chlorella* has also been used against infant malnutrition and neurosis (Yamaguchi, 1996), as well as being a food additive. Furthermore, algae are believed to have a positive effect on the reduction of cardio-circulatory and coronary diseases, atherosclerosis, gastric ulcers, wounds, constipation, anaemia, hypertension, and diabetes (Yamaguchi, 1996; Nuno et al., 2013).

The microalgae compounds, such as carotenoids have also been associated and claimed to reduce the risk of: (1) certain cancers (Gerster, 1993; Willet, 1994; Lupulescu, 1994; Tanaka et al., 2012), (2) cardiovascular diseases (Kohlmeier and Hastings, 1995; Giordano et al., 2012), (3) macular degeneration and cataract formation (Snodderly, 1995; Weikel et al., 2012) and possibly may have an effect on the immune system and may influence chronic diseases (Meydani et al., 1995; Park et al., 2010).

Besides nutritional, nutraceutical and therapeutic compounds, microalgae can also synthesize polysaccharides that can be used as an emulsion stabilizer or as biofloculants and polyhydroxyalkanoate, which are linear polyesters used in the production of bioplastics. Microalgae biomass has been demonstrated to improve the physical and thermal properties of plastic by replacing up to 25% of polymers, which increases the biodegradability of the final bioplastic. Microalgae can also produce isoprene, which is a key intermediate compound for the production of synthetic rubber and adhesives, including car and truck tires. It is also an important polymer building block for the chemical industry, such as for a wide variety of elastomers used in surgical gloves, rubber bands, golf balls, and shoes (Matos et al., 2013).

Furthermore, the amino acids produced by microalgae can be used as biofertilizers and therefore assist higher plant growth. Amino-acid based fertilization supplies plants with the necessary elements to develop their structures by adding nutrients through the natural processes of nitrogen fixation, solubilizing phosphorus, and stimulating plant growth through the synthesis of growth-promoting substances (eg., Painter, 1993; Dey, 2011; Sahu et al., 2012). Bio-fertilizers provide eco-friendly organic agro-input and are more cost-effective than chemical fertilizers.

Finally, regarding biofuels, they can be obtained from the microalgae biomass leftovers after the extraction of added-value compounds. According to the composition of the "waste" biomass, it can be used for the production of liquid biofuels (bioethanol, biodiesel, biobutano and bio-oil) (e.g., Gouveia and Oliveira, 2009; Miranda et al., 2012 or gaseous biofuels (biomethane, biohydrogen, syngas etc.) (Marques et al., 2011; Ferreira et al., 2013; Batista et al., 2014). The technology used to produce biofuels efficiently is not yet established, thus different biological and thermochemical processes still need to be studied and improved.

Unfortunately, the economic viability of algae-based biofuels is still unfeasible. However, the high metabolic versatility of microalgae and cyanobacteria metabolisms allow the production of the several mentioned non-fuel products, which have a very high-value and could play a major role in turning economic and energy balances more favorable. This versatility and huge potential of tiny microalgae could support a microalgae-based biorefinery and microalgae-based bioeconomy opening up vast opportunities in the global algae business.

The microalgae could play an important response to the worldwide biofuel demand, together with the production of high value-added products and assisting some other environmental issues such as water stream bioremediation and carbon dioxide mitigation.

Only the co-production of high added value products and environmental benefits could eventually off-set the high production costs of mass microalgae cultivation and support a microalgae-based bioeconomy. In fact, a microalgae-based biorefinery should integrate several processes and related industries, such as food, feed, energy, pharmaceutical, cosmetic, and chemical. Such an approach, in addition to the biomass, will take advantage of the various products synthesized by the microalgae. This adds value to the whole process which has a minimal environmental impact by recycling the nutrients and water, and by mitigating the CO_2 from the flue gases (Figure 1.)

This review highlights the potential of the tiny autotrophic microalgae for the production of several products in an experimental (lab scale) Biorefinery. The production contains biofuel(s) and other high value-added compounds which could be used for different applications and markets.

4.2 FROM (TINY) MICROALGAE TO (HUGE) BIOREFINERIES

The main bottleneck of the biorefinery approach is to separate the different fractions without damaging one or more of the product fractions. There is a need for mild, inexpensive and low energy consumption separation techniques to overcome these bottlenecks (Wijffels et al., 2010; Vanthoor-Koopmans et al., 2012). They should also be applicable for a variety of end products which have a sufficient quality but are also available in large quantities (Brennan and Owende, 2010; Lopes da Silva et al., 2014).

Some of the biorefinery techniques appropriate for metabolite separation and extraction are ionic liquids or surfactants (Vanthoor-Koopmans et al., 2013; Nobre et al., 2013). These techniques are relatively new and should therefore be studied thoroughly before commercial use will be possible.

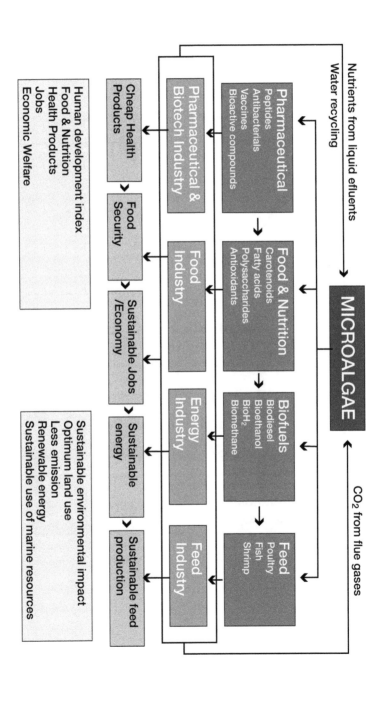

FIGURE 1: Example of a microalgae-based biorefinery and how it integrates several related industries (adapted from Subhadra, 2010).

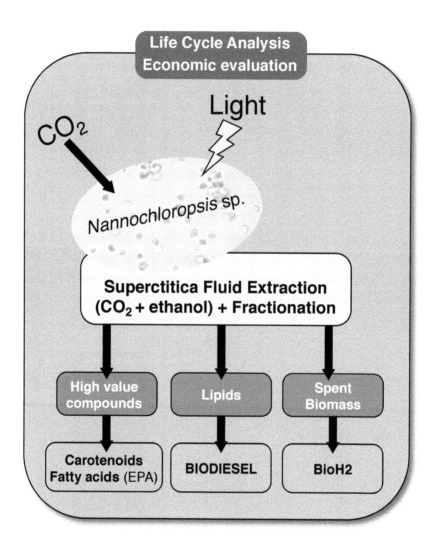

FIGURE 2: *Nannochloropsis* sp. Biorefinery.

4.2.1 NANNOCHLOROPSIS SP. BIOREFINERY

Nobre et al. (2013) used *Nannochloropsis* sp. microalga and developed a Biorefinery with the extraction of carotenoids and fatty acids (mainly EPA) for food and the feed industry as well as lipids for biodiesel production. The biomass composition is present in Table 1.

TABLE 1: *Nannochloropsis* sp. composition.

Composition	(%)
Crude fat	41
Total sugars	17
Total minerals	13
Others	29

The fractionated recovery of the different compounds was done by Supercritical Extraction using CO_2 and ethanol as an entrainer. From the biomass leftovers and using *Enterobacter aerogeneses* through dark fermentation, bioH_2 was also produced (Figure 2), yielding a maximum of 60.6 mL H_2/g alga (Nobre et al., 2013).

The energy consumption and CO_2 emissions emitted during the whole process (microalgae cultivation, harvesting, dewatering, milling, extraction and leftover biomass fermentation), as well as the economic factors were evaluated (Ferreira et al., 2013).

The authors showed five pathways and two biorefineries which were analysed (Figure 3):

1. Oil extraction by soxhlet (oil SE);
2. Oil and pigment extraction and fractionation through Supercritical Fluid Extraction (oil and pigment SFE);
3. Hydrogen production through dark fermentation of the leftover biomass after soxhlet extraction (bioH_2 via SE);
4. Hydrogen production by dark fermentation from the leftover biomass after Supercritical Fluid extraction (bioH_2 via SFE);

5. Hydrogen production from the whole biomass through dark fermentation (bioH$_2$ using the whole biomass).

Where path #1 and path #3 are the Biorefinery 1, path #2 and path #4 are the Biorefinery 2 and path #5 is the direct bioH$_2$ production. The analysis of pathways #1, #2 and #5 considers a system boundary from the *Nannochloropsis* sp. microalgal culture to the final product output (oil, pigments, or bioH$_2$, respectively). For pathways #3 and #4, the bioH$_2$ production from the leftover biomass from SE and SFE respectively was evaluated. The authors concluded that the oil production pathway by SE shows the lowest energy consumption, 176-244 MJ/MJ$_{prod}$, and CO$_2$ emissions, 13-15 kgCO$_2$/MJ$_{prod}$. However, economically the most favourable biorefinery was the one producing oil, pigments and H$_2$ via Supercritical Fluid Extraction (SFE).

From the net energy balance and the CO$_2$ emission analysis, Biorefinery 1 (biodiesel SE + bioH$_2$) presented the better results. Biorefinery 2 (biodiesel SFE + bioH$_2$) showed results in the same range of those in Biorefinery 1. However, the use of SFE produced high-value pigments in addition to the fact that it is a clean technology which does not use toxic organic solvents. Therefore, Biorefinery 2 was the best in terms if energy/CO$_2$/ and it being the most economically advantageous solution.

4.2.2 ANABAENA *SP. BIOREFINERY*

The experimental biohydrogen production by photoautrotophic cyanobacterium *Anabaena* sp. was studied by Marques et al. (2011). Hydrogen production from the *Anabaena* biomass leftovers was also achieved by fermentation through the *Enterobacter aerogenes* bacteria and was reported by Ferreira et al. (2012) (Figure 4).

Different culture conditions and gas atmospheres were tested in order to maximize the autotrophic bioH$_2$ yield versus the energy consumption and CO$_2$ emissions. The authors stated that the best conditions included an Ar+CO$_2$+20%N$_2$ gas atmosphere and medium light intensity (384 W) (Ferreira et al., 2012). The yielded H$_2$ could be increased using the biomass leftovers through a fermentative process, however this would mean a higher energy consumption as well as an increase in CO$_2$ emissions.

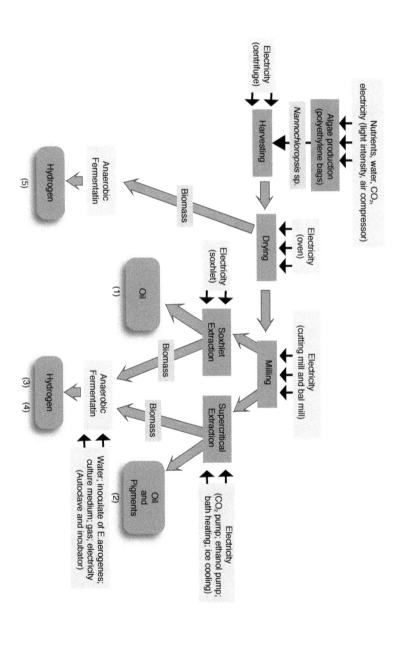

FIGURE 3: *Nannochloropsis* sp. biorefinery (including all steps, materials and energy, and different pathways) with the production of oils, pigments and bioHydrogen) (adapted from Ferreira et al., 2013).

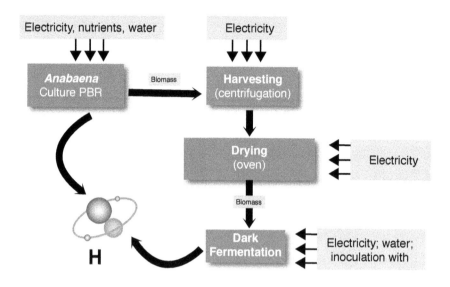

FIGURE 4: *Anabaena* sp. Biorefinery (including both H$_2$ production: autotrophically and by dark fermentation of its biomass (through Enterobacter aerogenes) (adapted from Ferreira et al., 2012).

4.2.3 CHLORELLA VULGARIS *BIOREFINERIES*

Quite a few reported works describe biorefineries from *Chlorella vulgaris:* these are stated below:

- Cv1: An integrating process for lipid recovery from the biomass of *Chlorella vulgaris* and methane production from the remaining biomass (after lipid extraction) was worked on by Collet et al. (2011). The authors demonstrated that, in terms of life cycle assessment (LCA), the methane from algae (algal methane) is the worst case, compared to algal biodiesel and diesel, in terms of abiotic depletion, ionizing radiation, human toxicity, and possible global warming. These negative results are mainly due to a strong demand for electricity. For the land use category, algal biodiesel also had a lesser impact compared to algal methane. However, algal methane is a much better option in terms of acidification and eutrophication.

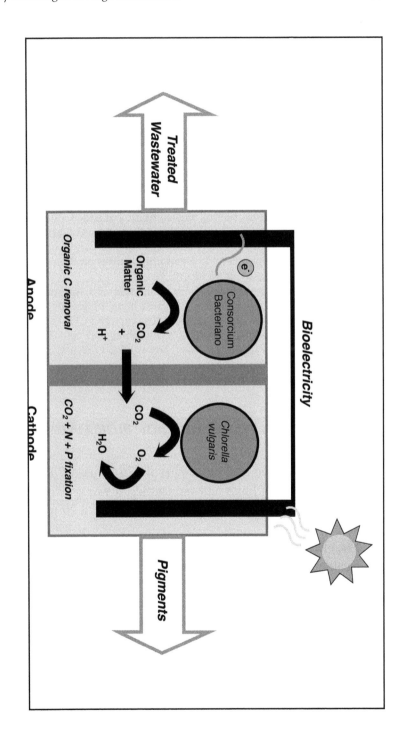

FIGURE 5: *Chlorella vulgaris* biorefinery (Photosynthetic Algal Microbial Fuel Cell) (adapted from Gouveia et al., 2014).

- Cv2: Another work concerning the simultaneous production of biodiesel and methane in a biorefinery concept was done by Ehimen et al. (2011). The authors obtained biodiesel from a direct transesterification on the *Chlorella* microalgal biomass, and from the biomass residues they obtained methane through anaerobic digestion. For a temperature of 40 °C and a C/N mass ratio of 8.53, a maximum methane concentration of 69 % (v/v) with a specific yield of 0.308 m^3 CH$_4$/kg VS was obtained. However, in this work the biodiesel yield was not reported.
- Cv3: In another work, the *Chlorella vulgaris* biorefinery approach was studied by Gouveia et al. (2014) and it included a Photosynthetic Algal Microbial Fuel Cell (PAMFC), where the microalga *Chlorella vulgaris* are present in the cathode compartment (Figure 5). The study demonstrated the simultaneous production of bioelectricity and added-value pigments, with possible wastewater treatment. The authors proved that the light intensity increases the PAMFC power and augments the carotenogenesis process in the cathode compartment. The maximum power produced was 62.7mW/m^2 with a light intensity of 96 μE/(m^2.s).
- Cv4: A bioethanol-biodiesel-microbial fuel cell was reported by Powel and Hill (2009) and basically consisted in an integration of photosynthetic *Chlorella vulgaris* (in the cathode) that captured CO$_2$ emitted by yeast (in the anode) fermenters, creating a microbial fuel cell. The study demonstrated the possibility of electrical power generation and oil for biodiesel, in a bioethanol production facility. The remaining biomass after oil extraction could also be used in animal feed supplement (Powel and Hill, 2009)

4.2.4 CHLORELLA PROTOTHECOIDS *BIOREFINERY*

The biorefinery stated by Campenni et al. (2013) used *Chlorella protothecoides* as a source of lipids and carotenoids, it was grown autotrophically and with nitrogen deprivation and the addition of a 20 g/l NaCl solution (Figure 6).

The total carotenoid content was 0.8% (w/w) (canthaxanthin (23.3%), echinenone (14.7%), free astaxanthin (7.1%) and lutein/zeaxanthin (4.1%)) which can be used for food applications. Furthermore, the total lipid content reached 43.4 % (w/w), with a fatty acid composition of C18:1 (33.6 %), C16:0 (23.3%), C18:2 (11.5 %), and C18:3 (less than 12 %), which is needed to fulfil the biodiesel EN 14214 quality specifications (EN 14214 2008) and can be used for the biofuel (biodiesel) industry.

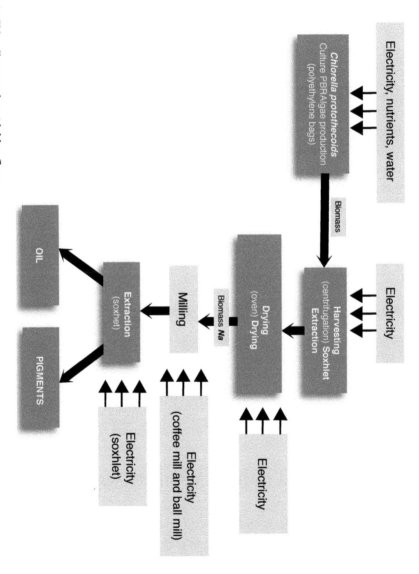

FIGURE 6: *Chlorella protothecoids* biorefinery

The leftover biomass is still available for hydrogen or bioethanol production in a biorefinery approach, as the residue still contains sugar taking advantage of all the *C. protothecoids* gross composition.

4.2.5 CHLORELLA REINHARDTII *BIOREFINERY*

The production of biohydrogen and the consequent biogas (methane) production by anaerobic fermentation of the residue of *Chlorella reinhardtii* biomass were achieved by Mussgnug et al. (2010).

The authors reported that using the biomass, after the hydrogen production cycle instead of using the fresh biomass, would increase the biogas production by 123 %.

The authors attributed these results to the storage compounds, such as starch and lipids with a high fermentative potential which is the key in the microalgae-based integrated process and could be used for more value-added applications.

4.2.6 DUNALIELLA SALINA *BIOREFINERY*

Sialve et al. (2009) attested the production of methane from the leftover biomass of *Dunaliella salina* after the oil extraction to make biodiesel. The authors found a much higher yield (around 50%) for a shorter hydraulic retention time (HRT, 18 days), than the corresponding values reported by Collet et al. (2011) using the *Chlorella vulgaris* biomass.

4.2.7 DUNALIELLA TERTIOLECTA *BIOREFINERY*

The chemoenzymatic saccharification and bioethanol fermentation of the residual biomass of *Dunaliella tertiolecta* after lipid extraction (for biodiesel production purposes) were investigated by Kim et al. (2013). The bioethanol was produced from the enzymatic hydrolysates without pretreatment by *S. cerevisiae*, resulting in yields of 0.14 g ethanol/g residual biomass and 0.44 g ethanol/g glucose produced from the residual biomass.

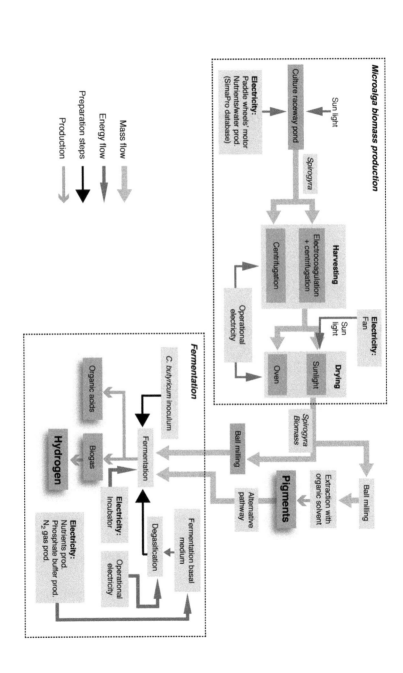

FIGURE 7: *Spirogyra* sp. biorefinery (adapted from Pacheco et al., 2014)

According to these authors, the residual biomass generated during mi-croalgal biodiesel production, could be used for bioethanol production in order to improve the economic feasibility of a microalgae-based integrated process.

4.2.8 ARTHROSPIRA (SPIRULINA) *BIOREFINERY*

Olguin (2012) highighted that the biorefinery strategy offers new opportunities for a cost-effective and competitive production of biofuels along with nonfuel compounds. The author studied an integrated system where the production of biogas, biodiesel, hydrogen and other valuable products (e.g. PUFAs, phycocyanin, and fish feed) could be possible.

4.2.9 SPIROGYRA *SP. BIOREFINERY*

Pacheco et al. (2014) pointed a biorefinery from *Spirogyra* sp., a sugar-rich microalga, for bioH$_2$ production as well as pigments (Figure 7). The economic and life cycle analysis of the whole process, allowed the authors to conclude that it is crucial to increase the sugar content of the microalgae to increase the bioH$_2$ yield. Furthermore, it is important to reduce the centrifugation needs and use alternative methods for pigment extraction other than using acetone solvents. The electrocoagulation and solar drying were used for harvesting and dewatering, respectively, and were able to reduce energy requirements by 90 %. Overall, centrifugation of the micro-algal biomass and heating of the fermentation vessel are still major energy consumers and CO$_2$ contributors to this process. Pigment production is necessary to improve the economic benefits of the biorefinery, but it is mandatory to reduce its extraction energy requirements that are demanding 62 % of the overall energy.

Mostafa et al. (2012) evaluated the growth and lipid, glycerol, and ca-rotenoid content of nine microalgae species (green and blue green microal-gae) grown in domestic wastewater obtained from the Zenein Wastewater

Treatment Plant in the Giza governorate in Egypt (Figure 8). The authors cultivated the different species under different conditions, such as without treatment after sterilization, with nutrients and sterilization, and with nutrients without sterilization, at 25±1°C, under continuous shaking (150 rpm) and illumination (2,000 lx), for 15 days. The highest biodiesel production from algal biomass cultivated in wastewater was obtained by *Nostoc humifusum* (11.80 %) when cultivated in wastewater without treatment and the lowest (3.8 %) was recorded by *Oscillatoria* sp. when cultivated on the sterilized domestic wastewater. The authors concluded that cultivating microalgae on domestic wastewater, combines nutrient removal and algal lipid production which has a high potential in terms of biodiesel feedstock. This methodology is suitable and non-expensive compared to the conventional cultivation methods for sustainable biodiesel and glycerol.

According to Subhadra and Edwards (2010) (Fig. 9), an integrated renewable energy park (IREP) approach can be envisaged by combining different renewable energy industries, in resource-specific regions, for synergetic electricity and liquid biofuel production, with zero net carbon emissions. Choosing the appropriate location, an IREP design, combining a wind power plant with solar panels and algal growth facilities to harness additional solar energy, could greatly optimize land. Biorefineries configured within these IREPs can produce about 50 million gallons of biofuel per year, providing many other value-added co-products and having almost no environmental impact (Subhadra, 2010) (Figure 9).

Clarens et al. (2011) suggested that the results from algae-to-energy systems can be either net energy positive or negative depending on the specific combination of cultivation and conversion processes used. The authors addressed the shortcoming "well-to-wheel", including the conversion of each biomass into transportation energy sources. The algal conversion pathway resulted in a combination of biodiesel and bioelectricity production for transportation, evaluated by vehicle kilometers traveled (VKT) per hectare. In this study, it was assumed that bioelectricity and biodiesel are used in commercially available battery electric vehicles (BEVs) and internal combustion vehicles (ICVs), respectively.

The authors depicted four pathways:

A. Methane-derived bioelectricity from the bulk algae biomass by an aerobic digestion
B. Biodiesel from the algae lipids and methane-derived bioelectricity from the residual biomass by anaerobic digestion
C. Biodiesel from the algae lipids and bioelectricity from the residual biomass by direct combustion
D. Bioelectricity from the bulk algae biomass by direct combustion

The four pathways follow various nutrient sources (e.g., virgin commercial CO_2, CO_2 from a coal-fired power plant, compressed CO_2 from flue gas, commercial fertilizers, and wastewater supplementation).

The authors found that algae-to-energy systems depend on the combination of cultivation and conversion processes used. They concluded that the conversion pathways involving direct combustion for the production of bioelectricity generally outperformed systems involving anaerobic digestion and biodiesel production. They ranked the four pathways as D > A > C > B in terms of energy return on investment.

The authors found an algae bioelectricity (D) generation of 1,402,689 MJ/km and algae biodiesel + bioelectricity (C) generation of 1,110 MJ/km. These algae-to-energy systems generate 4 and 15 times as VKT per hectare as switch grass or canola, respectively (Clarens et al. 2010).

Subhadra and Edwards (2011) analyzed the water footprint of two simulated algal biorefineries for the production of biodiesel, algal meal, and omega-3 fatty acids. The authors highlighted the advantages of multiproducts to attain a high operational profit with a clear return on investment. The energy return of algal biodiesel for different scenarios ranged between 0.016- 0.042 MJ.

Park et al. (2011) also studied algae which are grown as a by-product of high-rate algal ponds (HRAPs) operated for wastewater treatment. In addition to significantly better economics, algal biofuel production from wastewater treatment HRAPs has a much smaller environmental footprint compared to commercial algal production HRAPs which consume freshwater and fertilizers.

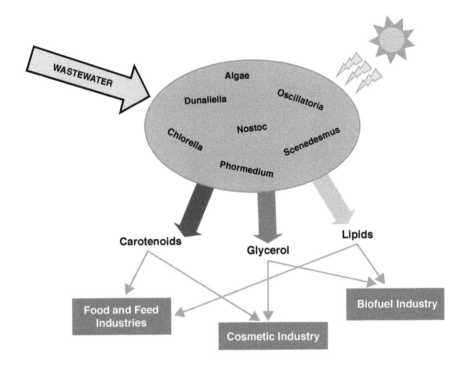

FIGURE 8: Biorefinery from several microalgae using wastewater with lipid, carotenoid and glycerol production (adapted from Mostafa at al., 2012).

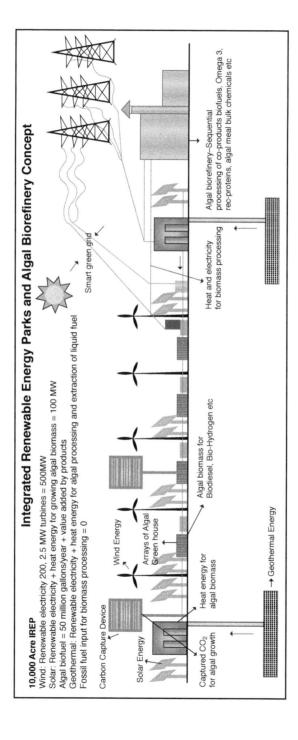

FIGURE 9: Integrated renewable energy and an Algal Biorefinery concept: a framework for the production of biofuel and high-value products with zero net carbon emissions (from Subhadra and Edwards, 2010).

4.3 CONCLUSIONS

Biomass, as a renewable source, is attracting worldwide attention to satisfy the so-called bioeconomy demand. Microalgae could be the appropriate feedstock as they did not compete with food and feed production, in terms of either land or water. Furthermore, microalgae remove/recycle nutrients from wastewater and flue-gases providing additional environmental benefits.

Due to their efficient sunlight utilization, microalgae are projected as living-cell factories with simple growth requirements. Their potential for energy and value-added products production is widely recognized. Nevertheless, to be economically sustainable the tiny microalgae should supply a huge biorefinery. Technical advances combined with the several advantages such as CO_2 capture, wastewater bioremediation and the extraction of value added-products will greatly increase algal bioproduct profitability. The versatility and the huge potential of the tiny microalgae could support a microalgae-biorefinery and microalgae-based bioeconomy, opening up a huge increase of opportunities in the global algae business.

REFERENCES

1. Batista, A.P., P. Moura, P.A.S.S. Marques, J. Ortigueira, L. Alves and L. Gouveia. 2014. Scenedesmus obliquus as a feedstock for bio-hydrogen production by Enterobacter aerogenes and Clostridium butyricum by dark fermentation. Fuel 117:537-543.
2. Brennan L. and P. Owende. 2010. Biofuels from microalgae - A review of technologies for production, processing, and extractions of biofuels and co-products. Ren Sustain Energy Rev 14:557–577.
3. Burja, A.M., B. Banaigs, E.B. Abou-Mansour, P.C. Wright. 2001. Marine cyanobacteria - a prolific source of natural products. Tetrahedron 57:9347-9377.
4. Campenni', L., B.P. Nobre, C.A. Santos, A.C. Oliveira, M.R. Aires-Barros, A.F. Palavra and L. Gouveia. 2013. Carotenoids and lipids production of autotrophic microalga Chlorella protothecoides under nutritional, salinity and luminosity stress conditions. Applied Microbiol Biotechnol 97:1383-1393.
5. Clarens, A.F., H. Nassau, E.P. Resurreccion, M.A. White and L.M. Colosi. 2011. Environmental Impacts of Algae-Derived Biodiesel and Bioelectricity for Transportation Environ Sci Technol 45:7554–7560.
6. Collet P., A. Hélias, L. Lardon, M. Ras, R.A. Goy and J.P. Steyer. 2011. Lifecycle assessment of microalgae culture coupled to biogas production. Bioresour Technol 102:207–214.

7. Dey, K. 2011. Production of Biofertilizer (Anabaena and Nostoc) using CO2. Presentation on Roll: DURJ BT No.2011/2. Regn.No: 660.

8. Ehimen E.A., Z.F. Sun, C.G. Carrington, E.J. Birch and J.J. Eaton-Rye. 2011. Anaerobic digestion of microalgae residues resulting from the biodiesel production process. Applied Energy 88:3454–3463

9. EN 14214. 2008. Automotive fuels—fatty acid methyl esters (FAME) for diesel engines—requirements and test methods.

10. Ferreira, A.F., A.C. Marques, A.P Batista, P.A.S.S. Marques, L. Gouveia and C. Silva. 2012. Biological hydrogen production by Anabaena sp. – yield, energy and CO2 analysis including fermentative biomass recovery. Internat Journal of Hydrogen Energy 37:179-190.

11. Ferreira, A.F., L. Ribeiro, A.P. Batista, P.A.S.S. Marques, B.P. Nobre, A.F. Palavra, P.P. Silva, L. Gouveia and C. Silva. 2013a. A Biorefinery from Nannochloropsis sp. microalga – Energy and CO2 emission and economic analyses. Bioresour Technol 138: 235-244.

12. Ferreira, A.F., J. Ortigueira, L. Alves, L. Gouveia, P. Moura and C. Silva. 2013b. Energy requirement and CO2 emissions of bioH2 production from microalgal biomass. Biomass & Bioenergy 49:249-259.

13. Gerster, H. 1993. Anticarcinogenic Effect of Common Carotenoids. Int Journal Vitamin Nutrition Res 63:93-121.

14. Giordano, P., P. Scicchitano, M. Locorotondo, C. Mandurino, G. Ricci, S. Carbonara, M. Gesualdo, A. Zito, A. Dachille, P. Caputo, R. Riccardi, G. Frasso, G. Lassandro, A.Di Mauro and M.M. Ciccone. 2012. Carotenoids and Cardiovascular Risk Curr Pharm Des 18:5577-5589.

15. Gouveia, L. and C. Oliveira. 2009. Microalgae as a raw material for biofuels production. Journal Industrial Microbiol and Biotechnol 36:269-274.

16. Gouveia, L., C. Neves, D. Sebastião, B.P. Nobre and C.T. Matos. 2014. Effect of light on the production of bioelectricity and pigments by a Photosynthetic Alga Microbial Fuel Cell. Bioresour Technol 154:171-177.

17. Kim, A.L., O.K. Lee, D.H. Seong, G.G. Lee, Y.T. Jung, J.W. Lee and E.Y. Lee. 2013. Chemo-enzymatic saccharification and bioethanol fermentation of lipid-extracted residual biomass of the microalga Dunaliella tertiolecta. Bioresour Technol 132:197–201.

18. Kohlmeier, L. and S.B.Hastings. 1995. Epidemiologic Evidence of a Role of Carotenoids in Cardiovascular-Disease Prevention. American Journal Clinical Nut 62:1370-1376.

19. Lopes da Silva, T., L. Gouveia and A. Reis. 2014. Integrated microbial processes for biofuels and high added value products: the way to improve the cost effectiveness of biofuel production. Applied Microbiol Biotechnol. doi:10.1007/s00253-013-5389-5.

20. Lupulescu, A. 1994. The role of vitamin-A, vitamin-beta-carotene, vitamin-E and vitamin-C in cancer cell biology. Int Journal Vitamin Nutrition Res 64:3–14.

21. Marques, A. E., T.A. Barbosa, J. Jotta, P. Tamagnini and L. Gouveia. 2011. Biohydrogen production by Anabaena sp. PCC 7120 wild-type and mutants under different conditions: Light, Nickel and CO2. Journal Biomass Bioenergy 35:4426-4434.

22. Matos, C.T., L. Gouveia, A.R. Morais, A. Reis and R. Bogel-Lukasik. 2013. Green metrics evaluation of isoprene production by microalgae and bacteria. Green Chemistry 15, 2854–2864

23. Meydani, S.N., D.Y.Wu, M.S.Santos and M.G.Hayek. 1995. Antioxidants and Immune-Response in Aged Persons - Overview of Present Evidence. Am J Clin Nut 62:1462-1476.

24. Miranda J.R., P.C. Passarinho and L. Gouveia 2012. Bioethanol production from Scenedesmus obliquus sugars: the influence of photobioreactors and culture conditions on biomass production. Applied Microbiol Biotechnol 96:555-564.

25. Mostafa, S.S.M., E.A. Shalaby and G.I. Mahmoud. 2012. Cultivating microalgae in domestic wastewater for biodiesel production. Nat Sci Biol 4:56–65.

26. Mussgnug J.H., V. Klassen, A. Schlüter and O. Kruse. 2010. Microalgae as substrates for fermentative biogas production in a combined biorefinery concept. J Biotechnol 150:51–56.

27. Nobre, B., F. Villalobos, B.E. Barragán, A.C. Oliveira, A.P Batista, P.A.S.S Marques, H. Sovotó, A.F. Palavra and L. Gouveia. 2013. A biorefinery from Nannochloropsis sp. microalga – Extraction of oils and pigments. Production of biohydrogen from the leftover biomass. Bioresour Technol Special Issue: Biorefinery 135:128-136.

28. Nuño, K., A.Villarruel-López, A.M.Puebla-Pérez, E.Romero-Velarde, A.G.Puebla-Mora and F.Ascencio. 2013. Effects of the marine microalgae Isochrysis galbana and Nannochloropsis oculata in diabetic rats. Journal Funct Foods 5:106-115.

29. Olguín, E.J. 2012. Dual purpose microalgae–bacteria-based systems that treat wastewater and produce biodiesel and chemical products within a biorefinery. Biotechnol Adv 30:1031–1046.

30. Pacheco, R., A.F. Ferreira, T. Pinto, B.T. Nobre, D. Loureiro, P. Moura, L. Gouveia and C.M. Silva. 2014. Life Cycle Assessment of a Spirogyra sp. biorefinery for the production of pigments, hydrogen and leftovers energy valorization. Applied Energy, under revision.

31. Painter, T.J. 1993. Carbohydrate polymers in desert reclamation. The potential of microalgal biofertilizers. Carbohydrate Polymers 20:77-86.

32. Park, J.S., J.H.Chyun, Y.K.Kim, L.L.Line and B.P.Chew. 2010. Astaxanthin decreased oxidative stress and inflammation and enhanced immune response in humans. Nutr Metab 7:18-28.

33. Park J.B.K., R.J. Craggs, A.N. Shilton. 2011. Wastewater treatment high rate algal ponds for biofuel production. Bioresour Technol 102:35–42.

34. Powel, E.E. and Hill, G.A. 2009. Economic assessment of an integrated bioethanol-biodiesel-microbial fuel cell facility utilizing yeast and photosynthetic algae. Chemical Eng Res Design 87:1340-1348.

35. Sahu, D., I. Priyadarshani and B. Rath. 2012. Cyanobacteria - as potential biofertilizer. CIBTech Journal Microbiol 1:20-26.

36. Shibata, S. and H. Sansawa. 2006. Preventive effects of heterotrophically cultured Chlorella regularis on lifestyle-associated diseases. Annu Rep Yakult Central Inst Microbiol Res 26:63-72.

37. Shibata, S., K. Hayakawa, Y. Egashira and H.Sanada. 2007. Hypocholesterolemic mechanism of Chlorella: Chlorella and its indigestible fraction enhance hepatic cho-

lesterol catabolism through up-regulation of cholesterol 7alpha-hydroxylase in rats. Biosc Biotechnol Biochem 71:916-925.

38. Shibata, S., Y. Natori, T. Nishihara, K. Tomisaka, K. Matsumoto, H. Sansawa and V.C. Nguyen. 2003. Antioxidant and anti-cataract effects of Chlorella on rats with streptozotocin-induced diabetes. Journal Nutr Sci Vitaminol (Tokyo) 49:334-339.

39. Sialve, B, N. Bernet and O. Bernard. 2009. Anaerobic digestion of microalgae as a necessary step to make microalgal biodiesel sustainable. Biotechnol Adv 27:409–416.

40. Singh, S., B.N. Kate, and U.C. Banerjee. 2005. Bioactive compounds from cyanobacteria and microalgae: an overview. Critical Reviews Biotechnol 25:73-95.

41. Snodderly, D.M. 1995. Evidence for Protection Against Age-Related Macular Degeneration by Carotenoids and Antioxidant Vitamins. Am Journal Clin Nut 62:1448-1461.

42. Subhadra, B. and M. Edwards. 2010. Algal biofuel production using integrated renewable energy park approach in United States. Energy Policy 38:4897–4902.

43. Subhadra, B. and Edwards, B 2011. Coproduct market analysis and water footprint of simulated commercial algal biorefineries. Applied Energy 88:3515-3523.

44. Subhadra, B.G. 2010. Sustainability of algal biofuel production using integrated renewable energy park (IREP) and algal biorefinery approach. Energy Policy 38:5892–901.

45. Tanaka, T., M. Shnimizu and H. Moriwaki. 2012. Cancer Chemoprevention by Carotenoids. Molecules 17:3202-3242.

46. Vanthoor-Koopmans, M., R.H. Wijffels, M.J. Barbosa and M.H.M. Eppink. 2013. Biorefinery of microalgae for food and fuel. Bioresour Technol 135:142-149

47. Weikel, K.A., C.J.Chiu and A.Taylor. 2012. Nutritional modulation of age-related macular degeneration. Mol. Aspects Med. 33:318-375.

48. Wijffels, R.H.,M.J. Barbosa and M.H.M. Eppink. 2010. Review: Microalgae for bulk chemicals and biofuels. Biofuels Bioprod Bioref 4:287–295.

49. Willett, W.C. 1994. Micronutrients and Cancer Risk. Am Journal Clin Nutr 59:1162-1165.

50. Yamaguchi, K. 1996. Recent advances in microalgal bioscience in Japan, with special reference to utilization of biomass and metabolites: A review. Journal Appl Phycol 8:487-502.

CHAPTER 5

Catalysis for Biomass and CO_2 Use Through Solar Energy: Opening New Scenarios for a Sustainable and Low-Carbon Chemical Production

PAOLA LANZAFAME, GABRIELE CENTI, AND SIGLINDA PERATHONER

5.1 INTRODUCTION

Many indications, from socio-economic to technological, point out that the chemical industry is moving to a new development cycle. This is characterized by global structural changes in the economy with a crucial reorganisation of the energy and resource infrastructure, in which the switch to renewable energies and sustainable issues is largely influencing the market and industrial objectives. [1,2a] The efficiency and use of renewable resources and energy, particularly in geographical regions poor of natural resources such as Europe, is becoming a driving element for competiveness and strategies in the chemical and process industry. The use of alternative "green/sustainable and renewable" resources such as biomass and

Catalysis for Biomass and CO₂ Use Through Solar Energy: Opening New Scenarios for a Sustainable and Low-Carbon Chemical Production. Lanzafame P, Centi G, and Perathoner S. Chemical Society Reviews **43** *(2014). DOI: 10.1039/C3CS60396B. Reproduced with permission from The Royal Society of Chemistry and the authors.*

solar energy for the development of new solutions to produce raw materials and energy vectors is thus becoming an industrial priority.

Catalysis is a key enabling factor to allow the development of new sustainable processes and technologies and thus plays a critical role to realize this transition. [2] Consequently, the development of catalysts and related catalytic technologies for the use of (i) biomass to produce chemicals and fuels, and (ii) renewable energy in chemical production is becoming a key area of the research. For the latter, we should clarify that the direct use of solar energy in photochemical organic syntheses is an area of minor interest from an industrial perspective (although potentially attractive) for the (still) low productivity/selectivity, while there is a growing interest in the use of renewable (solar) energy for producing H_2 and converting CO_2, either directly (photochemically) or indirectly. The latter refers to the use of the electrical energy generated by solar energy (photovoltaic, concentrated solar power – CSP, etc.) or of other related renewable energy (RE) sources (such as wind and hydro) to produce H_2 (used then to convert catalytically CO_2) or directly to convert (electrocatalytically) carbon dioxide. We refer here to this direct and indirect use of renewable (solar) energy for converting CO_2 when discussing this topic.

Many recent reviews have discussed these topics recently. A limited selection is represented by the reviews of Dumesic et al., [3] Kobayashi and Fukuoka, [4] Corma et al., [5] Miertus et al., [6] Stocker, [7] Rinaldi and Schüth, [8] and Gallezot [9] on different aspects of the catalytic chemistry in biomass conversion to chemicals and fuels, and the reviews of DuBois et al., [10] Centi et al., [11] Quadrelli et al., [12] Aresta and Dibenedetto, [13] Sakakura et al., [14] Wang et al., [15] and Leitner et al. [16] on different aspects of the catalytic chemistry of CO_2 utilization. Several books have also been published on these topics, amongst which may be cited those of Triantafyllidis et al. [17] on the role of catalysis for the sustainable conversion of biomass and of Aresta [18] on the use of CO_2 as a chemical feedstock.

The discussion on CO_2 catalytic chemistry, however, has been mainly centred on the use of CO_2, with limited aspects regarding the issue of incorporation of RE into the cycle of CO_2 conversion, particularly from an industrial perspective. [11b] Being carbon dioxide low on the thermodynamic energy scale, a sustainable use of CO_2 requires the energy to proceed uphill to the product of reactions (typically at a higher energy level than that of carbon dioxide) is provided by RE sources, directly (as photons, electrons) or indirectly, via high-energy molecules such as H_2 produced with the use of RE.

This concept is presented in Fig. 1. CO_2 could be converted to inorganic carbonate in an exo-energetic path or to organic carbonate (as example for the various routes to form CO_2-containing polymers or organic chemicals) by reaction with a high-energy molecule (ethene oxide in the example of Fig. 1). However, these routes do not incorporate RE in the process of carbon dioxide conversion. On the contrary, both direct routes of conversion of CO_2 using electrons produced using RE sources (electrocatalysis) or using photons (photo-catalysis), and indirect routes in which renewable H_2 (produced using RE sources) is first produced, and then this hydrogen is utilized in the catalytic processes of CO_2 conversion (to form CH_3OH, CH_4 or CO, for example) lead to incorporation of RE into the final product. [11b]

The direct route of conversion of CO_2 using photons (photo-catalysis) is apparently the more challenging and interesting. Various specific reviews have been published on the photoconversion of CO_2 on semiconductor materials, for example by Mao et al., [19] Habisreutinger et al., [20] Garcia et al. [21] and Fan et al. [22] In this case, the mechanism of light capture and energy transfer to a CO_2 molecule has been discussed, but the productivity is still too far from those necessary for exploitation (needing a three order of magnitude intensification). Industrialization still appears unrealistic, even taking into account the possible developments in the semiconductor materials. While typical aspects are discussed regarding the need of cocatalysts, the interfacial contact between semiconductor and other materials as well as the role of heterojunctions in promoting charge separation, and the design strategy in semiconductors to improve effectiveness in using the visible light portion of the solar spectrum, there are many aspects regarding the interaction of CO_2 and of the products of conversion with surface excited states which have been scarcely investigated. [23] The analysis of these aspects suggest the presence of intrinsic barriers to increase productivity in CO_2 direct photoconversion to the levels necessary for exploitation.

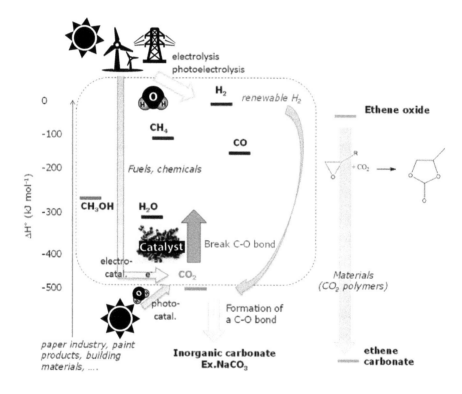

FIGURE 1: Energy scale in some of the products of CO_2 conversion, illustrating the different routes by which RE could be introduced in the carbon dioxide molecule to produce fuels or chemicals.

Also in the case of biomass conversion, the issue of the energy use in the production of the final products is often not analysed to evaluate the different possible paths, although in this case the analysis is more difficult and should consider the full cycle of production on a LCA (life-cycle assessment) basis.

5.2 SCENARIO ANALYSIS FOR A SUSTAINABLE AND LOW-CARBON CHEMICAL PRODUCTION

Current chemical production, limiting discussion here to petrochemistry, is mainly oil-centric (>90%) and based on the use of a few building blocks (light olefins, aromatics, few alkanes such as n-butane, and syngas). Other fossil fuel feedstocks still have limited use. Methane is used essentially to produce H$_2$ and syngas, the latter mainly used to produce methanol – about 65 Mt per year—and related value chain. Other alkanes present in natural gas (NG), such as ethane and propane, are used for the production of light olefins by steam cracking or direct dehydrogenation. Coal still finds limited usage for chemical production, even though its use to produce syngas converted then to methanol and finally light olefins is considerably promoted in China.

Shale gas is expected to change this panorama, due to the increased availability of low cost NG promoting an increased use also in chemical production. Several chemical companies are looking to the new opportunities created by shale gas to expand production capacity for ethylene, ethylene derivatives (i.e., polyethylene, polyvinyl chloride, etc.), ammonia, methanol, propylene, and chlorine. However, this situation cannot be generalized, particularly for geographical areas such as Europe where shale gas production will be limited due to the large environmental concerns.

Consequently, we may forecast that world chemical production will progressively move from an oil-centric common vision to different regional-based systems. US will promote chemical production centred on the use of shale-gas, China that based on coal utilization and the Middle East that based on oil use, due to the low local costs. For Europe, to remain competitive in this global competition, it needs to foster the use of alternative raw materials. The possibility for European chemical production, to remain competitive in this evolving scenario of raw materials, is to foster the use of its own resources, biomass (particularly, waste) and CO_2. The central role of these two raw materials in redefining the future scenario of chemical industry is recognized from important actions of the European chemical industry, such as the public–private partnership SPIRE (Sustainable Process Industry through Resource and Energy Efficiency).

We limit the discussion here, for conciseness, to the development of a sustainable and low-carbon chemical production, e.g. based on the use of biomass and CO_2, the latter through the use of renewable (solar) energy. As briefly mentioned above, this is a priority especially in Europe, but different geographical areas may have different priorities, particularly the use of shale gas in the US and coal in China. Although the chemical industry (and trade) is currently highly globalized, there is an evolution towards a deglobalization with a tendency to use and value local resources. The model of globalization of chemical production has shown its limits, and various economists have evidenced the role of the globalization approach in the current global crisis, often contrasted with governmental stimulus programs meant to rev up national markets. In the deglobalization approach, 24 production for the domestic market must again become the center of gravity of the economy rather than production for export markets. With respect to similar trends occurring at the beginning of the last century in response to the Great Depression, this time the "domestic" market is seen as an area market (Europe, for example, instead that limited to single European countries), and a larger social pressure exists (in terms of equitable income redistribution, deemphasizing growth to empower upgrading the quality of life and environmental preservation).

This change in approach also influences the priorities for chemical production moving beyond the key criterion of reduction of unit cost, to consider the integration in the social and environment value chain. Industrial and trade policies will be the driving element for this change, which reflect also in a change in future raw materials for chemical production. This is schematically shown in Fig. 2.

Moving to the use of alternative fossil fuel raw materials will be driven from economic (shale gas, in US) and in part geo-political (coal, in China) motivations. Although the increasing production trend of shale (or "tight") gas in the US has generated a wave of optimism, the actual EIA (US Energy Information Administration) data show that the total US gas production has not been growing for the past 1–2 years and that signs of decline are instead present. Drilling rigs for gas have been plummeting over the last two years. These data seem to confirm the interpretation of a financial "gas bubble", rather than a robust trend of development of new resources. In the near future, the decline in gas production in the US may lead to an increase in prices. In addition to these uncertainness about future shale gas costs, the environmental burdens (methane fugitive emissions; volumes of water and the chemicals used in fracking and their subsequent disposal; risk related to contaminating groundwater due to shale gas extraction; competing land-use requirements in densely populated areas; the physical effects of fracking in the form of increased seismic activity, etc.) have to be accounted for. For these reasons, shale gas use for chemical production is reasonably unrealistic in various areas of the world. As briefly outlined in Fig. 2, the contribution of shale gas to the future scenario is confined to (i) additional methane available for syngas/H$_2$ production or energetic uses associated to chemical industry, and (ii) additional light alkanes available for light olefins production (particularly, by steam reforming and to a lesser extent by dehydrogenation).

A different situation is present on the use of coal for chemical production in China, with motivations often being mixed with geopolitical reasons, and economics difficult to translate to different countries. Coal is essentially usable for chemical production only via the methanol/olefin route, except for some possible contribution to the energy needs of chemical industry.

It is thus clear that the relative weights of the three areas indicated in Fig. 2 (bottom) to the future scenario of chemical production will be different in different geographical areas and not easy to predict. We could note, however, that the use of new low-carbon raw materials (biomass, particularly biowaste, and CO$_2$/renewable energy) shows a better impact in terms of a (i) sustainable and low carbon economy, (ii) balanced use of local resources (integrated biorefineries) and synergy with other economic activities (agriculture and forestry, especially) and (iii) integration within the chemical production value chain (production of intermediates and high-added-value products, in addition to base raw materials).

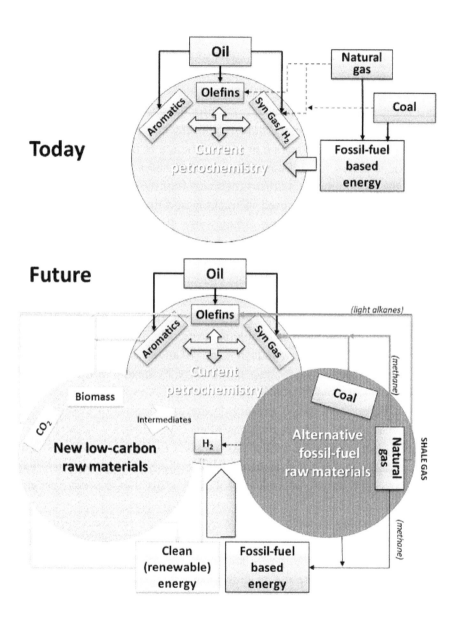

FIGURE 2: Evolution in the scenario of raw materials from current petrochemistry to the future scenario for a sustainable and low carbon chemical production.

The availability of these raw materials (biomass, CO_2) is clearly an important element for the evaluation of the effective impact of these routes. In 2010, the European chemicals industry, including pharmaceuticals, used a total of 54 million tonnes of oil equivalent (MTOE) of fuel and power consumption. The total biomass potential for EU27 (calibrated studies, e.g. compensated for geographical differences and biomass categories, for 2020–2030 decade) is between 4–20 EJ per year (96–478 MTOE), of which between 20 and 50% (depending on estimations) are derived from agricultural residues and organic biowaste, the remaining from energy crops, forestry and forestry residues. Considering 30% as the target for biomass substitution of fossil fuels in EU27 chemical production in year 2030, and 50% as the average yield to chemicals in converting biomass, the amount of available agricultural residues and organic biowaste is in excess to that necessary to cover the needs.

According to the European Environmental Agency (EEA), the European chemicals industry, including pharmaceuticals, emitted in 2010 a total of 166 million tonnes of CO_2 equivalent (down from a total of 330 million tonnes in 1990). Of these emissions, over 20% are easily recoverable. Many large-volume sources of rather pure CO_2 in refinery and chemical processes (ammonia production, ethylene oxide production, gas processing, H_2 production, liquefied natural gas, Fischer–Tropsch – synthesis) as well as from biorefineries (ethanol production) exist. The amount of CO_2 which could be easily recovered from chemical production alone could potentially account for about half of the light olefin production in EU27, while a reasonable target for year 2030 is about 10–15%.

Therefore, this brief analysis evidences how the availability of biowaste/agricultural residues and CO_2 will not be the limiting factor to develop a sustainable low-carbon chemical production in the next two decades. This statement is valid also for other geographical areas, such as Asia, US etc. where the degree of penetration of biowaste and CO_2 as alternative raw materials will be even lower than in Europe, as discussed before.

5.2.1 SELECTING THE FUTURE PATHS BASED ON THE USE OF BIOMASS AND SOLAR ENERGY

Sustainability is a major driver for the future scenario of the chemical industry, [1] but many other factors (internal to chemical industry, such as process economics, availability of raw materials, company strategies and synergies, etc., and external, for example, limitation on greenhouse gas emissions, environmental regulations, company visibility, etc.) determine the possibility that a potential route (technically feasible) becomes effectively a major industrial route.

This review, different from the others cited before as well as the many others present in the literature, will first analyse the future scenario of the chemical industry, in order to select the main routes having the possibility to become relevant production routes alternative to those actually in production (based mainly on the use of fossil fuels). The criteria for this selection are the following: [2a]

Economic drivers (ED). An important initial element for evaluation is whether the economic bases for the switch to new raw materials exist. We refer here only to considerations about estimations of raw material cost and product value (projection to future values), as well incidence of fixed costs based on the process complexity (in particular, regarding purification of both raw materials and products of reaction). It is thus limited to estimate when an economic potential exists to industrialize the process. The cost of production depends on many factors, which have also a large variability from country to country and company to company as well. In addition, estimation of process economics requires to have established in detail the process flowsheet, but this contrasts with a scenario analysis on processes often at an early stage of development. Therefore, evaluation should be limited to considering whether an economic potential to industrialize the process, and clear economic drivers to push the technological development of the process, exist.

There are also other important components in the evaluation of the economic drivers. Between these, the investment cost necessary to develop the process and the integration of the production within the existing infrastructure (drop-in products). In a rather uncertain future scenario of energy and raw material costs, processes requiring large investments, rather long

(>10 years) amortization times, and which not well integrate within the existing production infrastructure and value chain have lower possibilities to become major production routes.

Technological and strategic drivers (TSD). The scenario for the chemical production in the last half century was capital intensive, reflecting the large manufacturing facilities required to produce bulk quantities. This has given the industry a competitive advantage in terms of high barriers to entry, but the fast development in areas with low production cost such as Asia and globalization has broken down this model. Knowledge-based and high technology production is a major current driver for competiveness.

The development of new production routes based on alternative raw materials, particularly when using low-cost raw materials such as waste biomass and CO_2, offers clear incentives from this perspective, particularly for industrial newcomers that need to establish their business area. Time to market for these new players, however, is an important component of the possibility to success. The traditional approach based on lab-bench-pilot-demo sequence is time consuming. New models of production, based, for example, on parallel modules [2a] could reduce largely this time, bypass the scale-factor approach determining the industrial choices in the last half century, and could allow operating at full capacity of utilization even in the presence of a fluctuating market. An example is given from the results of the F3 (Fast, Future, Flexible)-Factory EU project, where major EU companies are collaborating to develop this new production model.

In terms of drivers for establishing new production routes, the possibility of a fast time to market, which is related to both the possibility to exploit new production approaches and to the presence of technologies not requiring costly development are relevant elements for the evaluation. Other elements such as flexibility of operations and capacity utilization rates are also important, as well as the innovation character of the process creating knowledge-based barriers to competitiveness rather than on other aspects.

Process complexity index is an aspect in part related to those discussed above, and which is another relevant aspect to consider in evaluating the possible scenario for the future chemical production. A reduced number of process and separation steps, simple separation units, high productivity and reduced number of byproducts, limited need of special materials and safety measures/devices are important elements determining the possibility of success for new production routes.

Between the strategic drivers, it must also be considered the trend towards a de-globalization of the chemical production. There is a need to realize stronger synergies with downstream industries and user, as well as symbiosis with other productions on a regional basis. This is the clear trend observed in biorefineries. [25] The need to use local biomass resources, of their full utilization to produce also high-added value products (integrated chemical and energy production), the necessary strong link with the territory are all elements driving towards new models of biorefineries (with respect to the traditional ones), with clear relevant impact also on the future routes of chemical production.

Environmental and sustainability drivers (ESD). The efficiency in using energy, resources and in limiting the impact on the environment are elements of increasing relevance for the chemicals industry and to determine the successful rate of new chemical productions. The efficiency includes the possibility of symbiosis with other productions.

However, a careful analysis has to be made, based on LCA or related methodologies. The use of biomass as raw material, for example, does not imply that a process shows a better sustainability than the analogous based on the use of fossil fuels. Often only using waste biomass the process shows a lower impact on the environment. On the other hand, it has to remark that the application of LCA methodology to a chemical process shows still several limits, both in terms of a lack of reliable data for the analysis, and in terms of categories of analysis that do not well adapt to evaluate the chemical production.

It is also to important to remark that still many of the actual chemical production routes (based on fossil fuels) suffer from significant drawbacks. [26] If we consider a combined parameter of efficiency reflecting both the feed efficiency (C lost per C built into the product) and the energy efficiency (C lost per C built into the product, after converting the consumption of fuel, power and steam into methane equivalent), there are various processes showing a high value of this parameter (higher than about 1), for example: (i) cyclohexane to caprolactam, (ii) adipic acid synthesis from benzene, (iii) methane to HCN and to NH_3, (iv) dinitrotoluene to toluene diisocyanate, (v) methanol to dimethylcarbonate, etc. There are also still many processes with a high inorganic waste production, for example (a) toluene to dinitrotoluene, (b) acetone to methyl-methacrylate, (c) cyclo-hexane to caprolactam, and (d) propylene to epichloridrin. Integration of the manufacturing line up to the final polymer (from raw materials) is also important.

Most efficient processes are those for polyolefins and polystyrene. Energy-consuming routes are the production of PVC (polyvinyl chloride) and Nylons, and large amounts of wastes are produced in Nylon, PMMA (polymethyl-methacrylate), polyurethanes, epoxy resins, and polycarbonates processes. The sustainability of these routes does not imply necessarily the use of alternative raw materials to fossil fuels, but can be a possibility to reduce the environmental impact.

Socio-political drivers (SPD). There is low public esteem for the chemicals industry. Public opinion must not be underestimated—it can be a powerful 'driver' of the business environment. In addition, public opinion can act as a strong catalyst for regulatory initiatives, which do not necessarily create a favourable business environment. Therefore, socio-political visibility of the different routes is an element for their possible success.

Establishing the possible routes for the future scenario of the chemical industry will thus require a multifactor evaluation. Several of the discussed aspects cannot be fully quantified, at least in a scenario analysis where several of the elements necessary for the evaluation are missing. Scenario analyses thus contain some arbitrary elements that may be questioned. Nevertheless, it is a useful exercise, particularly in a tutorial review, to discriminate between the many possible routes in using biomass and solar energy (the latter, through CO_2 conversion) which can confuse researchers approaching this field. The aim of this introductory scenario analysis is thus to pose the discussion of the status of the development of the catalysts in these different selected routes on more solid bases.

5.2.2 DEFINING THE NEW SCENARIO FOR A SUSTAINABLE AND LOW CARBON CHEMICAL PRODUCTION

In a recent paper dedicated to the challenge of introducing green energy in the chemical production chain [1] we have anticipated a possible new scenario for a sustainable chemical production based on the reuse of CO_2 and of biomass to produce both raw materials (building blocks) and specific intermediates, even though discussion was limited to a few aspects of CO_2 reuse. Fig. 3 illustrates schematically the concept of how CO_2 as a carbon-source in integration with platform molecules derived from biomass, and lignin to produce aromatics, make it potentially possible to avoid the use of fossil fuels for chemical production (limited to organic products).

In a more realistic target, about 30% of the use of fossil fuels (in year 2030) could be substituted by use of biomass, CO_2 and renewable energy (for the part concerning the use of fossil fuels as energy vector). This is the target of the cited SPIRE initiative promoted by the European Chemical Industry Council (Cefic) and thus shared by many chemical companies at least for the European scenario (which covers about one third of the chemical production worldwide). This percentage of 30% is expected to increase further in subsequent years, an objective which can be reached even with a faster rate. History teaches us that about six decades ago, the introduction of olefins as alternative raw materials to acetylene and other chemicals led in about two decades to a complete change in the chemical production. There are many aspects indicating the existence of a similar situation that may lead to a fast transition to the new scenario for chemical production. It is thus important and relevant to prepare this transition and highlight the routes having the higher possibilities to become major future routes of chemical production.

Fig. 3 does not include the use of alternative fossil fuels (coal and shale gas, see Fig. 2), because it is focused on new low-carbon raw materials. At the beginning of this Section 2 the possible role of these alternative raw materials in a future scenario has been discussed, but from a sustainability and low-carbon economy perspective both these raw materials show significant drawbacks. This is especially true for coal, due to the high impact on the environment associated with the production and use of coal.

Another note regards the fact that only some main routes can be discussed here to focus the discussion. The aim is to provide some relevant examples, rather than a systematic and complete analysis of all the possibilities and new routes, not compatible with the objectives of a tutorial review.

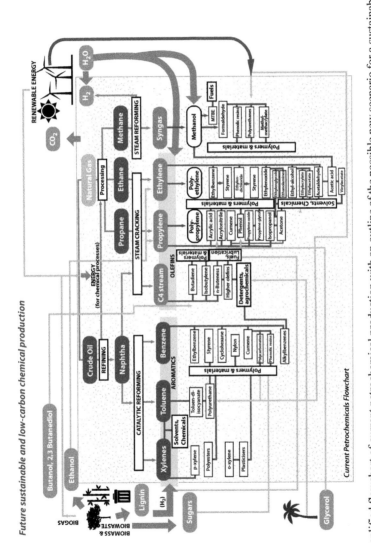

FIGURE 3: Simplified flowchart of current petrochemical production with an outline of the possible new scenario for a sustainable chemical production based on the reuse of CO_2 and of biomass to produce both raw materials (building blocks) and specific intermediates. Modified from ref. 1, © 2014.

5.2.2.1 NOVEL SUSTAINABLE LIGHT
OLEFIN PRODUCTION ROUTES

Light olefins (ethylene and propylene) are produced worldwide in an amount of about 200 Mtons per year. Their synthesis is the single most energy-consuming process in the chemical industry. The largest part of ethylene and propylene is used to produce polymers by direct routes (polyethylene and polypropylene; polypropylene production, for example, accounts for more than 60% of the total world propylene consumption) or by indirect routes. For example, the main products of propylene conversion (acrylonitrile, propylene-oxide, acrylic acid and cumene) are also mainly used to produce polymers (see Fig. 3).

Currently, light olefins are produced principally by steam cracking and this process accounts for about 3×10^{18} J of primary energy use, not counting the energy content of the products. The pyrolysis section of a naphtha steam cracker alone consumes approximately 65% of the total process energy and accounts for approximately 75% of the total energy loss. The specific emission factor (CO_2 Mt/Mt light olefin) depends on the starting feedstock, but ranges between 1.2 and 1.8. About 300 Mtons per year of CO_2 derive from the production of these building blocks of the chemical production chain.

Light olefins can be produced from different sources (crude oil, natural gas, coal, biomass and bio-waste such as recycled plastics, and CO_2), [27,28] as summarized in Fig. 4. The main current processes (indicated with bold black arrows in Fig. 4) are the steam cracking of oil and NG fractions, with minor production by direct dehydrogenation of alkanes. Olefins are also a side product of the fluid catalytic cracking (FCC) process in the refinery, but are usually utilized inside the refinery for alkylation or oligomerization processes.

New process routes, already at an industrial level, include the dehydration of ethanol produced from biomass fermentation and the production via syngas (through the intermediate synthesis of methanol), with the syngas deriving from coal combustion or biomass pyrolysis/gasification. Methanol can be converted to olefins (MTO – methanol to olefins) or even selectively to propylene (MTP – methanol to propylene). These processes are indicated briefly in Fig. 4 with the acronym DH (dehydration process-es) which comprises different types of processes, e.g. methanol to olefins, methanol to propylene and ethanol dehydration.

New routes under development are based on the direct Fischer–Tropsch (FT) reaction, e.g. direct use of syngas or even of CO_2–H_2 mixtures to selectively synthesize light olefins. To make sustainable the process, H_2 should be produced using RE sources (for example, by electrolysis using the electrical energy produced by hydropower or solar energy).

As indicated in Fig. 4, we have selected seven possible alternative routes to form olefins starting from biomass, biowaste and CO_2 (arrows, indicated by a number such as 1). Scenario analysis refers to the European case. The first two routes are thermochemical, and consider the formation of gas (by gasification, or pyrolysis followed by gasification) and then conversion of syngas (for conciseness, we have omitted in Fig. 4 the need of a purification step, but this is one of the critical elements) either directly to olefins by FTO (Fischer–Tropsch to olefin) process (although catalysts for this reaction have to be further improved) or to methanol followed by MTO (methanol to olefin) or MTP (methanol to propylene) processes. The methanol synthesis is well established commercially, while MTO–MTP are industrial processes, even if there are still some critical aspects regard-ing productivity/deactivation.

The third route starts instead from CO_2 and renewable H_2, e.g. pro-duced using RE and PEM electrolyzers (current preferable technology; high-temperature electrolysis has some advantages, but needs to be further developed). CO_2 may be directly hydrogenated to methanol, or instead CO_2 may be first converted to CO via reverse water gas shift reaction (RWGS). In the second case, the process is slightly more complex, but productivity is higher. Methanol can then be converted as above. This route depends considerably on the cost of production of renewable H_2, which in turn de-pends on the cost of electrical energy. We consider here the use of unused electrical energy sources (electrical energy which cannot be introduced to the grid, for example produced by wind during night, or in remote areas not connected to the grid). For CO_2, we consider the recovery from con-centred streams (for example, in some chemical and refinery processes or in biorefineries from fermentation) and that the reuse of CO_2 allows for a net introduction of RE in the process. [11b]

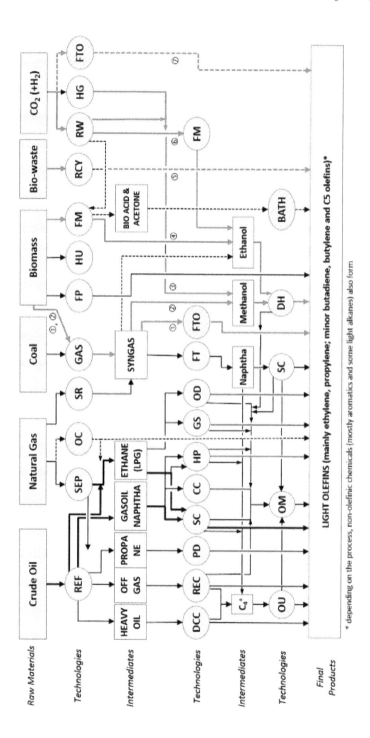

FIGURE 4: Raw material and technology map in the production of light olefin. The paths in bold (black) are those mainly used today, while those in red indicate possible future routes starting from biomass, biowaste and CO_2, the latter with H_2 generated using renewable energy sources. Paths with a dashed line and technologies indicated with a background formed by parallel lines indicate paths/technologies still under development. The numbers in the red lines indicate the paths starting from biomass, biowaste and CO_2 that are discussed in the text. Re-elaboration from Fig. 3 of ref. 27. Acronyms: BATH: Bio-acid acetone to hydrocarbons (e.g. olefins); CC: Catalytic Cracking or Catalytic Pyrolysis; DCC: Deep Catalytic Cracking, etc.; DH: De-hydration process (e.g. methanol to olefins, methanol to propylene and ethanol dehydration); FM: Fermentation; FP: Flash pyrolysis; FT: Fischer–Tropsch synthesis; FTO: FT to olefin; GAS: Gasification and liquefaction; GS: Gas stream reactor technologies, e.g. shockwave reactors; HG: Hydrogenation; HP: Hydro-Pyrolysis; HU: Hydro-Thermal Upgrading Liquefaction which produces naphtha from biomass feedstock; OC: Oxidative coupling of methane; OD: Oxidative Dehydrogenation of ethane; OM: Olefin Metathesis, e.g. ABB-Lummus Olefin Conversion Technology, IFP-CPC meta-4; OU: Olefins Upgrading (conversion of C4–C10) to light olefins, e.g. Superflex, Propylur and Olefins Cracking; PD: Propane dehydrogenation; RCY: Re-cycling pyrolysis using organic waste such as discarded plastics, used rubber, etc.; REC: Recovery of refinery off gases, which contains ethylene, propylene, propane, etc.; REF: Refinery processes (distillation, catalytic cracking, cryogenic separation and absorption produces ethane and LPG, etc.); RW: reverse water gas shift; SC: Steam cracking (conventional); SEP: Gas separation process which produces methane, ethane and propane; SR: Steam Reforming of natural gas.

The fourth path is based on the production of ethanol by fermentation, followed by dehydration to form ethylene. Production of other olefins would require to convert further ethylene, for example by olefin metathesis. This route depends greatly on the cost of production of ethanol. In Brazil, where cheap ethanol is available by sugar fermentation, already a couple of industrial plants (by Braskem, Dow and Solvay Indupa) produce ethylene from bioethanol, but the cost of ethanol production in Europe is greater. However, ethanol could be easily transported by ship, for example. Van Haveren et al. [29] indicated that bio-based ethylene production will significantly increase in the short and medium term, first in Brazil and then extending to regions such as USA and Europe. They suggested that this route would lead also to the production of bio-based vinyl chloride, with thus the two most dominant thermoplastic materials (polyethylene and polyvinylchloride) produced to a significant extent from biomass.

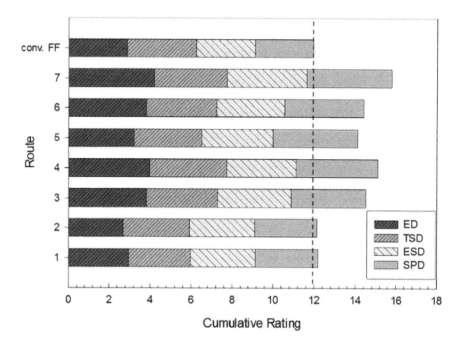

FIGURE 5: Cumulative rating for the four drivers analysed to discuss the role of different routes (alternative to the use of fossil fuels) to produce olefins (from biomass, biowaste and CO_2) in relation to the future scenario for sustainable chemical production.

The fifth route considers the re-cycling pyrolysis using organic waste such as discarded plastics, used rubber, etc. The potential advantage is to avoid the costs of disposal of these wastes, but at the same time the use of waste does not guarantee a constant feed composition. There are also problems in terms of process and separation costs, as well purification. The sixth route is based on a first step of RWGS from CO–ren.H_2 mixtures, followed by a fermentation of the CO–H_2–CO_2 mixture to produce ethanol. LanzaTech has already developed some semi-commercial units for the second step of ethanol production, although productivity is still limited and ethanol has to be recovered from solution. Ethanol could then be dehydrated to ethylene as mentioned before. Renewable H_2 could derive also from processes using micro-organisms, such as cyanobacteria, able to produce H_2 using sunlight.

Finally, the last process considered is the direct conversion of CO_2–ren. H_2 mixtures by FT to the olefin process. This case assumes (as in route 1) that a yield of C_2–C_4 olefins >75% could be obtained. Although current data are still lower, this yield seems a reasonable target that can be reached. [28] These seven routes are compared with steam cracking of naphtha, the current most used process in Europe to produce light olefins. This route is considered as a reference for the olefin production by conversion of fossil fuel (conv. FF).

In order to evaluate the different routes, each of the drivers discussed before was sub-divided in various specific factors that were considered separately. These specific aspects are the following:

- ED: (i) cost of raw material versus product value; (ii) process complexity; (iii) investment necessary; (iv) integration with other processes.
- TSD: (i) technological barriers to develop the process; (ii) time to market; (iii) flexibility of the process; (iv) requirements for reaction and separation steps; (v) synergy with other process units.
- ESD: (i) energy efficiency; (ii) resource efficiency; (iii) environmental impact; (iv) GHG impact.
- SPD: (i) social acceptance; (ii) political drivers; (iii) public visibility.

A value between 1 (low) and 5 (high) was given to each of these specific factors, estimating then the value of ED, TSD, ESD and SPD by averaging the value of the relative specific aspects considered. The cumulative rating was the sum of the estimated value of each of these drivers (ED, TSD, etc.).

The results are reported in Fig. 5 that evidences how incentives (e.g. a higher cumulative rating) exist to use alternative raw materials to produce light olefins. It may be discussed the specific values assigned to each of the discussed drivers for the various routes analysed and in turn how this affects the relative ranking. However, this is not relevant. The result indicated from this discussion about the future scenario for the sustainable chemical production is that a multi-parameter analysis (rather than single aspects as typically used) indicates that there are motivations to develop novel sustainable olefin production routes.

5.2.2.2 SUSTAINABLE PRODUCTION OF OTHER OLEFINS AND POLYOLEFINS

Other olefins could be produced by alternative routes (to those based on fossil fuels as raw materials), but scenario analysis is less favourable that in the case discussed above. For example, various microorganism strains could produce propane and propylene from glucose media (together with butane, butene, pentene, etc.), but productivity is low and separation costs high. Propylene could be produced from bio-chemical production of 1-propanol or 2-propanol via fermentation, but again actual productivity is still too low. Probably, the preferable route involves the production of 1,2-propanediol, converted to 2-propanol and finally dehydrated to propylene. The alternative is the ABE fermentation process leading to ethanol, butanol and acetone, the latter reduced to propylene. Also these routes, however, do not appear to be competitive. In fact, it would be preferable to use directly the glycols (1,2-propanediol, for example) obtained by catalytic conversion of sugars or other platform molecules (glycerol, lactic acid) as a substitute of those derived from fossil fuels. The latter are obtained by catalysed ring-opening of propylene oxide, for example. The glycol is then used to produce humectant, antifreeze or brake fluids and as a component of polyesters and alkyd resins. Polyol Chemical Inc., for example, is producing propylene and ethylene glycols (together with other products such as glycerine and butanediol) starting from sorbitol/glucose. The process economics, however, are positive only using waste sugar streams.

Butadiene is another interesting olefin that can be produced along different sustainable alternative routes (Fig. 6) with respect to the current process based on either the recovery from naphtha steam cracking fractions, or by dehydrogenation of butane. Butadiene could be produced by dehydrogenation of ethanol to acetaldehyde, followed by aldol condensation and dehydration. A one-pot process over MgO–SiO$_2$ catalysts (Lebedew process) is possible with an overall yield of 70% or more. Hüls has been using this process over three decades, but butadiene could only be economic using cheap bioethanol. Butadiene could be used to produce various rubbers, or converted (via epoxidation) to tetrahydrofuran and 1,4-butanediol.

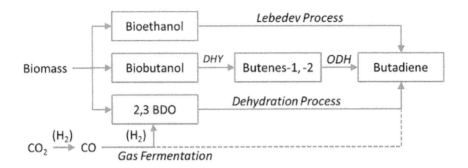

FIGURE 6: Sustainable alternative routes to produce bio-butadiene. DHY: dehydration. ODH: oxidative dehydrogenation. 2,3-BDO: 2,3-butanediol.

Alternatively, 2,3-butanediol (2,3-BDO) produced by fermentation of sugars could be dehydrated to butadiene. LanzaTech has developed on a pilot-scale a fermentation process to produce 2,3-BDO (using $CO-H_2$ as feed). In a joint venture with other companies (Invista, a global nylon producer interested in converting butadiene to adiponitrile, an intermediate in the manufacture of Nylon 6,6) LanzaTech is also developing a process to convert 2,3-BDO to butadiene via fermentation (leading potentially to the single-step production of butadiene via gas fermentation). Versalis in a joint venture with Genomatica is also developing the production of butadiene via 2,3-BDO obtained by fermentation. There is a high potential for these routes (by sugar or gas fermentation) to become commercially attractive, but still productivity and the cost of separation are critical elements.

Other olefins interesting for rubbers are also developed using biomass. Genecor in a joint venture with Goodyear is developing a bio-isoprene production process, and Glycos Biotechnologies plan also to commercialize the production of isoprene from crude glycerine. Global Bioenergies and Gevo/Lanxess are developing processes to produce isobutene from glucose or from isobutanol, respectively. The latter is produced by fermentation, as demonstrated on an already relatively large scale by Gevo and claimed to be competitive to the fossil-fuel based route to C_4 olefins.

However, butenes and butadiene can be produced from n-butane dehydrogenation or oxidative dehydrogenation. n-Butane cost, similar to other light alkanes in natural gas, has been decoupled from the oil price, due to the abundance of the market related to shale gas and the discovery of new gas fields. n-Butane can also be easily transported being a liquid under mild pressure. This fact, associated with the lower productivity in biobutanol or 2,3-BDO production by fermentation and lower selectivity as well (thus higher separation costs) with respect to bioethanol production by fermentation, make the bio-routes to C_4 olefins less attractive in comparison to the alternative path from n-butane. Notwithstanding the presence of various companies interested in developing bio-based intermediates for rubber manufacture, as discussed above, this scenario analysis thus indicates a less favourable outlook for producing biobutadiene with respect to the case of bioethylene. However, the use of cheap raw biomass and higher productivity micro-organisms could improve the process economics. As shown in Fig. 6, gas fermentation (from CO and in principle from CO_2) could be an alternative route based on waste streams (CO-rich emissions, for example).

5.2.2.3 SUSTAINABLE PRODUCTION OF DROP-IN INTERMEDIATES

Various new routes are under development for the production of intermediates starting from biomass, biowaste or CO_2 as an alternative to current fossil-fuel based processes. An illustrative example is given from the production of 1,4-BDO. Current market is about 1.5 Mtons to produce polymers such as polybutylene terephthalate, copolyester ethers and thermoplastic polyurethanes. It can be converted to tetrahydrofuran (THF, used in some polymers synthesis or as a solvent) and to γ-butyrolactone (GBA), used in various syntheses for fine and speciality chemicals. Current production is a multi-step process, via synthesis of maleic anhydride (MA) by catalytic selective oxidation of n-butane, esterification of MA with ethanol followed by hydrogenation to diethylsuccinate converted then to GBA and finally to 1,4-BDO with THF as co-product. It is thus a complex multistep route, and thus a key advantage of alternative routes is the possibility to reduce the number of steps.

The current main alternative passes through the production of biosuccinic acid from sugars, eventually in the presence of CO_2. Various companies (BASF, DSM, Rochette, BioAmber, Purac, etc.) are working on pilot/demo scale production of biosuccinic acid. The latter can then be hydrogenated to BDO–THF. BDO can be alternatively produced by direct fermentation of sugar (Genomatica), or via formation of PHA (polyhydroxyalkanoate) (Metabolix). The biosuccinic acid route is currently the most advanced, with a good potential to become a main route to produce BDO–THF, mainly because it reduces the process complexity and improves energy & resource efficiency with respect to the current path starting from n-butane.

Table 1 reports a list of processes to prepare drop-in intermediates for the chemical industry starting from biomass or biowastes. These processes are expected to be introduced commercially soon and become significant production routes. Table 1 also indicates the main companies involved in the development of these novel (sustainable) routes. [30]

TABLE 1: Processes to prepare drop-in intermediates for the chemical industry starting from bio-mass/-waste [30]

Product	Main companies involved in developing these routes
BTX	– Anellotech's Bio-BTX, Virent's BioFormPX
	– Paraxylene from Gevo's isobutanol
Adipic acid	– Verdezyne, Ronnovia, DSM, Genomatica, BioAmber
Acrylic acid	– OPX Biotechn./Dow Chem., Myriant, Novozyme/Cargill/BASF, ADM, Novomer, Metabolix
BDO	– Genomatica, Myriant, BioAmber, LanzaTech
Rubber feedstocks (butadiene, isoprene, isobutene)	– Butadiene – Amyris/Kuraray, LanzaTech/Invista, Versalis/Genomatica, Global Bioenergies/Synthos, Cobalt Biotechnologies
	– Isoprene – Amyris/Michelin, Ajinomoto/Bridgestone, DuPont/Goodyear, Aemetis, Glycos Biotechn.
	– Isobutylene – Global Bioenergies/LanzaTech, Gevo/Lanxess

BTX: benzene, toluene and xylenes. BDO: butanediol

Table 2 reports instead already commercial or semi-commercial production routes of bio-based intermediates for the chemical industry.30 The introduction of these routes (mainly focused at producing bioplastics) in the market shows impressive growth rates, as reported by European Bioplastics e.v. The worldwide production capacity for bioplastics will increase from around 1.2 million tonnes in 2011 to approximately 5.8 million tonnes by 2016, although still remaining a few percentage points of the global plastic market.

TABLE 2: Commercial or semi-commercial production routes of bio-based intermediates for the chemical industry [30]

Product	Process (company and starting raw material)
1,3-Propane-diol	– Dupont from corn sugar
	– In China from glycerol
Butanol	– From corn by Cathay Industrial Biotech, Laihe Rockley and other small Chinese firms
Isobutanol	– Gevo from corn
Propylene glycol	– From glycerol by Oleon
	– From sorbitol by ADM
	– From corn glucose by Global Bio-Chem
Ethylene glycol	– From sugar-cane ethanol by Greencol and India Glycols
Epichlorohy-drin	– From glycerol by Vinythai
Farnesenea	– From sugarcane by Amyris
Polyamides	– From castor-oil by many companies, such as Arkema, Evonik, BASF, Solvay, DSM, Radici Group, etc.

a Farnesene refers to six closely related sesquiterpenes.

By far the strongest growth will be in the biobased, non-biodegradable bioplastics group, especially drop-in solutions, i.e. biobased versions of bulk plastics like PE (polyethylene) and PET (polyethylene terephthalate) deriving from fossil fuels. Leading the field is partially biobased PET, which is already accounting for approximately 40% of the global bioplastics production capacity. Partially biobased PET will continue to extend this lead to more than 4.6 million tonnes by 2016. That would correspond to 80% of the total bioplastics production capacity. Following PET is biobased PE with 250[thin space (1/6-em)]000 tonnes, constituting more than 4% of the total production capacity. Biodegradable plastic market, particularly PLA (polylactic acid) and PHA (polyhydroxyalkanoates) each of them accounting for 298[thin space (1/6-em)]000 tonnes (+60%) and 142[thin space (1/6-em)]000 tonnes (+700%) respectively, also show rather high growth rates, but the market is much smaller in size. This analysis evidences that bio-based, non-biodegradable commodity plastics (drop-in products such as PE, PET, or PP-polypropylene), are the main future route, being their possible use and recycling along their conventional counterparts.

5.2.2.4. SUSTAINABLE PRODUCTION OF NEW INTERMEDIATES AND PLATFORM MOLECULES

The previous section has emphasized the concept of drop-in products, e.g. which are produced by alternative raw materials without changing the final product already in commerce (however, quality may be different in terms of impurities). Biodegradable bioplastics will remain instead an interesting, but niche market, in part due to the fact that they are not drop-in products. Drop-in products can be faster commercialized, as it is not necessary to create a market and overcome the many barriers (authorization, REACH, etc.) necessary to introduce new chemicals/products.

There are, however, some relevant examples of non-drop-in chemicals which have a good potential to become large-scale products, because the final product has better properties and existing specific incentives (on the market) to products not derived from fossil fuels.

An example is offered by polyethylene furanoate (PEF) and its use as a new material for packaging. It must first be recalled that bottles and other packaging are the main areas of the use of bioplastic. These applications are thus expected to be the main areas for possible development of alternative bio-based chemicals to conventional ones. Between these, it can be cited especially the production of alternatives to PET, used mainly to produce bottles and food containers for the consumer markets. This market is sensitive to the green image of the product and thus there is interest from leading companies in this field (Coca Cola, Danone, for example) to use bio-based containers which reduce, if not eliminate, the use of fossil fuels.

A route of growing interest in this direction is that developed by Avantium for the synthesis of 2,5-FDCA (furan dicarboxylic acid) by oxidation of 5-HMF (5-hydroxymethyl-furfural), which can be obtained from fructose by dehydration. [31] FDCA and MEG (mono ethylene-glycol) are then used to produce polyethylene furanoate (PEF), a valid alternative to PET with better O_2-barrier capacity. MEG can also be produced from bio-resources, for example by one-pot catalytic conversion of cellulose in the presence of H_2. [32] The reaction is quite selective and could be an alternative to the two-step current process involving the synthesis of ethylene oxide (EO) by ethylene epoxidation, followed by catalytic hydratation of EO. It is thus possible to produce 100% biobased PEF which can substitute commercial PET (Fig. 6). The driving force for the development is from one side the social push to have more sustainable commodities and from the other the formation of a final product with superior performance (5-times higher resistance barrier to O_2 permeation).

Eerhart et al. [33] evaluated the energy and greenhouse gas (GHG) balance in the production of the bioplastic polyethylene furandicarboxylate (PEF) starting from corn based fructose and compared to its petrochemical counterpart polyethylene terephthalate (PET). The production of PEF can reduce the non-renewable energy use (NREU) approximately 40% to 50% while GHG emissions can be reduced approximately 45% to 55%, compared to PET for the system cradle to grave. These reductions are higher than for other biobased plastics, such as polylactic acid (PLA) or polyethylene (PE). With an annual market size of approximately 15 million metric tonnes (Mt) of PET bottles produced worldwide, the complete bottle substitution of PEF for PET would allow us to save between 440 and 520 PJ of NREU and to reduce GHG emissions by 20 to 35 Mt of CO_2 equivalents. If also substantial substitution takes place in the PET fibres and film industry, the savings increase accordingly. The GHG emissions could be further reduced by a switch to lignocellulosic feedstocks, such as straw, but this requires additional research.

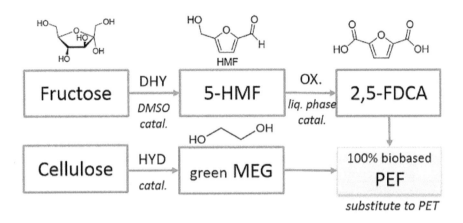

FIGURE 7: Block-diagram for the production of 100% biobased PEF, as a substitute for oil-derived PET. DHY: dehydration. OX: selective catalytic oxidation.

The development of this route also opens the market to the use of the platform intermediate HMF (particular, when improved routes for the direct synthesis from cheaper raw materials such as cellulose can be developed) as well as of FDCA in a number of other applications. [31] FDCA could be used to produce polyesters (for example, PEF for bottles, flexible packaging and carpets/textiles), co-polyesters (engineering resins), polyamides (for engineering plastics, nylons and fibers and bullet proof vests), polyurethanes (for footwear), thermosets (polyester resin for powder coating) and plasticizers (esters, for PVC cables).

HMF also finds use in other applications. Some of them are highlighted in Fig. 8 which reports some of the possible conversion routes to valuable high-volume chemicals. Between these routes, the following can be evidenced:

- Caprolactam: commercially produced in a multistep process from benzene; it is the monomer for Nylon-6.

- 1,6-Hexanediol: commercially prepared by hydrogenation of adipic acid; it is widely used for industrial polyester and polyurethane production.
- Adipic acid: commercially produced in a multistep process from benzene; it is the monomer for Nylon-66.

Fig. 8 reports, as an example, the scheme of the catalytic synthesis of adipic acid from 2,5-FDCA, which can be made with an overall yield >85%. Also the synthesis of the other two selected large-volume monomers could be realized with a high yield.

Table 3 reports other interesting routes under development, in addition to those discussed above. Also these routes are potentially relevant for large-volume products and are based on alternative raw materials to fossil fuels (in part or full).

TABLE 3: Production routes under development based on alternative raw materials to fossil fuels (in part or full) [30]

Product	Process under development
2,5-FDCA	– FDCA + MEG → PEF as alternative to PET
	– Nylons and Aramids using adipic acid derived from FDCA
CO/CO_2	– CO_2 + propylene oxide → polypropylene carbonate (PPC)
	– CO_2 + ethylene oxide → polyethylene carbonate (PEC)
	– CO_2 → polyether polycarbonate polyols (PPP)
	– CO + ethylene oxide → propiolactone
	– CO/CO_2 → C2–C5 products (via FTO, GTL, or fermentation)
Levulinic acid (LA)	– LA → β-acetacrylic acid (new acrylate polymer)
	– LA → diphenolic acid (replacement for bisphenol-A)
	– LA → 1,4-pentanediol (new polyesters)
	– LA-derives lactones for solvent use

5.3 ROLE OF CATALYSIS IN ESTABLISHING THE NEW SCENARIO FOR THE CHEMICAL INDUSTRY

The previous section evidenced that there are many novel routes based on alternative raw materials (biomass, biowaste and CO_2) that are actively developed both at academic and industrial levels. As commented, these routes are at a different stage of development, and may have different chances to substitute current processes based on fossil fuel derived raw materials. A common aspect, however, is the need to improve process performances, typically by developing improved catalysts (a definition which includes bio-catalysis).

The discussion on the new scenario for the chemical industry has evidenced that some priority areas could be identified. This section will analyse the status and perspectives of catalysis in these areas, to identify the expected targets and possible breakthrough challenges, which in turn determines how fast and effective the transition to the new sustainable economy could be. Discussion is focused here on the use of solid catalysts, being in general their use preferable for more sustainable (resource and energy efficient) processes.

5.3.1 CATALYSIS FOR ALTERNATIVE ROUTES OF LIGHT OLEFIN PRODUCTION

Light olefins are the building blocks for most of the petrochemistry routes and polymeric materials. The production of light olefins is thus one of the largest productions (mainly by steam cracking of naphtha in Europe), but is also very energy-intensive as commented before. The panorama for light olefin production is fast changing. There are two main challenges for their production (and thus for the petrochemical industry) in Europe. The first is the differing cost level of feedstocks and raw materials throughout the regions. The second is to cope with the age and the size of the existing plants in Europe. To maintain competiveness, it is thus necessary to convert these challenges into new opportunities.

Regarding the cost of feedstocks, the last few years have seen a large capacity addition of petrochemical plants in the Middle East mainly based on associated gas from oil production leading to a superior cost position. In addition to that, the extensive shale gas exploration in the US leading to a substantial reduction of prices for natural gas and ethane was observed. As a result most cracker operators in the US have increased their share of lighter feedstocks and thus have experienced a significant improvement in their cost position. In Europe and Asia this change was not present, since availability of these NG feedstocks is and will probably remain limited.

But the advantage of lighter feedstocks comes at a price as these feedstocks result in a lower production of higher olefins. Propylene and crude C4 are, meanwhile, an important factor for the competitiveness of naphtha crackers compared to ethane or light feed crackers. The existing supply shortage and the high demand for C3 and C4 are increasing their prices and thus the profitability of naphtha crackers.

Another important element regards the size of the plants for light olefin production. Steam-cracking plants have a typical size between 1.5 and 3.0 Mtons per year and they are not profitable for smaller sizes or at lower operation capacity. It is thus necessary to transport the olefins at high distances, with associated costs and risks, and there is a low flexibility in operations. There is thus interest to develop dedicated, small size olefin production plants (on-purpose). Due to the shortage of propylene (mostly driven by increased consumption of PP, which is expected to expand by an average 5–6% for the next several years), new plants for propane dehydrogenation and olefin metathesis have been constructed, but the cost of propane (or n-butane for C4 olefins–diolefins) is still high in Europe. Here there is the opportunity for the production of light olefins from alternative routes to fossil-fuel, because it can use (potentially) cheaper and more sustainable raw materials, and is suited for on-purpose production. Ethylene from bioethanol or CO_2 to olefins (FTO) are two of the interesting routes from this perspective.

5.3.1.1 CATALYTIC DEHYDRATION OF ETHANOL

The simplified reaction mechanism of ethanol catalytic conversion over mixed oxides is presented in Fig. 9. Although this is a reaction apparently simple and acid-catalysed, it is necessary to tune the catalyst properties to maximize the selectivity, because it is necessary to avoid the (i) consecutive reaction of ethylene with surface acid sites, and (ii) redox reaction leading to dehydrogenation rather than dehydration. An acido-base concerted mechanism with formation of a surface ethoxy species occurs. This is an easy reaction occurring typically in very mild conditions and the rate limiting step of the reaction is thus water desorption to regenerate the active site.

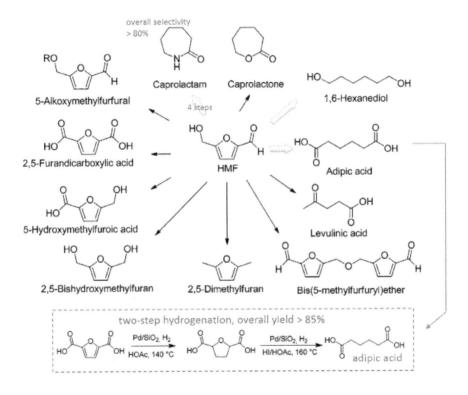

FIGURE 8: HMF as platform molecule to produce chemicals and fuel components. The larger, light arrow indicates the three routes discussed in the text. In the bottom, the scheme of catalytic synthesis of adipic acid from 2,5-FDCA (one of the key products from HMF) is reported. Adapted from ref. 31, © ACS 2013.

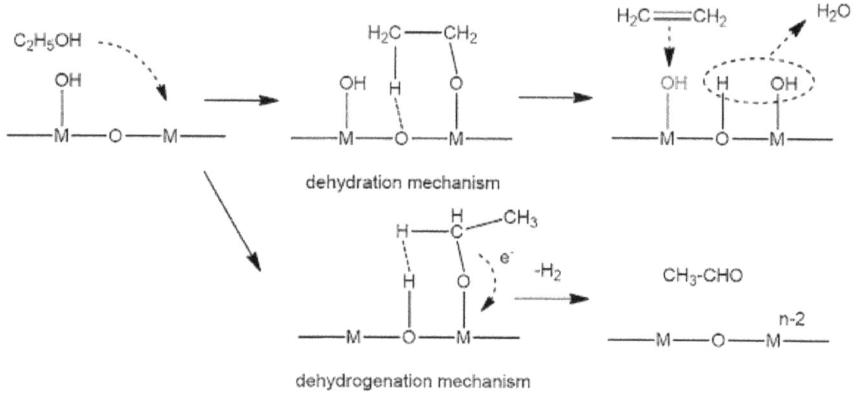

FIGURE 9: Simplified reaction mechanism of ethanol conversion over mixed oxide catalysts.

γ-Alumina has been one of the first used catalysts for this reaction, but requires a high reaction temperature (450 °C) resulting in a relatively low ethylene yield (about 80%). Doping of the alumina with KOH and/or ZnO or the preparation of MgO–Al_2O_3/SiO_2 mixed oxides (Syndol catalyst) were used by companies such as Phillips Oil Co. and Halcon SD, respectively, to increase the selectivity. Further modifications with other dopants resulted in selectivities of over 98–99%, although high reaction temperatures were still necessary.

Zeolites, particularly H-ZSM-5, were a second class of catalysts used for ethanol dehydration. The main advantage is the activity at lower temperatures (200 to 300 °C). At 300 °C, HZSM-5 can reach an ethanol conversion level of 98% and 95% ethylene selectivity. The main disadvantage of HZSM-5 is its acidity, which reduces its stability and coking resistance. Modification with phosphorus to reduce acidity improves both selectivity (over 99%) and stability. Modification with La (eventually as co-dopant with P) leads also to interesting results. With almost 100% ethylene selectivity and ethanol conversion and low temperatures (about 240 °C), 0.5% La-2% P-HZSM-5 is currently one of the best catalysts for industrial use.

Due to diffusional limitations, the use of nano-scale zeolites leads to better results. SAPO zeolites, such as SAPO-34 which is one of the best catalysts for methanol to olefin (MTO) reaction together with H-ZSM-5, also shows good performance in ethanol dehydration.

A third class of catalysts investigated was heteropolyacids. Particularly, $Ag_3PW_{12}O_{40}$ has demonstrated high catalytic ability, making it a promising catalyst for the dehydration of ethanol to ethylene, but its high acidity reduces its stability. This catalyst gives 99.2% selectivity at 100% ethanol conversion in rather mild conditions (220 °C, GHSV = 6000 h^{-1}),34a but long term stability has to be demonstrated. As a comparison, an industrial catalyst such as the cited SynDol (Halcon) gives comparable performance (96.8% selectivity at 99% conversion), but requires higher temperatures (450 °C, LHSV = 26–234 h^{-1}).34b

Therefore, recent developments in catalysts, [34b,c] particularly nanoscale HZSM-5, which has a 99% ethylene yield at 240 °C and a lifespan of over 630 h before ethylene selectivity decreased to below 98%, and in heteropolyacids such as $Ag_3PW_{12}O_{40}$, achieving over 99% ethylene yield at temperatures as low as 220 °C, have significantly improved performances over currently used catalysts in industrial plants for ethanol dehydration. We have to remark, however, that operations at low temperatures are not necessarily better (due to less efficient heat recovery), if not higher productivities and stability are achieved at the same time.

The profitability of the process depends essentially on the cost of production of bioethanol, which may vary considerably depending on the raw materials and technology of production. Energy integration of the process is also critical. Actual ethanol to ethylene plants have a production capacity about one order of magnitude lower than that typical of steam cracking plants, but this aspect could be an advantage in terms of on-purpose plants. Process intensification is one of the ways to make profitable also small-medium size plants.

5.3.1.2 CO_2 TO OLEFINS VIA FTO

There are different possible routes to produce light olefins from CO_2 and renewable H_2 (Fig. 10). RWGS reaction is typically promoted from the same catalysts of the consecutive steps (methanol synthesis or FT reaction) and thus a single reactor/catalyst could be used. However, a direct route converting CO_2 without involving the RWGS step is preferable, because reversibility of the latter limits the performance. There are two main paths to light olefins: (a) a direct route from syngas (CO + H_2) using modified Fischer–Tropsch (FT) catalysts and (b) an indirect two-step route via the intermediate formation of methanol. In this indirect route, a conventional commercial methanol catalyst is used for the first step and small-pore zeolites (CHA or MFI-type) for the second methanol-to-olefin (MTO) step. In the presence of an acid catalyst, two methanol molecules could be dehydrated to dimethyl ether (DME), which can also be converted to light olefins (it is an intermediate in the process).

Being current methanol catalysts active both in RWGS reaction and in the methanol synthesis, it may be seem the same to start from CO–H_2 or CO_2–H_2 mixtures. However, the productivities in the second case are typically one third of those using syngas (CO–H_2), even if the addition of small amounts of CO_2 (less than 3–4%) to syngas promotes the methanol synthesis rate. There are two main motivations. CO_2 is a better oxidant than CO and thus in large amounts alters the surface active state of the catalysts (Cu–ZnO–Al_2O_3 based materials). The water formed in the RWGS reaction inhibits the reaction. It is thus convenient to use two reactors in series, with intermediate removal of water from the stream. The alternative is to use a reactor approach with in situ removal of water (catalytic distillation, membrane reactor). It is also possible to combine the catalysts for methanol to the zeolite for MTO to have in one-step the direct formation of light olefins from CO_2 and H_2.

In terms of R&D there are two main objectives. The first is to develop more productive catalysts for the direct use of CO_2–H_2 mixtures, which probably should be not active in the RWGS reversible reaction. It is necessary to remark that most of the catalysts tested up to now are based on catalysts for syngas (CO–H_2) adapted to operate with CO_2 and H_2, but not specifically developed to work with carbon dioxide. It would be desirable, for example, to have a novel FT catalyst able to directly and selectively convert CO_2 and H_2 to light olefins, or novel methanol catalysts able to directly convert CO_2 without the presence of the RWGS reaction. Therefore, even if the methanol, MTO and FT catalysts (from syngas) are well established and current methanol and FT processes operate in the presence of some carbon dioxide, converting pure CO$_2$ would require the development of novel or improved catalysts. Maximizing selectivity to light olefins, and possibly also their relative ratio of formation (currently it is preferable to form ≥C3= over ethylene) is another challenge.

We focus here the discussion on the catalysts for the direct FTO conversion of CO$_2$ and ren.H$_2$ to light olefins, because this reaction would be preferable for a resource and energy efficient process.

The probability for the selective formation of lower olefins increases with temperature (in the 200–400 °C temperature range, the typical one for FT reaction) and decreases at higher pressures and H2[thin space (1/6-em)]:[thin space (1/6-em)]CO ratios in the feed. Olefins can also be incorporated into the growing chain involving a metallo-cyclobutane transition state followed by β-H transfer to form a α-olefin. It is thus necessary to prevent the re-adsorption of olefins which increases the formation of longer-chain compounds. Shorter contact times are preferable, but also the

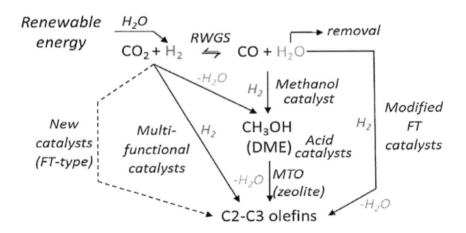

FIGURE 10: Different routes to synthesize light olefins from CO$_2$ and renewable H$_2$. Reproduced with permission from ref. 11b. © RCS 2013.

choice of the reactor is important. Operations in liquid phase (slurry-type reactors) allow limiting olefin readsorption and surface overheating due to the exothermic reaction. Operations in a slurry reactor lead to maximizing the yields of lower olefins.

Fig. 11 reports selected relevant literature results [11b,28] in the FTO reaction starting from syngas, because up to now the studies using CO_2–H_2 instead of CO–H_2 (eventually with small amounts of CO_2) are limited. Fig. 10 also indicates the target area from an industrial perspective. In general, overall, yields up to over 55% in C_2–C_4 olefins have been observed, but together with C_2–C_4 alkanes, methane and C_5+ products. [31] These conditions are still not satisfactory and a further improvement would be necessary, to start from CO_2 instead of that from CO. There is thus the need for further R&D on catalysis to develop and exploit this route.

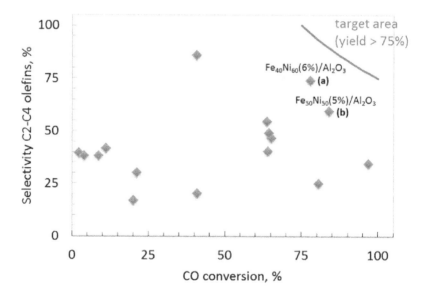

FIGURE 11: Relevant literature results in FTO reaction (see ref. 11b and 28 for references) starting from syngas, with an indication of the target yields for possible industrial exploitation. Composition of two of the better catalysts formulations is also indicated (data refer to a temperature of 360 °C): (a) ref. 35a. (b) ref. 35b.

5.3.1.3 CATALYTIC SYNTHESIS OF BUTADIENE

Butadiene is predominantly sourced by extraction from the mixed C_4 stream produced in steam crackers, particularly of naphtha sources. Historically the revenue from these valuable co-products maintained margins for cracking naphtha at a premium to cracking lighter natural gas liquids. This situation, however, is changing quickly, and there is thus interest in bio-routes to produce butadiene, particularly for on-purpose applications. The catalytic dehydrogenation or oxidative dehydrogenation of n-butane, however, is a strong competitive route.

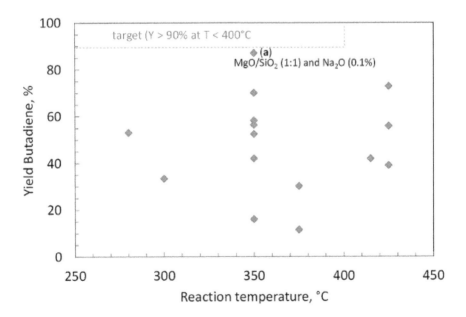

FIGURE 12: Catalytic synthesis of butadiene from bioethanol. Composition of the better catalyst is also indicated. (a) ref. 37. Based on the data reported in Table 7 of ref. 36. © Wiley 2013.

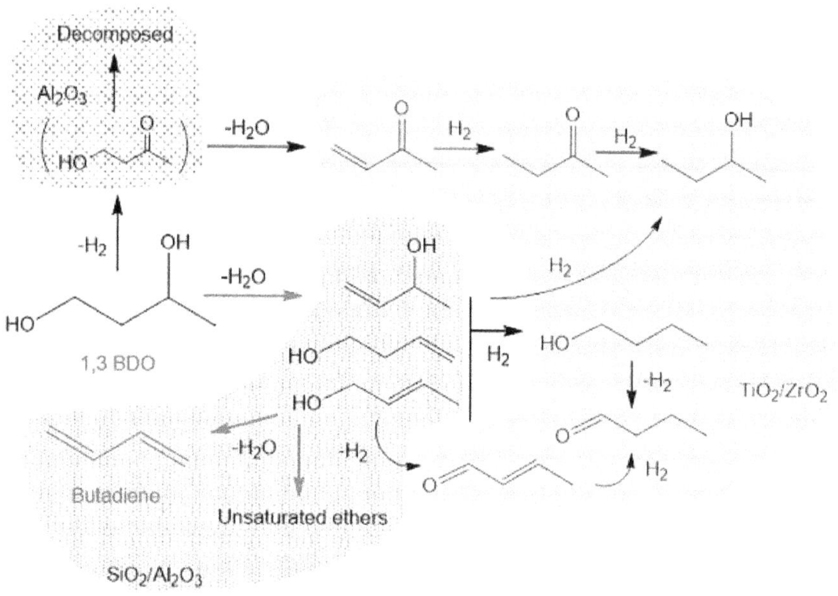

FIGURE 13: Reaction network in the conversion of 1,3-BDO to butadiene and other products on mixed oxides. The coloured areas indicate the prevailing mechanism of transformation over the oxides indicated in the coloured area. Adapted from ref. 40. © Elsevier 2006.

There are some main possibilities to produce butadiene by alternative, non-fossil-fuel based routes (Fig. 6). The process to convert ethanol into butadiene is not new, but still rather inefficient. Ethanol is converted into acetaldehyde after which an aldolization is performed followed by dehydration. A target yield of over 90%, at temperatures below about 450 °C, is required for industrial exploitability, and good stability as well. A selection of catalytic results [36] is reported in Fig. 12.

While in general yields were unsatisfactory (below 70%), a catalyst based on magnesia-silica, doped with Na_2O, has been reported having performances close to the target. [37] However, the results have not been reproduced later by other authors. There is thus the need to develop improved catalysts for this reaction, as well as to understand a number of fundamental aspects (reaction mechanism, structure–activity–selectivity relationship, etc.). [36]

In the evaluation (using a multi-criteria approach) of different routes for using bioethanol as a chemical building block for biorefineries, Posada et al. [38] indicates the conversion of ethanol to ethylene or 1,3-butadiene as promising routes for an integrated biorefinery concept, in contrast to other possibilities (for example, ethanol conversion to acetic acid, n-butanol, isobutylene, hydrogen and acetone).

Butanol dehydration to butenes and butadiene has been presented as a valuable route, [39] but we would instead suggest that it is necessary to produce the raw material at a more competitive cost. In addition, the further catalytic conversion to butadiene is also difficult. While ethanol to ethylene conversion is a dehydration reaction, butanol conversion to butadiene requires an oxidative dehydration–dehydrogenation mechanism. Data are quite limited on this reaction and related catalysts.

The third alternative is the conversion of butanediol to butadiene, which is investigated by companies such as LanzaTech and Genomatica/Versalis. 1,3-, 1,4- or 2,3-butanediol could be dehydrated to butadiene over acid catalysts, but various byproducts (unsaturated alcohol, ketone, etc.) form and the reaction is thus more challenging with respect to ethanol dehydration.

ABE fermentation (acetone–butanol–ethanol) with wild and genetically modified strains (from the Clostridium family) has been known for a long time, but has received renewed interest recently. However, there are still many aspects to improve in order to produce n-butanol at commercially attractive prices, such as (i) improve yields of butanol, (ii) expanding substrate utilization and (iii) minimizing energy consumption during separation and purification. The cost of n-butanol is thus still high. It is necessary to develop micro-organisms able to give the selective fermentation to butanol to make competitive the synthesis.

Solid acids such as SiO_2–Al_2O_3, Al_2O_3, ZrO_2 and TiO_2 convert 1,3-butanediol (1,3-BDO) depending on their acid properties (Fig. 13). [40] Strong acid catalysts (SiO_2–Al_2O_3) catalyse the dehydration of 1,3-butanediol at reaction temperatures below 250 °C, while weak acid catalysts (ZrO_2 and TiO_2) require temperatures above 325 °C. SiO_2–Al_2O_3 catalyses the dehydration of 1,3-butanediol into unsaturated alcohols. The latter are then dehydrated into 1,3-butadiene. Alumina alone instead forms 4-methyl-1,3-dioxane, which is the acetal compound of 1,3-butanediol and formaldehyde. Several compounds were produced over TiO_2 and ZrO_2 owing to the side reactions such as dehydrogenation and hydrogenation. On strong solid acids, the butadiene selectivity is still unsatisfactory and thus more research is needed.

On the other hand, the method could be interesting for the synthesis of unsaturated alcohols (raw materials for the synthesis of various fine and speciality chemicals for applications such as medicines, perfumes, agricultural products). Over weak basic oxides such as CeO_2 at 325 °C, 3-buten-2-ol and trans-2-buten-1-ol are produced with selectivities of about 58% and 36%, respectively. [41]

5.3.2 CATALYSIS FOR PLATFORM MOLECULES

Section 2.2.4 evidenced how HMF and its derivatives (Fig. 7 and 8) are good platform molecules [31] to produce important chemicals according to alternative routes to commercial production starting from fossil fuels.

5.3.2.1 CATALYTIC SYNTHESIS OF 5-HYDROXYMETHYLFURFURAL (HMF)

Among the platform chemicals which can be obtained from biomass (for example, levulinic acid or bioethanol), HMF shows some distinctive advantages: (i) it has retained all six carbon atoms present in the hexoses and (ii) high selectivities are possible in its preparation, in particular from fructose. Current (year 2013) cost for fructose syrup is about 0.6 \$ kg^{-1}, making it possible to produce HMF at a cost of around 0.9–1.0 \$ kg^{-1}, which is suitable for the chemicals discussed below, but not for fuels, except some booster additives.

HMF forms by acid-catalysed dehydration of hexoses. Fructose is much more reactive and selective toward HMF than glucose, because the latter has a more stable ring structure, which hinders its ability to form the cyclic reaction intermediate. It is thus necessary to isomerize glucose to fructose, if starting from the former. The solvent influences the tautomeric forms present in solution and thus reactivity (even if there is a fast interconversion between them). Therefore, the solvent plays a key role in both reactivity and selectivity.

Fructose dehydration to HMF has been known for a long time, but only relatively recently have high HMF yields been obtained working in organic solvents. Both homogeneous and heterogeneous acid catalysts could be used, but the latter are preferable for an industrial process. Fig. 14 summarizes selected catalytic results (using solid catalysts) obtained in aqueous systems or in organic solvents. [31] Target yield ($\geq$95%) is also indicated. As shown from the figure there are some catalysts meeting the requirements. The table in Fig. 14 shows some of the catalyst compositions and relative operating conditions, which meet the target HMF yield. Relatively mild reaction conditions are necessary (about 120 °C, 2 h of reaction), but it is important that good yields are obtained also with high fructose concentrations. Dimethyl sulfoxide (DMSO) is the best solvent, but also acetone–DMSO mixtures as well as dimethylformamide (DMF) give good results. Very high yields (>90%) are found for a number of catalysts, such as zeolites, heteropolyacids and acidic resins in DMSO, but in general a continuous water removal is necessary. Amberlyst 15 is claimed to yield 100% HMF, even after recycling (three times) and at fructose concentrations as high as 50%. [42] In the case of $FePW_{12}O_{40}$ and H-BEA zeolite, the HMF yields decreased to below 50% at 50 wt% fructose concentration, while yields > 90% at low fructose concentration.

The same catalysts giving excellent yield in HMF from fructose show lower performances starting from glucose, and even worse starting from polysaccharides. The HMF yields from cellulose are generally very low (<10%). There is thus the need to develop novel multifunctional catalysts for the selective HMF synthesis starting from polysaccharides. Taking into account the price differential with respect to fructose, a target yield over 60% has to be obtained.

The use of ionic liquids (IL) to overcome the limits of solubility in water–organic solvents of some polysaccharides, and to improve performances starting from the latter is an active research line. [31] However, to apply IL as reaction media for HMF production from biomass, highly efficient recycling is required because of their high cost. Another challenge is related to progressive contamination of IL when using untreated biomass feedstock, because these contain many inorganic (ash) and organic impurities that should be periodically removed from the IL.

5.3.2.2 CATALYTIC CONVERSION OF HMF

As discussed before, HFM could be converted to a variety of valuable chemicals/intermediates, but the synthesis of 2,5-FDCA, caprolactam, caprolactone and 1,6-hexanediol are more interesting.

HMF oxidation into FDCA was achieved in the past using different stoichiometric oxidants like N_2O_4, HNO_3 and $KMnO_4$, but today a cleaner catalytic synthesis is necessary. The oxidation of HMF using air or O_2 could be achieved by different catalysts. Catalysts such as those currently used for terephthalic acid production (Co/Mn/Br) could be used at high pressure (70 bar air), but operations using heterogeneous catalysts would be preferable. Between the most active/selective catalysts reported recently, it may be cited.

- Pt supported on carbon or Al_2O_3 gives 98% FDCA yield at complete HMF conversion in the liquid phase oxidation with oxygen in the presence of a base (need to keep the FDCA formed in aqueous solution as dialkaline salt); [42a] reaction conditions are 100 °C, 40 and 260 min for C and alumina supports respectively, and 150 psi O_2.
- Au supported on hydrotalcite, which under similar reaction conditions as above, gives nearly 100% FDCA yield. [42b] This catalyst, due to the use of a basic support, may avoid the use of a base, the base being the support itself. However, strong chemisorption of FDCA on the support may cause incomplete recovery and possible catalyst deactivation over time. In addition, Pt-based catalysts oxidize HMF to FDCA an order of magnitude faster than Au-based catalysts.

The use of bimetallic catalysts, for example $Au–Cu/TiO_2$ giving 99% FDCA yield at 110 °C for 4 h under 20 bar O_2, [43] improves catalyst stability, but remains a critical aspect in these catalysts.

Direct synthesis routes of FDCA from fructose by combining dehydration and oxidation have also been reported using Pt–Bi/C in combination with a solid acid in water–MIBK. However, an FDCA yield of only 25% was obtained. Direct (base-free) oxidative esterification of HMF into the diester (which may be directly used to produce polymers) was instead demonstrated by Casanova et al. [44] using Au/CeO_2 as catalyst in methanol under 10 bar oxygen in an autoclave reactor (yield about 99% at about 65 °C).

Therefore, although some improvements in catalysts could still be necessary, for example to avoid the use of a base and to have more stable operations, catalysts for the selective oxidation of HMF to FDCA are available.

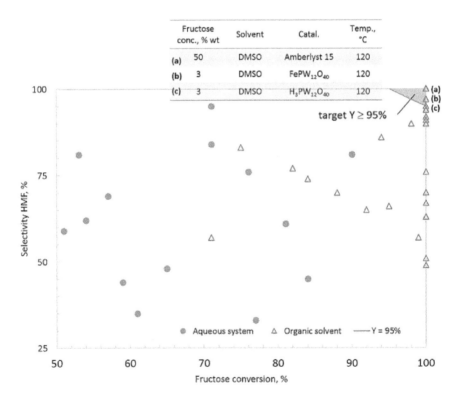

FIGURE 14: Catalytic synthesis of HMF from fructose in aqueous systems or organic solvent. Composition of the better catalyst is also indicated. (a–c) ref. 42. Based on the data reported in Tables 9 and 12 of ref. 31. © ACS 2013.

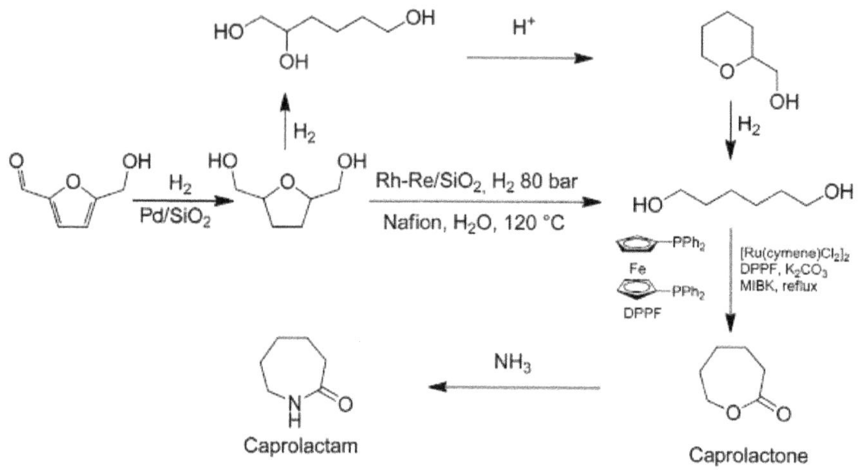

FIGURE 15: Catalytic conversion of HMF into caprolactam or 1,6-hexanediol. Adapted from Scheme 42 of ref. 31. © ACS 2013.

As outlined in Fig. 8, several other compounds can be made from HMF, eventually via intermediate FDCA formation. Adipic acid, used in the synthesis of nylon 6,6, synthesis from HMF is one of the most attractive routes. The hydrogenation of FDCA to adipic acid occurs in a two-step process. [45] In the first step, 2,5-tetrahydrofuran-dicarboxylic acid is produced in 88% yield by hydrogenating FDCA at 140 °C for 3 h in acetic acid, catalyzed by Pd on silica. Yields up to 99% of adipic acid were claimed by reacting this product under hydrogen at 160 °C for 3 h in acetic acid in the presence of 0.2 M of HI and 5% Pd on silica catalyst.

A multistep route to produce caprolactam (the monomer for nylon-6) or 1,6-hexanediol (monomer for high performance polyesters, polyurethane resins, and adhesives) has been proposed by de Vries and co-workers (Fig. 14).46 HMF is hydrogenated in >99% yield to 2,5-bishydroxymethyl-tetrahydrofuran which can then be further hydrogenated to 1,6-hexanediol in 86% yield, by using a Rh–Re/SiO$_2$ catalyst in the presence of Nafion. This is a tandem three step process proceeding through formation of 1,2,6-hexanetriol, which is cyclized under the influence of the acid to tetrahydropyran-2-methanol, which in turn is hydrogenated to 1,6-hexanediol using the same catalyst. The diol can then be converted into caprolactone using a ruthenium catalyzed Oppenauer oxidation in virtually quantitative yield. Overall selectivity from HMF to caprolactone was 86%. Conversion of ε-caprolactone into caprolactam can be made selectively using ammonia at 170 bar and 300–400 °C (UCC process), but an alternative catalytic route would be preferable (Fig. 15).

HMF could thus be converted to caprolactam in a 4 steps process, whereas the current industrial process, which is based on benzene, contains seven steps. Using this technology, producing 1 kg of caprolactam would require 1.44 kg of HMF, 0.11 kg of H_2, and 0.17 kg of NH_3. These results indicate the interest in this novel process using renewable resources, but it is necessary to improve the productivity of the steps, particularly of the diol to caprolactone step for which the development of heterogeneous catalysts appears necessary. Combining this step with the consecutive one, e.g. developing direct catalysts for 1,6-hexanediol to caprolactam, would be also useful.

There are thus various attractive routes by which important monomers can be synthesized in alternative routes to those currently based on fossil fuels. For conciseness, we have limited the discussion here to the use of HMF as a renewable platform molecule, but there are other important possibilities. One of these is glycerol, produced in large amounts as a by-product during bio-diesel production.47

Glycerol can be catalytically converted to C3 diols (alternative method to the production by fermentation, or from propylene). 1,2-Propanediol (1,2-PDO or propylene glycol; it is currently produced from propylene via propene oxide) is used as a chemical feedstock for the production of un-saturated polyester resins. Other relevant uses are as humectant (E1520), solvent, and preservative in food and for tobacco products, including in electronic cigarettes. 1,3-Propanediol (1,3-PDO) has a wide range of applications from carpets and textiles to cosmetics, personal, and home care industry. Polytrimethylene terephthalate (PTT, mainly used in the carpet industry) is the largest application of 1,3-PDO. It also has the potential to substitute propylene glycol, 1,4-butanediol (BDO), butylenes glycol, and nylon in a number of applications. Glycerol can be also dehydrated to acrolein and then oxidized to acrylic acid and acrylate esters, [48] which are also important monomers for polymer industry. The main current limitation to industrial development of these processes is the large uncertainty in defining the long-term trend in the price of glycerol. This is one of the main motivations having inhibited the industrial exploitation of these processes.

Acrylic acid can be also produced from CO_2 and ethylene. [49] Although important progress has been made recently in particular by BASF researchers, to realize this reaction on an industrial scale, it is still a challenge. Using ethylene from bioethanol, a full alternative renewable route to acrylic acid and derivatives could be realized.

5.4 CONCLUSIONS

The use of biomass, bio-waste and CO_2 derived raw materials, the latter synthesized using H_2 produced using renewable (solar) energy sources, open up new scenarios to developing a sustainable and low carbon chemical production, particularly in regions such as Europe lacking in other resources. There are many motivations for this transition to a new economy and some of them are also discussed in the Introduction. Typically in the transition to new economies, there is the start of many R&D initiatives to explore new possibilities and new synthetic routes, even though only a minimal part of them will be effectively exploited as main synthetic routes.

It is thus necessary to have some guidelines to analyse these alternative routes, and determine those having better possibilities to become future main routes of production. This tutorial review discusses first this new scenario, in particular from a sustainability perspective, with the aim to point out, between the different possible options, those more relevant to enable this new future scenario for chemical production. In particular the different drivers (economic, technological and strategic, environmental and sustainability and socio-political) guiding the selection are commented upon.

The case of the use of non-fossil fuel based raw materials for the sustainable production of light olefins is discussed in a little more detail, but the production of other olefins and polyolefins, of drop-in intermediates and other platform molecules are also analysed. These examples will not systematically cover the entire possibilities, it not being possible given the limits of a tutorial review, but are valuable cases of the aspects to consider.

The final part discusses the role of catalysis in establishing this new scenario, summarizing the development of catalysts with respect to industrial targets, for (i) the production of light olefins by catalytic dehydration of ethanol and by CO_2 conversion via FTO process, (ii) the catalytic synthesis of butadiene from ethanol, butanol and butanediols, and (iii) the catalytic synthesis of HMF and its conversion to 2,5-FDCA, adipic acid, caprolactam and 1,6-hexanediol. Also in this case, other valuable platform molecules are possible (the case of glycerol is briefly discussed).

We have also limited the discussion about CO_2 conversion routes (except for the case of CO_2 to FTO), because these have been discussed in detail elsewhere. [1,11b,12] To mention, however, that the conversion of CO_2 over novel designed electrocatalysts for photoelectro-catalytic (artificial-leaf like) devices [50] opens new perspectives to move to what is defined as Economy 3.0, e.g. where a distributed energy and chemical production will exist.

In conclusion, the panorama for chemical production is fast evolving and new raw materials, substituting fossil fuels for sustainability motivations and to develop a low carbon society, will become a main driver to develop novel production routes. Catalysis is a key element to enable this possibility. Starting from scenario analysis to selecting the key routes, we have examined the status of development and some of the remaining critical aspects on which to focus the research.

REFERENCES

1. G. Centi and S. Perathoner, ChemSusChem, 2014 DOI:cssc.201300926 .
2. (a) F. Cavani, G. Centi, S. Perathoner and F. Trifirò, Sustainable Industrial Chemistry: Principles, Tools and Industrial Examples, Wiley-VCH, Germany, 2009 ; (b) G. Centi and R. A. van Santen, Catalysis for Renewables, Wiley-VCH, Germany, 2007 .
3. (a) J. N. Chheda, G. W. Huber and J. A. Dumesic, Angew. Chem., Int. Ed., 2007, 46, 7164 ; (b) D. M. Alonso, S. G. Wettstein and J. A. Dumesic, Chem. Soc. Rev., 2012, 41, 8075 RSC .
4. H. Kobayashi and A. Fukuoka, Green Chem., 2013, 15, 1740 RSC .
5. (a) G. W. Huber, S. Iborra and A. Corma, Chem. Rev., 2006, 106, 4044 ; (b) A. Corma, S. Iborra and A. Velty, Chem. Rev., 2007, 107, 2411 .
6. S. Zinoviev, F. Müller-Langer, P. Das, N. Bertero, P. Fornasiero, M. Kaltschmitt, G. Centi and S. Miertus, ChemSusChem, 2010, 3, 1106 .
7. M. Stocker, Angew. Chem., Int. Ed., 2008, 47, 9200 .

8.   R. Rinaldi and F. Schüth, Energy Environ. Sci., 2009, 2, 610 CAS .

9.   P. Gallezot, ChemSusChem, 2008, 1, 734 .

10. A. M. Appel, J. E. Bercaw, A. B. Bocarsly, H. Dobbek, D. L. DuBois, M. Dupuis,
     J. G. Ferry, E. Fujita, R. Hille, P. J. A. Kenis, C. A. Kerfeld, R. H. Morris, C. H. F.
     Peden, A. R. Portis, S. W. Ragsdale, T. B. Rauchfuss, J. N. H. Reek, L. C. Seefeldt,
     R. K. Thauer and G. L. Waldrop, Chem. Rev., 2013, 113, 6621 .

11. (a) G. Centi and S. Perathoner, Catal. Today, 2009, 148, 191 CAS ; (b) G. Centi, E.
     A. Quadrelli and S. Perathoner, Energy Environ. Sci., 2013, 6, 1711 RSC .

12. E. A. Quadrelli, G. Centi, J.-L. Duplan and S. Perathoner, ChemSusChem, 2011, 4,
     1194 .

13. M. Aresta and A. Dibenedetto, Dalton Trans., 2007, 2975 RSC .

14. T. Sakakura, J.-C. Choi and H. Yasuda, Chem. Rev., 2007, 107, 2365 .

15. W. Wang, S. Wang, X. Ma and J. Gong, Chem. Soc. Rev., 2011, 40, 3703 RSC .

16. M. Peters, B. Kohler, W. Kuckshinrichs, W. Leitner, P. Markewitz and T. E. Muller,
     ChemSusChem, 2011, 4, 1216 .

17. K. Triantafyllidis, A. Lappas and M. Stöcker, The Role of Catalysis for the Sustain-
     able Production of Bio-fuels and Bio-chemicals, Elsevier, The Netherlands, 2013 .

18. M. Aresta, Carbon Dioxide as Chemical Feedstock, Wiley-VCH, Germany 2010 .

19. J. Mao, K. Li and T. Peng, Catal. Sci. Technol., 2013, 3, 248 .

20. S. N. Habisreutinger, L. Schmidt-Mende and J. K. Stolarczyk, Angew. Chem., Int.
     Ed., 2013, 52, 7372 .

21. S. Navalon, A. Dhakshinamoorthy, M. Alvaro and H. Garcia, ChemSusChem, 2013,
     6, 562 .

22. W. Fan, Q. Zhang and Y. Wang, Phys. Chem. Chem. Phys., 2013, 15, 2632 RSC .

23. V. P. Indrakanti, J. D. Kubicki and H. H. Schobert, Energy Environ. Sci., 2009, 2,
     745 CAS .

24. W. Bello, De-Globalization: Ideas for a New World Economy, Zed Books, London,
     UK, 2005 .

25. (a) G. Centi, P. Lanzafame and S. Perathoner, Catal. Today, 2011, 167, 14 ; (b) G.
     Centi, P. Lanzafame and S. Perathoner, Catal. Today, 2013 , submitted.

26. F. Cavani and J. H. Teles, ChemSusChem, 2009, 2, 508 .

27. T. Ren, M. Patel and K. Blok, Energy, 2006, 31, 425 .

28. G. Centi, G. Iaquaniello and S. Perathoner, ChemSusChem, 2011, 4, 1265 .

29. J. van Haveren, E. L. Scott and J. Sanders, Biofuels, Bioprod. Biorefin., 2008, 2, 41
     CrossRef CAS .

30. D. de Guzman, Bio-Based Chemical Feedstocks, presented at Tecnon OrbiChem
     Marketing Seminar at APIC 2013, Taipei, 9 May 2013.

31. R.-J. van Putten, J. C. van der Waal, E. de Jong, C. B. Rasrendra, H. J. Heeres and J.
     G. de Vries, Chem. Rev., 2013, 113, 1499 .

32. A. Wang and T. Zhang, Acc. Chem. Res., 2013, 46, 1377 .

33. A. J. J. E. Eerhart, A. P. C. Faaij and M. K. Patel, Energy Environ. Sci., 2012, 5,
     6407 CAS .

34. (a) J. Gurgul, M. Zimowska, D. Mucha, R. P. Socha and L. Matachowski, J. Mol.
     Catal. A: Chem., 2011, 351, 1 ; (b) D. Fan, D.-J. Dai and H.-S. Wu, Materials, 2013,
     6, 101 CrossRef CAS ; (c) M. Zhang and Y. Yu, Ind. Eng. Chem. Res., 2013, 52,
     9505 CrossRef CAS .

35. (a) M. Feyzi, A. A. Mirzaei and H. R. Bozorgzadeh, J. Nat. Gas Chem., 2010, 19, 341 CrossRef CAS ; (b) C. Wang, L. Xu and Q. Wang, J. Nat. Gas Chem., 2003, 12, 10 CAS .

36. C. Angelici, B. M. Weckhuysen and P. C. A. Bruijnincx, ChemSusChem, 2013, 6, 1595 CAS .

37. R. Ohnishi, T. Akimoto and K. Tanabe, J. Chem. Soc., Chem. Commun., 1985, 1613 RSC .

38. J. A. Posada, A. D. Patel, A. Roes, K. Blok, A. P. C. Faaij and M. K. Patel, Bioresour. Technol., 2013, 135, 490 .

39. M. Mascal, Biofuels, Bioprod. Biorefin., 2012, 6, 483 CrossRef CAS .

40. N. Ichikawa, S. Sato, R. Takahashi and T. Sodesawa, J. Mol. Catal. A: Chem., 2006, 256, 106 .

41. S. Sato, F. Sato, H. Gotoh and Y. Yamada, ACS Catal., 2013, 3, 721 CrossRef CAS .

42. (a) M. A. Lilga, R. T. Hallen and M. Gray, Top. Catal., 2010, 53, 1264 CrossRef CAS ; (b) N. K. Gupta, S. Nishimura, A. Takagaki and K. Ebitani, Green Chem., 2011, 13, 824 RSC ; (c) K.-I. Shimizu, R. Uozumi and A. Satsuma, Catal. Commun., 2009, 10, 1849 .

43. T. Pasini, M. Piccinini, M. Blosi, R. Bonelli, S. Albonetti, N. Dimitratos, J. A. Lopez-Sanchez, M. Sankar, Q. He, C. J. Kiely, G. J. Hutchings and F. Cavani, Green Chem., 2011, 13, 2091 RSC .

44. O. Casanova, S. Iborra and A. Corma, J. Catal., 2009, 265, 109 .

45. T. R. Boussie, E. L. Dias, Z. M. Fresco and V. J. Murphy, Int. Patent WO, 2010144873, 2010, (assigned to Rennovia, Inc.) .

46. T. Buntara, S. Noel, P. H. Phua, I. Melián-Cabrera, J. G. de Vries and H. J. Heeres, Angew. Chem., Int. Ed., 2011, 50, 7083 .

47. C. H. Zhou, H. Zhao, D. S. Tong, L. M. Wu and W. H. Yu, Catal. Rev. Sci. Eng., 2013, 55, 369 CAS .

48. B. Katryniok, S. Paul and F. Dumeignil, ACS Catal., 2013, 3, 1819 CrossRef CAS .

49. M. L. Lejkowski, R. Lindner, T. Kageyama, G. E. Bodizs, P. N. Plessow, I. B. Müller, A. Schaefer, F. Rominger, P. Hofmann, C. Futter, S. A. Schunk and M. Limbach, Chem.–Eur. J., 2012, 18, 14017 .

50. (a) S. Bensaid, G. Centi, E. Garrone, S. Perathoner and G. Saracco, ChemSusChem, 2012, 5, 500 ; (b) C. Genovese, C. Ampelli, S. Perathoner and G. Centi, J. Energy Chem., 2013, 22, 202 CrossRef CAS ; (c) C. Ampelli, R. Passalacqua, C. Genovese, S. Perathoner and G. Centi, Chem. Eng. Trans., 2011, 25, 683 ; (d) C. Ampelli, R. Passalacqua, S. Perathoner and G. Centi, Chem. Eng. Trans., 2009, 17, 1011 .

CHAPTER 6

Quantifying the Climate Impacts of Albedo Changes Due to Biofuel Production: A Comparison with Biogeochemical Effects

FABIO CAIAZZO, ROBERT MALINA, MARK D. STAPLES,
PHILIP J. WOLFE, STEVE H. L. YIM, AND STEVEN R. H. BARRETT

6.1 INTRODUCTION

Biofuels may hold promise to promote energy security, reduce the environmental impact of transportation and foster economic development. For these reasons, many countries have enacted policies to encourage their production (EU 2009, US EPA 2013). In the US, biofuel production for transportation aims to replace 30% of petroleum consumption by 2030 (Perlack et al 2005, US Department of Energy 2011). Targets are also set for the EU (10% replacement of diesel and gasoline by 2020; EU 2009) and other countries such as China (2 million tons of biodiesel by 2020; Koizumi 2011) and Indonesia (20% replacement of diesel and gasoline by 2025; Zhou and Thomson 2009).

Quantifying the Climate Impacts of Albedo Changes Due to Biofuel Production: A Comparison with Biogeochemical Effects. © *Caiazzo F, Malina R, Staples MD, Wolfe PJ, Yim SHL, and Barrett SRH.* Environmental Research Letters *9,2 (2014). doi:10.1088/1748-9326/9/2/024015. Licensed under a Creative Commons Attribution 3.0 Unported License, http://creativecommons.org/licenses/by/3.0/.*

Historically, environmental assessments of biofuels have focused on biogeochemical effects (i.e. greenhouse gas (GHG) emissions) directly or indirectly attributable to the lifecycle of the fuel. Emissions are considered for all relevant lifecycle steps, including feedstock cultivation, extraction, and transportation as well as fuel production, distribution, and combustion (Kim and Dale 2005, Larson 2006, Lardon et al 2009, Yee et al 2009, Stratton et al 2010, Van der Voet et al 2010, Guinée et al 2011). Land-use change (LUC) to cultivate biomass feedstock for biofuel production may lead to GHG emissions if it changes the amount of carbon stored in vegetation and soil (Stratton et al 2010).

Distinct from assessing the biogeochemical effects, there is limited research focused on the biogeophysical effects of LUC for biomass feedstock cultivation. Biogeophysical effects include changes in surface albedo (Betts 2000, Lee et al 2011), evapotranspiration (Pitman et al 2009, Georgescu et al 2011), surface roughness/canopy resistance (Lean and Rowntree 1993, Betts 2007, Georgescu et al 2009), leaf area index and rooting depth of the vegetation (Georgescu et al 2009). Of these, the LUC-induced change in surface albedo is considered the dominant biogeophysical effect at the global scale (Betts 2000, 2001, Claussen et al 2001, Bala et al 2007). A change in albedo alters the surface reflectivity of sunlight (the incoming shortwave radiation), thus changing the Earth's radiative balance. Albedo changes can be quantified in terms of global radiative forcing (RF) (Betts 2000, Georgescu et al 2011, Bright et al 2012, Cherubini et al 2012), which can be expressed in terms of GHG equivalent emissions (Betts 2001, Bird et al 2008). This allows for a direct comparison against the biogeochemical effects calculated by traditional LCA.

In contrast, additional biogeophysical effects such as evapotranspiration and surface roughness cannot be adequately expressed in terms of global RF (Davin et al 2007, Betts 2011, Cherubini et al 2012), although they may be relevant at a local scale (Bounoua et al 2002, Georgescu et al 2011). In previous work, the climate impact of albedo changes has been assessed to describe the effect of forestation policies (Rautiainen et al 2009, Lohila et al 2010, Rautiainen et al 2011). Recent studies have also attempted to evaluate the albedo effect of biomass feedstock cultivation, using either numerical models (Georgescu et al 2011, Anderson-Teixeira et al

2012, Hallgren et al 2013, Anderson et al 2013) or satellite measurements (Bright et al 2011, Loarie et al 2011, Cherubini et al 2012). The results of those analyses suggest that albedo effects are potentially as important as the biogeochemical effects assessed by traditional LCA (Georgescu et al 2011, Anderson-Teixeira et al 2012). The assessments that are available in the literature often focus only on a single feedstock (Georgescu et al 2011, Bright et al 2011, Loarie et al 2011) and are based on different methodologies, making cross-study comparison difficult.

In this study we perform an assessment of the LUC-induced albedo effects of a range of LUC scenarios by considering the cultivation of five different biomass feedstocks (switchgrass, soybean, palm, rapeseed and salicornia) and compare these effects to the biogeochemical effects quantified by traditional LCA. To the best of our knowledge, this is the first study to consider the albedo effects from a broad range of feedstocks using direct satellite measurements, and the first to quantify and compare the albedo effect of replacing an original land use with the cultivation of palm, rapeseed or salicornia, in particular. The LUC scenarios considered are derived from Stratton et al (2010, 2011a) and the albedo effects are presented in terms of gCO_2e/MJ of renewable middle distillate (MD) fuel, which is the fuel considered in the Stratton et al (2011a) lifecycle analysis. This enables a consistent comparison of the LUC-induced albedo effect, the biogeochemical effects from LUC, and the GHG emissions from the production of renewable MD fuels.

6.2 METHODOLOGY

In this study, we evaluate the induced albedo effect of a number of discrete LUC scenarios. Each of these scenarios is evaluated at multiple geographic locations in order to account for variability in surface and meteorological conditions within the same land types involved in the LUC. Satellite measurements of albedo and transmittance parameters are retrieved for each geographic location of interest, and an analytical radiative balance model is used to convert the albedo changes into a RF, then into equivalent GHG emissions.

TABLE 1: Biomass feedstock types (first column) and original land uses from this study (second column), compared to the reference LCA from Stratton et al (2010) (fourth column). Each LUC scenario for which the albedo effect is calculated is associated to a LUC code (third column). Geographic regions (fifth column) are consistent with the reference LCA (Stratton et al 2010) for each LUC scenario, in order to enable comparison between albedo and biogeochemical effects.

Biomass feedstock type	Original land use (this study)	LUC code	Original land use (reference LCA)a	Region
Switchgrass	Corn cultivation	B1	Carbon-depleted soil	Central US (Midwest-North-east states)
	Soybean cultivation	B2		
	Barren land	B3		
Soy	Cerrado grassland	S1	Cerrado grassland	Brazil (Central and Southern regions)
	Tropical rainforest	S2	Tropical rainforest	
Palm	Previously logged-over forest	P1	Previously logged-over forest	Southeast Asia (Malaysia and Indonesia)
	Tropical rainforest	P2	Tropical rainforest	
	Peat land rainforest	P3	Peat land rainforest	
Rapeseed	Corn cultivation	R1	Set-aside land	Europe (United Kingdom, France and Denmark)
	Uncultivated land	R2		
Salicornia	Desert	H1	Desert	Mexico (Sonora desert) and US (Southern states)

6.2.1 LUC SCENARIOS

We consider eleven LUC scenarios, comprised of five different biomass feedstocks for up to three original land uses each, as shown in table 1. Each scenario is restricted to one geographic region. The scenarios are consistent between this study and the traditional LCA study that we use as a reference (Stratton et al 2010; LUC combinations S1, S2, P1, P2, P3, H1), wherever possible. The switchgrass and rapeseed scenarios are redefined due to ambiguities in the reference LCA study. In Stratton et al (2010), switchgrass cultivation is assumed to take place on generic carbon-depleted soils. In this study we consider three possible LUC scenarios associated with carbon-depleted soils (McLaughlin et al 2002, Adler et al 2007): corn cultivation (LUC B1), soybean cultivation (LUC B2), and barren land (LUC B3) replaced by switchgrass cultivation. Furthermore, in the reference LCA rapeseed cultivation in Europe is assumed to take place on set-aside land, i.e. land areas temporarily removed from agricultural production (Stratton et al 2010). In this case we consider two LUC scenarios: corn cultivation (LUC R1) and uncultivated land (LUC R2) replaced by rapeseed cultivation. Table 1 also indicates the geographic region in which each LUC scenario is assumed to take place in the reference LCA (Stratton et al 2011a).

A minimum of four geographic locations are selected to describe each original land use type, and a minimum of eight combinations of biomass feedstock and original land use locations are used to define each of the 11 LUC scenarios shown in table 1. This multi-location approach allows for a more complete picture of the potential land conversions and the associated natural variability. The latitude and longitude of these locations are retrieved using current literature (e.g., Mosali et al 2013), satellite observations (e.g., Rhines 2008) and farming databases (e.g., FIC 2013), and are confirmed using the Moderate Resolution Imaging Spectroradiometer (MODIS) Land Cover Type database (MCD12Q1) (NASA MODIS 2013a).

6.2.2 ALBEDO AND TRANSMITTANCE DATA RETRIEVAL AND CALCULATION

Black-sky shortwave (BSW) broadband albedo coefficients (encompassing both the near infrared and visible spectra) are retrieved for each biomass feedstock and original land use pairing, i, that represents a specific LUC scenario. BSW albedo coefficients are obtained from the MODIS satellite database MCD43A3 (NASA MODIS 2013b) and vetted using a separate MODIS database (MCD43A2; NASA MODIS 2013c) Albedo data from the MODIS database are produced every 8 days and are linearly interpolated to obtain daily albedo evaluations for a full year. In case of missing or low quality data, the time interpolation is performed between the two closest acceptable observations. The average daily BSW albedo for each i is obtained for a full year by averaging the daily values retrieved for three reference years (2009, 2010 and 2011), in order to account for annual variability in local conditions. Biases in each individual albedo observation are reduced by taking the space average across the 500 m × 500 m cell where the location under investigation is found and the eight cells surrounding it. Consistency between the land-use type of these cells is verified using the MODIS Land Cover Type database (NASA MODIS 2013a). Table 2 shows the yearly-averaged BSW albedo, retrieved and processed as described, for each land use type considered in the LUC scenarios from table 1. The BSW albedos in table 2 are given as mean, minimum and maximum values among the yearly-averaged albedos retrieved for all the sample locations representing each specific land use type.

For each biomass feedstock type and original land use pairing i, representing a LUC scenario from table 1 we evaluate the albedo effect as the difference in RF induced by the conversion of the original land use to biomass feedstock cultivation. The geographical location and conditions of radiative transmittance are kept the same as that of the original land use; i.e. feedbacks between albedo changes and local weather/cloudiness conditions are not accounted for. For each i, the planetary albedo change is computed as a function of the day of the year d:

$$\Delta\alpha_i(d) = K_{T\,orig,i}(d)T_\alpha\big[\alpha_{bio,i}(d) - \alpha_{orig,i}(d)\big] \tag{1}$$

where $\alpha_{bio,\,i}$ and $\alpha_{orig,\,i}$ are the daily cloud-free shortwave albedo coefficients for biomass feedstock cultivation and the original land use, respectively, obtained from the MODIS database (NASA MODIS 2013b) and averaged in time and space as previously described. $K_{T\,orig,\,i}$ is the mean daily all-sky clearness index for the original land use point, and T_a is the transmittance factor, as in Bright et al (2012). The calculation of planetary albedo differences from clear-sky surface albedo values in equation (1) follows a procedure widely reported in the literature (Lenton and Vaughan 2009,

Muñoz et al 2010, Bright et al 2012, Cherubini et al 2012). Local mean daily values of KTorig, i are constant for each month and are retrieved from the NASA Atmospheric Science and Data Center (ASDC) database, which provides monthly 22-year averages of the all-sky clearness index (including maximum and minimum bounds) (NASA ASDC 2013). The transmittance factor Ta is chosen as a global annual average of 0.854, consistent with previous findings and modeling comparisons (Lenton and Vaughan 2009, Muñoz et al 2010, Cherubini et al 2012, Bright and Kvalevåg 2013).

TABLE 2: Black-sky shortwave albedo for each land use type considered in this study. Albedo values are given as mean, minimum and maximum values among yearly-averaged BSW albedo coefficients retrieved for the sample locations representing a specific land use type. The number of sample locations for each land use type is indicated in the Table. Yearly-averaged BSW albedos at each location are obtained from the MODIS satellite database MCD43A3 (NASA MODIS 2013b), and treated as described in section 2.2.

Land type	Number of samples	BSW albedo		
		Mean	Min	Max
Switchgrass field (US)	4	0.177	0.156	0.191
Soybean cultivation (Brazil)	6	0.175	0.139	0.204
Palm plantation (SE Asia)	14	0.088	0.045	0.134
Rapeseed field (Europe)	11	0.162	0.147	0.179
Salicornia cultivation (Mexico)	5	0.149	0.140	0.156
Corn field (US)	5	0.165	0.151	0.178
Soybean cultivation (US)	8	0.163	0.148	0.200
Barren land (US)	4	0.175	0.135	0.222
Cerrado grassland (Brazil)	4	0.133	0.122	0.148
Tropical rainforest (Brazil)	4	0.120	0.111	0.128
Previously logged-over forest (SE Asia)	9	0.125	0.093	0.138
Tropical rainforest (SE Asia)	4	0.066	0.033	0.103
Peat land rainforest (SE Asia)	4	0.091	0.046	0.129
Corn field (Europe)	11	0.156	0.145	0.174
Uncultivated land (Europe)	8	0.151	0.132	0.184
Desert land (South US/Mexico)	10	0.326	0.251	0.535

6.2.3 RADIATIVE FORCING (RF) MODEL

The planetary albedo change due to biomass feedstock cultivation on land originally used for some other purpose, found in (1), alters the radiative balance of the Earth which can be quantified as a radiative forcing. This is equal to the time integral of the product of daily albedo variation (1) and daily radiative flux at the top of the atmosphere $R_{TOA,i}(d)$, calculated at the original land use location (Bright et al 2012, Cherubini et al 2012). For each biomass feedstock and original land use pairing, i, the yearly global RF (measured in W m^{-2}) is therefore:

$$\Delta RF_{global,i} = \left\{ -\frac{1}{365} \sum_{d=1}^{365} [R_{TOA,i}(d) \cdot \Delta \alpha_i(d)] \right\} \cdot \frac{A_a}{A_{earth}} \quad (2)$$

where A_a is the reference area subject to the albedo change, and A_{earth} is the total area of the earth. The RF associated with each of the LUC scenarios in table 1, ΔRF_{LUC}, is calculated as the average of all the global yearly radiative forcings $\Delta RF_{global,i}$ found for all of the pairings, i representing the same LUC case.

6.2.4 CO$_2$-EQUIVALENT EMISSION CONVERSION

CO$_2$e emissions per unit energy of biofuel produced (gCO$_2$e/MJ) is a common metric adopted in LCA studies (Larson 2006, Adler et al 2007, Stratton et al 2010, 2011a). To establish direct comparison between albedo change effects and biogeophysical effects for the same LUC scenario, the global RF associated with each one of the LUC scenarios in table 1 is converted into CO$_2$e. The correspondence between the RF induced by albedo changes and CO$_2$e emissions is well established in the literature (Betts

2000, Bird et al 2008, Muñoz et al 2010, Joos et al 2013). First, RF is converted into a change in atmospheric carbon concentration ΔC by using a logarithmic relation with the background carbon concentration (Betts 2000) (linearized for small perturbations). Positive RF (induced by a decrease in the land albedo) corresponds to carbon emissions, while negative RF corresponds to carbon sequestrations. The concentration ΔC (in parts per million, ppm) is then converted into an equivalent carbon emission ΔC_T per unit area subject to albedo change:

$$\Delta C_T = \left(\frac{1}{AF_{TH=100}}\right) \Delta Cm_{atm} \left(\frac{M_C}{M_{air}}\right) \times \left(\frac{1}{A_a}\right) \frac{1}{10^6} \text{ (tC ha}^{-1}) \tag{3}$$

where m_{atm} is the total mass of the atmosphere (in tons), M_C and M_{air} are the molecular weights of carbon and air respectively (in g mol^{-1}), A_a is expressed in hectares and $AF_{TH}=100$ is the airborne emission fraction of CO_2 for a time horizon (TH) of 100 years, consistent with the reference biogeochemical impacts assessment by Stratton et al (2010). Finally, using data about the biomass yield, mass conversion factor, and specific energy conversion efficiencies, the carbon emission per unit area in (3) is converted into a CO_2e emission per unit energy of the fuel produced. In order to evaluate the albedo effects under the same assumptions as the biogeochemical impacts, the resulting emissions are distributed over 30 years, as in the reference LCA (Stratton et al 2010).

In order to represent the magnitude of natural variability, we examine low, baseline, and high cases for each LUC scenario in table 1. The baseline cases utilize the mean of the RFs calculated for all the biomass feedstock and original land use pairings representing the same LUC scenario. Low and high cases utilize the maximum and minimum RFs calculated among the pairings used to simulate the same LUC scenario, and the upper and lower estimates of the relevant all-sky clearness index from the ASDC database (NASA ASDC 2013). Variability in meteorological conditions is therefore accounted for.

The geographic locations and relevant physical parameters for all sample locations are given in the Supporting Information (SI available at stacks.iop.org/ERL/9/024015/mmedia). A more detailed derivation of the

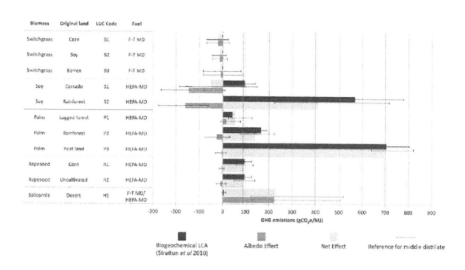

FIGURE 1: Climate impacts of biofuel production and use for different LUC scenarios. Each row of the table on the left contains a biomass feedstock type and original land use pairing, corresponding to a particular MD fuel, indicated in the last column. In the histogram, one set of bars indicates the impact of the albedo variations due to each LUC, in terms of CO_2e per unit energy of fuel produced. The high and low cases include variability in the geographical locations and local meteorological conditions (as described in section 2.4). The other bars show the biogeochemical effects (i.e., the direct GHG emissions) as calculated in the reference LCA (Stratton et al 2010). The related whisker bars account for the variability of process efficiency and biomass feedstock yield (Stratton et al 2010). The bars in the background show the net impact of considering albedo and biogeochemical effects in the baseline, low and high emission scenarios. Both albedo and biogeochemical effects are distributed over a time span of 30 years, consistent with Stratton et al (2010). The dashed black line indicates the results for conventional MD fuel from the reference LCA (Stratton et al 2010).

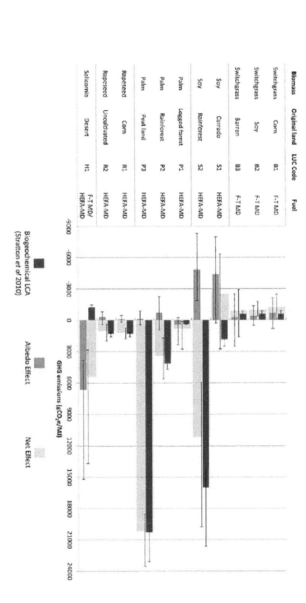

Figure 2. Climate impacts of direct LUC for different LUC scenarios. Each row of the table on the left contains a biomass feedstock type and original land use pairing, corresponding to a particular MD fuel, indicated in the last column. In the histogram, the blue bars indicate the instantaneous effect of changes in albedo due to LUC, in terms of CO_2 equivalent emissions per unit energy of fuel. The high and low cases include variability in the geographical locations and local meteorological conditions (as described in section 2.4). The red bars show the biogeochemical effects (i.e., LUC-induced GHG emissions) related exclusively to LUC, as calculated by Stratton et al (2010). The related whisker bars account for the variability in biomass feedstock yield (Stratton et al 2010). The green bars in the background show the aggregate climate impact of LUC considering albedo and biogeochemical effects in the baseline, low and high emission scenarios.

albedo-emission conversion model and the yields and energy efficiencies of each biomass feedstock type are also discussed in the SI (available at stacks.iop.org/ERL/9/024015/mmedia).

6.3 RESULTS AND DISCUSSION

The albedo effect for the eleven LUC scenarios is shown in figure 1 in terms of emissions or sequestrations of CO_2e per unit energy of biofuel produced (gCO$_2$e/MJ). The whisker bars represent low and high cases, corresponding to the minimum and maximum RF from the albedo changes induced by each LUC scenario, taking into account the variability of the albedo and meteorological conditions among the locations representative of the same LUC scenario.

The bars in figure 1 show the biogeochemical impacts calculated in the reference LCA by Stratton et al (2010) for the same biomass feedstock cultivation and original land use pairings. The biogeochemical effects calculated in the reference study include GHG emissions from cultivation, harvesting, extraction and transportation of the biomass; processing of biomass into MD fuels; and transportation, distribution, and combustion of the finished fuel product. Non-CO_2 GHGs and emissions species from direct fuel combustion are not considered in the reference LCA. In the case of aviation, these effects can result in a doubling of CO_2e direct fuel combustion emissions for a 100 year time horizon (Stratton et al 2011b). Only GHG emissions associated with direct LUC are considered in the reference LCA (Stratton et al 2010). Emissions from indirect LUC, which occurs if direct LUC disrupts the equilibrium between supply and demand for the displaced crop, and for downstream products relying on this crop (Plevin et al 2010), are not taken into account. The low and high ranges for the red bars in figure 1 reflect the variability of parameters used for LCA, such as process efficiency and biomass feedstock yield (Stratton et al 2010). The green bars in figure 1 represent the sum of albedo and biogeochemical effects. This net effect can be compared to the reference lifecycle emissions for conventional MD (90 gCO$_2$e/MJ, dashed black line in figure 1), assumed to be equal to the results for conventional diesel from Stratton et al (2011a). We do not consider albedo effects attributable to conventional middle distillate fuels, since in this case land use change per unit energy of finished fuel is estimated to be two to three orders of magnitude lower than for biomass-based fuels (Yeh et al 2010).

6.3.1 SWITCHGRASS

The albedo change due to replacement of corn cultivation, soy cultivation and barren land with switchgrass (scenarios B1–B3) leads to a negative RF in the baseline results, the equivalent of a sequestration of CO_2e. The effect is stronger when switchgrass replaces corn (LUC B1, −22 gCO_2e/MJ) or soy (LUC B2, −13 gCO_2e/MJ) than it is for barren land (LUC B3, −9 gCO_2e/MJ). A negative RF (or equivalently a cooling effect) for the conversion of land to switchgrass cultivation has also been computed by Georgescu et al (2011) (−0.0053 W m^{-2}) and Anderson-Teixeira et al (2012) (− 50 Mg CO_2e ha^{-1} 50 yr^{-1}) however the disparate methodologies, metrics and LUC assumptions used in those studies do not allow a quantitative comparison with this work. The biogeochemical lifecycle impacts of renewable MD fuel production and use from switchgrass replacing carbon-depleted soils are estimated by Stratton et al (2010) to be nearly carbon neutral (−1.6 gCO_2e/MJ). In absolute terms, the albedo impact of switchgrass cultivation for MD production is therefore 1380%, 813% and 563% greater than the biogeochemical lifecycle effects of corn, soybean, and barren land conversion, respectively. The net sum of albedo and biogeochemical effects is −23 gCO_2e/MJ, −15gCO_2e/MJ and −11 gCO_2e/MJ for corn replacement (LUC B1), soy replacement (LUC B2) and barren land replacement (LUC B3), respectively. It should be noted that the high emission cases (high whisker bar limits) demonstrate that a net positive RF is also possible for these three LUC scenarios.

6.3.2 SOY

In figure 1, LUC S1 and S2 show that the albedo effect of replacing Brazilian Cerrado and tropical rainforest with soybean cultivation results, in both cases, in a negative RF, equivalent to a cooling effect (−146 and −161

gCO_2e/MJ, respectively). Soybean cultivation in Brazil generally exhibits a higher reflectivity (average albedo of 0.175, see table 2) than the savannah land of the Brazilian Cerrado (average albedo of 0.133) or than the dense dark tropical rainforest in the Amazon region (average albedo of 0.120). An albedo-induced CO_2e reduction due to the establishment of soy cultivation in Brazil has also been found by Anderson-Teixeira et al (2012) (-70 $MgCO_2e$ $ha^{-1}50$ yr^{-1}). The bar in figure 1 for LUC S1 shows that, considering both the effects of albedo change and of biogeochemical lifecycle GHG emissions calculated by the reference LCA (Stratton et al 2010), the aggregate climate impact of the conversion of cerrado grassland to soybean cultivation is equivalent to a sequestration of 50 gCO_2e/MJ. By adding the albedo effect to the biogeochemical results from the reference LCA, renewable MD from soybean cultivated on land that was previously Cerrado grassland exhibits a decrease in climate impact of 156% with respect to conventional MD (from 90 to -50 gCO_2e/MJ). This reverts the results of the reference LCA which suggests that soybean derived MD has greater climate impact, in terms of lifecycle GHG emissions, than conventional MD (97 gCO_2e/MJ versus 90 gCO_2e/MJ) (Stratton et al 2010). In the case of tropical rainforest replacement with soybean cultivation (LUC S2), inclusion of the albedo effects does not revert the findings of the reference LCA: the direct GHG emission calculated by the reference LCA (569 gCO_2e/MJ) (Stratton et al 2010) is 453% larger than the impact attributable to the increase in albedo (-161 gCO_2e/MJ).

6.3.3 PALM

Due to their leaf characteristics and plantation density, palms are characterized by the lowest average shortwave albedo (0.088, see table 2) among the biomass feedstock types considered in this study. Furthermore, palm is not harvested, meaning that the low albedo coefficient is nearly constant throughout the year. This accounts for the smaller cooling effect shown in figure 1 for the case of tropical rainforest replaced by palm plantations (LUC P2, -25 gCO_2e/MJ) compared to the cooling effect occurring for soybean cultivation replacement of tropical rainforest (LUC S2, -161 gCO_2e/MJ). The low albedo of palm also accounts for the positive RF induced by conversion of previously logged-over forest (LUC P1), equivalent to a GHG emission of 14 gCO_2e/MJ. Palm replacement of peat land rainforest (LUC P3) yields a relatively small sequestration of -4 gCO_2e/MJ. For the aggregate climate impact of LUC P1, indicated by the green bar in figure 1, the baseline case of 55 gCO_2/MJ remains below the conventional jet fuel baseline even if the albedo effect is included. If the albedo effect is included, the high emission case (high whisker bar limit, 129 gCO_2e/MJ) is 43% larger than conventional MD, whereas the high emission case for the biogeochemical effects does not exceed the conventional MD reference. For LUC P2, while the baseline warming effect remains higher than that of conventional MD even if the albedo effect is included, the aggregate climate impact in the low emission case (77 gCO_2e/MJ) is below that of conventional MD, which was not the case in the reference LCA. For LUC P3, the biogeochemical effects from the reference LCA are so large (705 gCO_2e/MJ) (Stratton et al 2010), that the inclusion of albedo effects does not change the total climate impact significantly.

6.3.4 RAPESEED

LUC R1 and R2 show the albedo effect of replacing corn or uncultivated land with rapeseed cultivation in Europe. In both cases a small cooling effect is found, $-3gCO_2e/MJ$ for LUC R1 and -10 gCO_2e/MJ for LUC R2. According to the reference LCA, the biogeochemical effects of renewable MD fuel from rapeseed cultivated on previously set-aside land, yields a biogeochemical effect of 96 gCO_2e/MJ (Stratton et al 2010). The contribution of the albedo effect is negligible for this set of LUC scenarios.

6.3.5 SALICORNIA

LUC H1 considers salicornia cultivation on land that was previously desert. The albedo of the desert (average of 0.326 from table 2) is much larger than the albedo of the salicornia cultivations (average of 0.149), since deserts are composed of smooth, clear sand, and are highly reflective of incoming solar radiation (Pielke and Avissar 1990). The large desert albedo accounts for the relatively large positive RF induced by LUC H1, equivalent to 222 gCO_2e/MJ. In comparison, the direct GHG emissions computed by the reference LCA are only 6 gCO_2e/MJ (Stratton et al 2010). This result shows that production and use of renewable MD using salicornia grown on desert lands can have a larger warming effect than producing and using conventional MD.

Figure 2 shows a comparison between albedo effects and GHG emissions from LUC, excluding all the other stages accounted for in the reference LCA (Stratton et al 2010) (i.e. biomass cultivation and transport; feedstock to fuel conversion; and biofuel transport and combustion). This comparison is instructive since it shows the relative magnitude of the biogeochemical and biogeophysical effects that exclusively stem from (direct) alterations in land use. Both effects are evaluated as instantaneous CO_2e emissions, not distributed across the reference 30-years time span as in figure 1 (see equation S19 in the SI available at stacks.iop.org/ERL/9/024015/mmedia) but only across the first year of land use change. This is because the albedo effect is obtained using albedo differences and transmittance parameters averaged over a full year of variation, and carbon sequestration due to LUC is evaluated after a full year of vegetation replacement (Stratton et al 2010). Therefore, the results for albedo effects are directly proportional to the ones shown in figure 1. Results for LUC emissions are instead proportionally smaller than the ones reported in figure 1, since they exclude steady-state transport, production and combustion emissions. The results indicate that albedo effects are on the same order of magnitude as the traditional direct biogeochemical LUC emissions for most of the LUC scenarios considered.

6.4 CONCLUSIONS

This study shows that changes to surface albedo due to biomass cultivation can have a significant impact on the aggregate climate impact of biofuels. The albedo effects of LUC related to biomass feedstock cultivation for biofuel production, shown in figure 1, are on the same order of magnitude as the biogeochemical effects calculated by traditional LCA for the same LUC scenarios. The largest effects are calculated for LUC scenarios S1 and H1. Renewable MD production from soybean cultivated on land that was previously cerrado grassland (LUC S1) is found to yield a net cooling effect, equivalent to -50 gCO_2e/MJ. This makes renewable MD derived from soy oil a potentially viable alternative to conventional MD, a result that was not apparent when the albedo effect was not included in the reference LCA (Stratton et al 2010). Conversely, renewable MD production from salicornia cultivated on land that was previously desert yields a net warming effect, corresponding to 228 gCO_2e/MJ of MD fuel. This is the first evidence that salicornia-derived biofuel obtained by converting desert land could be potentially detrimental from a climate impact standpoint when compared to conventional fuels. Our results give support for further evaluating the consideration of LUC-induced surface albedo changes in global biofuels policies (Betts 2000, 2001, Claussen et al 2001, Bala et al 2007).

Some limitations of this study warrant acknowledgment. First, our analysis is restricted to changes in surface albedo, and other biogeophysical impacts such as evapotranspiration, surface roughness and rooting depth are not quantified here. Second, the albedo effects shown in figure 1 are dependent on the sample geographical locations chosen, and should not be interpreted as characteristics of the feedstocks considered, but rather as a function of the biomass feedstock and original land use pairing investigated. Third, the use of equivalent emissions based on RF has a theoretical weakness (Davin et al 2007) because albedo effects and biogeochemical effects act on different spatial and temporal scales. Nevertheless, equivalent emissions of CO_2 per unit energy of combusted fuel is a widely accepted metric used to compare both effects (Betts 2001, Bird et al 2008, Georgescu et al 2011, Cherubini et al 2012). When RF is used to compare different climate change mechanisms, there is an implicit assumption that the climate response is proportional to forcing. However, research by Hansen and Nazarenko (2004) show that surface forcing may be twice as effective at high latitudes as at low latitudes in generating surface temperature change. Further discussion is available in Betts et al (2007), Bird et al (2008), and in Cherubini et al (2013). With respect to time scales, the global RFs (and equivalent CO_2 emissions) evaluated in this study reflect the impacts of albedo variations averaged over a whole year of LUC in order to compare the albedo effect with the long-term biogeochemical impacts. However, albedo coefficients are dependent upon transient surface conditions (Song 1999), and these variations may lead to significant seasonal impacts on the local climate. These impacts may be offsetting when averaged over a whole year, as found by Georgescu et al (2013) for the biogeophysical effects of savannah to sugarcane conversion. Finally, the albedo impacts of other variables such as snow cover variation due to LUC, and climate-meteorology feedbacks potentially affecting local cloudiness, are not accounted for in this study.

REFERENCES

1. Adler P R, Grosso S J D and Parton W J 2007 Life-cycle assessment of net greenhouse-gas flux for bioenergy cropping systems Ecol. Appl. 17 675–91
2. Anderson C J, Anex R P, Arritt R W, Gelder B K, Khanal S, Herzmann D E and Gassman P W 2013 Regional climate impacts of a biofuels policy projection Geophys. Res. Lett. 40 1217–22
3. Anderson-Teixeira K J, Snyder P K, Twine T E, Cuadra S V, Costa M H and DeLucia E H 2012 Climate-regulation services of natural and agricultural ecoregions of the Americas Nature Clim. Change 2 177–81
4. Bala G, Caldeira K, Wickett M, Phillips T J, Lobell D B, Delire C and Mirin A 2007 Combined climate and carbon-cycle effects of large-scale deforestation Proc. Natl Acad. Sci. 104 6550–5
5. Betts R A 2000 Offset of the potential carbon sink from boreal forestation by decreases in surface albedo Nature 408 187–90
6. Betts R A 2001 Biogeophysical impacts of land use on present-day climate: near-surface temperature change and radiative forcing Atmos. Sci. Lett. 2 39–51
7. Betts R A 2007 Implications of land ecosystem-atmosphere interactions for strategies for climate change adaptation and mitigation Tellus B 59 602–15
8. Betts R A 2011 Mitigation: a sweetener for biofuels Nature Clim. Change 1 99–101
9. Betts R A, Falloon P D, Goldewijk K K and Ramankutty N 2007 Biogeophysical effects of land use on climate: model simulations of radiative forcing and large-scale temperature change Agricult. Forest Meteorol. 142 216–33
10. Bird D N, Kunda M, Mayer A, Schlamadinger B, Canella L and Johnston M 2008 Incorporating changes in albedo in estimating the climate mitigation benefits of land use change projects Biogeosci. Discuss. 5 1511–43
11. Bounoua L, DeFries R, Collatz G J, Sellers P and Khan H 2002 Effects of land cover conversion on surface climate Clim. Change 52 29–64
12. Bright R M, Cherubini F and Strømman A H 2012 Climate impacts of bioenergy: inclusion of carbon cycle and albedo dynamics in life cycle impact assessment Environ. Impact Assess. Rev. 37 2–11

13. Bright R M and Kvalevåg M M 2013 Technical note: evaluating a simple parameterization of radiative shortwave forcing from surface albedo change Atmos. Chem. Phys. Discuss. 13 18951–67

14. Bright R M, Strømman A H and Peters G P 2011 Radiative forcing impacts of boreal forest biofuels: a scenario study for norway in light of albedo Environ. Sci. Technol. 45 7570–80

15. Cherubini F, Bright R M and Strømman A H 2012 Site-specific global warming potentials of biogenic CO2 for bioenergy: contributions from carbon fluxes and albedo dynamics Environ. Res. Lett. 7 045902

16. Cherubini F, Bright R M and Strømman A H 2013 Global climate impacts of forest bioenergy: what, when and how to measure? Environ. Res. Lett 8 014049

17. Claussen M, Brovkin V and Ganopolski A 2001 Biogeophysical versus biogeochemical feedbacks of large-scale land cover change Geophys. Res. Lett. 28 1011–4

18. Davin E L, De Noblet-Ducoudré N and Friedlingstein P 2007 Impact of land cover change on surface climate: Relevance of the radiative forcing concept Geophys. Res. Lett. 34 L13702

19. 2009 Renewable Energy Directive (Dir 2009/28/EC) http://eur-lex.europa.eu/LexUriServ/LexUriServ.do?uri=Oj:L:2009:140:0016:0062:en:PDF

20. FIC 2013 Farmland Information Center, www.farmlandinfo.org/, accessed in May 2013

21. Georgescu M, Lobell D B and Field C B 2009 Potential impact of US biofuels on regional climate Geophys. Res. Lett. 36 L21806

22. Georgescu M, Lobell D B and Field C B 2011 Direct climate effects of perennial bioenergy crops in the United States Proc. Natl. Acad. Sci. 108 4307–12

23. Georgescu M, Lobell D B, Field C B and Mahalov A 2013 Simulated hydroclimaticimpacts of projected Brazilian sugarcane expansion Geophys. Res. Lett. 40 972–7

24. Glenn E P, O'leary J W, Watson M C, Thompson T L and Kuehl R O 1991 Salicornia bigelovii Torr.: an oilseed halophyte for seawater irrigation Science 251 1065–7

25. Guinée J B, Heijungs R, Huppes G, Zamagni A, Masoni P, Buonamici R, Ekvall T and Rydberg T 2011 Life cycle assessment: past, present, and future Environ. Sci. Technol. 45 90–6

26. Hallgren W, Schlosser C A, Monier E, Kicklighter D, Sokolov A and Melillo J 2013 Climate impacts of a large-scale biofuels expansion Geophys. Res. Lett. 40 1624–30

27. Hansen J and Nazarenko L 2004 Soot climate forcing via snow and ice albedos Proc. Natl Acad. Sci. 101 423–8

28. Hill J, Nelson E, Tilman D, Polasky S and Tiffany D 2006 Environmental, economic, and energetic costs and benefits of biodiesel and ethanol biofuels Proc. Natl Acad. Sci. 103 11206–10

29. Hungate B A and Hampton H M 2012 Ecosystem services: valuing ecosystems for climate Nature Clim. Change 2 151–2

30. Joos F et al 2013 Carbon dioxide and climate impulse response functions for the computation of greenhouse gas metrics: a multi-model analysis Atmos. Chem. Phys. 13 2793–825

31. Kim S and Dale B E 2005 Life cycle assessment of various cropping systems utilized for producing biofuels: Bioethanol and biodiesel Biomass Bioenergy 29 426–39

32. Koizumi T 2011 Biofuel programs in East Asia: developments, perspectives, and sustainability Environmental Impact of Biofuels ed M Aurelio Dos and S Bernardes (Rijeka: InTech Europe)

33. Lardon L, Hélias A, Sialve B, Steyer J-P and Bernard O 2009 Life-cycle assessment of biodiesel production from microalgae Environ. Sci. Technol. 43 6475–81

34. Larson E D 2006 A review of life-cycle analysis studies on liquid biofuel systems for the transport sector Energy Sust. Dev. 10 109–26

35. Lean J and Rowntree P R 1993 A GCM simulation of the impact of Amazonian deforestation on climate using an improved canopy representation Q. J. R. Meteorol. Soc. 119 509–30

36. Lee X et al 2011 Observed increase in local cooling effect of deforestation at higher latitudes Nature 479 384–7

37. Lenton T M and Vaughan N E 2009 The radiative forcing potential of different climate geoengineering options Atmos. Chem. Phys. 9 5539–61

38. Loarie S R, Lobell D B, Asner G P, Mu Q and Field C B 2011 Direct impacts on local climate of sugar-cane expansion in Brazil Nature Clim. Change 1 105–9

39. Lohila A, Minkkinen K, Laine J, Savolainen I, Tuovinen J-P, Korhonen L, Laurila T, Tietäväinen H and Laaksonen A 2010 Forestation of boreal peatlands: impacts of changing albedo and greenhouse gas fluxes on radiative forcing J. Geophys. Res.: Biogeosci. 115 G4

40. McLaughlin S B, De la Torre Ugarte D G, Garten C T Jr, Lynd L R, Sanderson M A, Tolbert V R and Wolf D D 2002 High-value renewable energy from prairie grasses Environ. Sci. Technol. 36 2122–9

41. Mosali J, Biermacher J T, Cook B and Blanton J 2013 Bioenergy for cattle and cars: a switchgrass production system that engages cattle producers Agron. J. 105 960

42. Muñoz I, Campra P and Fernández-Alba A R 2010 Including CO2-emission equivalence of changes in land surface albedo in life cycle assessment. Methodology and case study on greenhouse agriculture Int. J. Life Cycle Assess 15 672–81

43. NASA ASDC 2013 NASA Surface Meteorology and Solar Energy—Location available at https://eosweb.larc.nasa.gov/cgi-bin/sse/grid.cgi? accessed in July 2013

44. NASA MODIS 2013a Index of /MOTA/MCD12Q1.051 available at http://e4ftl01.cr.usgs.gov/MOTA/MCD12Q1.051/ accessed in July 2013

45. NASA MODIS 2013b Index of /MOTA/ MCD43A3.005 available at http://e4ftl01.cr.usgs.gov/MOTA/MCD43A3.005/ accessed in July 2013

46. NASA MODIS 2013c Index of /MOTA/ MCD43A2.005 available at http://e4ftl01.cr.usgs.gov/MOTA/MCD43A2.005/ accessed in July 2013

47. OECD/IEA 2004 World Energy Outlook 2004

48. Perlack R D, Wright L L, Turhollow A F, Graham R L, Stokes B J and Erbach D C 2005 Biomass as feedstock for a bioenergy and bioproducts industry: the technical feasibility of a billion-ton annual supply US Department of Energy and US Department of Agriculture Report ORNL/TM-2005/66

49. Pielke R A and Avissar R 1990 Influence of landscape structure on local and regional climate Landscape Ecol. 4 133–55

50. Pitman A J et al 2009 Uncertainties in climate responses to past land cover change: First results from the LUCID intercomparison study Geophys. Res. Lett. 36 L14814

51. Plevin R J, Michael O'Hare, Jones A D, Torn M S and Gibbs H K 2010 Greenhouse gas emissions from biofuels' indirect land use change are uncertain but may be much greater than previously estimated Environ. Sci. Technol. 44 8015–21

52. Rautiainen M, Nilson T and Lükk T 2009 Seasonal reflectance trends of hemiboreal birch forests Remote Sens. Environ. 113 805–15

53. Rautiainen M, Stenberg P, Mottus M and Manninen T 2011 Radiative transfer simulations link boreal forest structure and shortwave albedo Boreal Environ. Res. 16 91–100

54. Rhines S 2008 Oklahoma's Bioenergy Future http://www.ces.ncsu.edu/nreos/forest/feop/biomass-south/proceedings/pdf/0923-1515c-Rhines.pdf

55. Song J 1999 Phenological influences on the albedo of prairie grassland and crop fields Int. J. Biometeorl. 42 153–7

56. Stratton R W, Wong H M and Hileman J I 2010 Lifecycle greenhouse gas emissions from alternative jet fuels PARTNER Project 28 Report, PARTNER-COE-2010-001

57. Stratton R W, Wong H M and Hileman J I 2011a Quantifying variability in life cycle greenhouse gas inventories of alternative middle distillate transportation fuels Environ. Sci. Technol. 45 4637–44

58. Stratton R W, Wolfe P J and Hileman J I 2011b Impact of aviation Non-CO2 combustion effects on the environmental feasibility of alternative jet fuels Environ. Sci. Technol. 45 10736–43

59. US Department of Energy 2011 U.S. billion-ton update: biomass supply for a bioenergy and bioproducts industry R D Perlack, B J Stokes (Leads), ORNL/TM-2011/224

60. US Environmental Protection Agency 2013 Renewable Fuels Standard (Washington, DC) www.epa.gov/otaq/fuels/renewablefuels/index.htm

61. Van der Voet E, Lifset R J and Luo L 2010 Life-cycle assessment of biofuels, convergence and divergence Biofuels 1 435–49

62. Wicke B, Dornburg V, Faaij A P C and Junginger H M 2007 A Greenhouse Gas Balance of Electricity Production from Co-Firing Palm Oil Products from Malaysia NWS-E-2007-33

63. Yee K F, Tan K T, Abdullah A Z and Lee K T 2009 Life cycle assessment of palm biodiesel: Revealing facts and benefits for sustainability Appl. Energy 86 S189–96

64. Yeh S, Jordaan S M, Brandt A R, Turetsky M R, Spatari S and Keith D W 2010 Land use greenhouse gas emissions from conventional oil production and oil sands Environ. Sci. Technol. 44 8766–72

65. Zhou A and Thomson E 2009 The development of biofuels in Asia Appl. Energy 86 S11–20

CHAPTER 7

Biofuel for Energy Security: An Examination on Pyrolysis Systems with Emissions from Fertilizer and Land-Use Change

CHIH-CHUN KUNG, HUALIN XIE, TAO WU, AND SHIH-CHIH CHEN

7.1 INTRODUCTION

Taiwan is vulnerable to high energy prices and market distortions in the world energy market because only a small fossil fuel stock is found in Taiwan and most of Taiwan's energy is imported [1]. To enhance Taiwan's energy security, there is interest for the Taiwanese to produce energy on its own. In addition to the energy insecurity, another serious challenge facing Taiwan is climate change. According to the 2007 report by the Intergovernmental Panel on Climate Change [2], the Earth is warming due to anthropogenic emissions of greenhouse gases (GHGs) and its temperature is very likely to increase in the next decades. Such warming would have consequences ranging from increased desertification, a rise in the ocean

Biofuel for Energy Security: An Examination on Pyrolysis Systems with Emissions from Fertilizer and Land-Use Change. © Kung C-C, Xie H, Wu T, and Chen S-C. Sustainability **6**,2 (2014), doi:10.3390/su6020571. Licensed under Creative Commons Attribution 3.0 Unported License, http://creativecommons.org/licenses/by/3.0/.

level to the possible increased occurrences of hurricanes, which may bring potential significant damages to Taiwan. As the 25th largest CO_2 emissions country [3], Taiwan is willing to reduce CO_2 emissions and mitigate global climate shift to avoid unwelcome climate impacts, once the energy security issue is resolved. Renewable energy sources that can potentially substitute fossil fuels and provide some of the domestic energy supply include wind and solar energy, hydro-power, geothermal energy and bio-energy [4]. Among these renewable energy alternatives, Taiwan has been developing bioenergy for several years. Geographically, Taiwan's land area is about 14,000 square miles with 67% of that land being mountainous. Land is a scarce resource in Taiwan and has been intensively utilized in various ways. From this point of view, Taiwan would not be able to produce bioenergy because a substantial amount of land is required for bioenergy production. However, participation in the World Trade Organization (WTO) offers a possibility of development of bioenergy in Taiwan because Taiwan's agricultural sector is less competitive and part of Taiwan's agricultural land has been idle. Net idled cropland has increased from 68,000 hectares to 280,000 hectares, which provides a potential stock of land for bioenergy feedstock production [1].

Although bioenergy can potentially enhance Taiwan's energy security and reduce GHG emissions [5,6], two important factors, the GHG emissions from land use change and fertilizer use, have been ignored. When agricultural land is converted into other uses, NOx emissions will change and result in different CO_2 equivalent (CO_2e hereafter) emissions [7,8,9]. If the change in NOx is small, neglecting to consider this factor may not significantly affect the result. However, this change is usually large [10,11]. Snyder et al. [12] also point out that the most important GHG issue from agriculture is N_2O, mainly from soils and N inputs to crop and soil systems. They show that, from the global warming potential (GWP) point of view, even though N_2O is a small part of the overall GHG issue, agriculture is considered to be the main source that is linked to soil management and fertilizer use. Therefore, examining bioenergy production and GHG emissions offset without considering associated GHG emissions from land use change and fertilizer use may result in the disaster. This study aims to examine the GHG emissions from various bioenergy production levels under different gasoline, coal and GHG prices. The work makes contributions by integrating multiple bioenergy technologies (ethanol, co-fire and pyrolysis), energy crops, energy and GHG prices and emissions from land use change and fertilizer use into a single study, which provides information about potential enhancement of Taiwan's energy security and GHG emissions offset to the Taiwanese government.

7.2 LITERATURE REVIEW

Among available bioenergy techniques, Taiwan can produce bioenergy in the forms of ethanol, direct combustion biopower (conventional bioelectricity) and biopower through pyrolysis (pyrolysis-based electricity). Because these technologies are not mutually exclusive and can be employed at the same time, it is necessary for us to consider all combinations. Bioenergy involving ethanol and conventional electricity have been examined and applied for more than a decade, but bioenergy produced from pyrolysis is intensively studied only in recent years. Pyrolysis involves heating biomass in the absence of oxygen and results in the decomposition of biomass into biooil, biogas and biochar. Biooil and biogas are used to generate electricity in the pyrolysis plant while biochar was also used as an energy source but many studies found that biochar can bring significant environmental and associated economic benefits when it is used as a soil amendment [13,14,15,16,17]. In general, pyrolysis can be categorized as fast pyrolysis, medium pyrolysis, slow pyrolysis and gasification. In this study, we examine the two popular types of pyrolysis techniques (fast and slow pyrolysis), and two uses of biochar (burn biochar in the pyrolysis plant or haul biochar back to the cropland) are incorporated in our bioenergy production framework.

The reason that we would like to examine biochar is because it has been shown to improve agricultural productivity and the environment in several ways. Specifically, biochar is stable in the soil [13] and has nutrient-retention properties that lead to increases in crop yields [18]. Moreover, biochar offers a chance to sequester carbon [13]. As pyrolysis can provide a significant amount of renewable energy and offset more GHG emissions [5,6,16], it is a potential bioenergy technique that Taiwan is interested in.

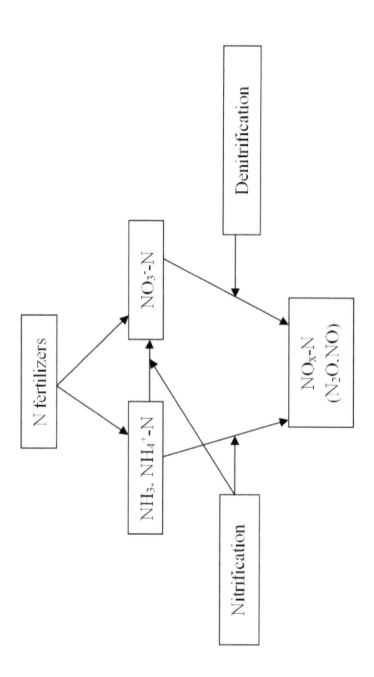

FIGURE 1: N_2O emissions due to N fertilizer application (Source: [28]).

Lifecycle analysis of GHG emissions has been examined widely in bioenergy production [7,13,16,19] and land use changes [10,11,20,21,22]. Some studies examined the impacts of emissions from global land use changes on the lifecycle emissions of corn ethanol [7,9] and other studies focused on the land use change emissions when specific land types are culti-vated for cropland use [22,23,24]. Land use change has also been examined in large scale. For example, Timilsina et al. [25] analyzed the long-term im-pacts of large-scale expansion of biofuels on land use change, food supply and prices, and the overall economy in various countries or regions, while Kwon et al. [26] examined the state-level soil carbon emissions for direct land use change in the United States. Land use change also has various GHG effects. Schaufler et al. [21] found that different land-use strongly affected GHG fluxes in cropland, grassland, forests and wetland. N_2O and CO_2 emis-sions are highest in grassland soils while NO emissions are highest in forest soils, which are also positively correlated with N input. Moreover, Baldos [20] utilized the Land Use Change Emissions (LUCE) modules and found that the direct lifecycle GHG emissions of corn ethanol fuel can exceed the 20% GHG reduction requirement in the US EISA given the data and as-sumptions during corn farming and ethanol production. Wang et al. [8] also developed a widely applied modeling approach (including GREET model and CCLUB model (Carbon Calculator for Land Use Change from Biofuels Production)) in the US to study biofuel life cycle. Baggs et al. [19] found that zero tillage resulted in higher N_2O emissions than conventional tillage and N_2O emissions were generally correlated with CO_2 emissions. How-ever, this result must be adopted very carefully when different land types are examined. Grover et al. [11] indicated that soil-based GHG emissions will increase from 53 to 70 t CO_2-equivalents after land use change. Far-quharson and Baldock [27] indicated that adding N fertilizers will increase N_2O emissions due to nitrification and denitrification process where the am-monium (NH_4^+) is available for nitrification. Synder et al. [12] also showed that fertilizer induced N_2O emissions from soil equates to a GWP of 4.65 kg CO_2 kg^{-1} of N applied. Therefore, when we change crops and associated N fertilizer application, it will lead to different NOx emissions. Qin et al. [28] developed an agroecosystem model (AgTEM) that was incorporated with biogeochemical and ecophysiological processes. They found that N_2O emitted from croplands with high N application rates is mostly larger than those with lower N input levels. The average N_2O flux is 1.8 kg N ha^{-1} and most of the simulation results are within a reasonable range (e.g., less than 5 kg N ha^{-1}). Their recent work [29] also examined carbon and nitrogen dynamics for maize and cellulosic crops. They indicated that for maize the global warming potential (GWP) amounts to 1–2 Mg CO_2eq ha^{-1} yr^{-1}, with a dominant contribution of over 90% from N_2O emissions while cellulosic crops contribute to the GWP of less than 0.3 Mg CO_2eq ha^{-1} yr^{-1}. The general process of NOx emissions is depicted as Figure 1.

However, they also point out that this estimate will vary depending on local climates since NH_3 volatilization and NO_3^- leaching are heavily affected by the climate. These studies focus on how many GHG emissions are produced during bioenergy production or the mechanism of GHG emissions from land use change, all of which have provided significant contributions to the literature, but the relationship between bioenergy production strategies and GHG emissions from land use change are usually not linked. Therefore, in order to know how land use GHG emissions may affect the benefits from bioenergy production, it is necessary to incorporate the emissions from fertilizer and change on land use into the bioenergy production to present a more general lifecycle analysis on bioenergy.

7.3 MODEL STRUCTURE

The model used herein is based on price endogenous mathematical programming, which is originally illustrated by Samuelson [30]. Samuelson shows the equilibrium in the perfect competition market can be derived from the optimization model that maximizes the consumer surplus and producer surplus. Takayama and Judge [31] establish a mathematical programming model on spatial model based on Samuelson's idea while McCarl and Spreen [32] point out that this model is useful in policy analysis, especially in its property of price endogeneity. In addition, McCarl and Spreen [32] compare the linear programming models used by other planned economic systems to the price endogenous model, and the results showed that the price endogenous model can represent the economic system in a perfectly competitive market and thus, can be useful in policy analysis including soil conservation policy [33], global climate change [34,35,36], and climate change mitigation [37]. It has also been applied extensively for research evaluation [38,39].

Chen and Chang [40] develop the Taiwan Agricultural Sector Model (TASM) to analyze the Taiwanese agricultural policy in terms of production and market issues. The TASM is a multi-product partial equilibrium model based on the previous work [32,33,39,41]. This empirical structure has been adapted to Taiwan and used in many policy-related studies [40,42,43]. The current version of TASM accommodates more than 110 commodities in 15 subregions aggregated into four major production and processing regions. We extended the TASM to evaluate the potential economic and GHG implications of bioenergy crop production plus competition with other land uses. GHG emissions from land and fertilizer use are also incorporated into the modified TASM. The modified TASM simulates market operations under assumptions of perfect competition with individual producers and consumers as price-taker. It also incorporates price-dependent product demand and input supply curves.

7.3.1 MODIFIED TAIWAN AGRICULTURAL SECTOR MODEL

TASM was constructed by Chen and Chang [41] under above theory and for this analysis we extend this model by adding features related to bioenergy and N_2O emissions. Specifically, to get a version for use herein, we have to address how energy crops and GHG emissions are incorporated in the modified TASM. We illustrate the algebraic form of the objective function of the modified TASM and its constraints. The objective function and constraints of modified TASM are shown as follows:

$$Max \sum_i \int \psi(Q_i) dQ_i$$

$$- \sum_i \sum_k C_{ik} X_{ik}$$

$$- \sum_k \int \alpha_k(L_k) dL_k - \sum_k \int \beta_K(R_k) dR_k + \sum_i P_i^G * Q_i^G + \sum_k P^L * AL_K$$

$$+ \sum_j \sum_k SUB_j * EC_{jk} + \sum_i \int ED(Q_i^M) dQ_i^M - \sum_i \int ES(Q_i^X) dQ_i^X$$

$$+ \sum_i \int EXED\ (TRQ_i) dTRQ_i + \sum_i [tax_i * Q_i^M + outtax_i * TRQ_i]$$

$$- P_{GHG} * \sum_g GWP_g * GHG_g$$

(1)

Subject to:

$$Q_i + Q_i^X + Q_i^G - \Sigma_k Y_{ik} X_{ik} - \Sigma_j EC_{jk} X_{jk} - (Q_i^M + TRQ_i) \leq 0 \qquad \text{for all i,}$$

(2)

$$\Sigma_i X_{ik} + AL_k + \Sigma_j EC_{jk} - L_k \leq 0 \qquad \text{for all k,} \tag{3}$$

$$\Sigma_i f_{ik} X_{ik} - \Sigma_j f_{jk} X_{jk} - R_k \leq 0 \qquad \text{for all k,} \tag{4}$$

$$\Sigma_{i,k} E_{gik} X_{ik} - Baseline_g = GHG_g \qquad \text{for all g} \tag{5}$$

Table 1 details the variables using in the objective function and constraints.

The objective function of the modified TASM model incorporates the domestic and trade policies where the first term is the area under the domestic demand curve and the second, third and fourth terms stand for input costs, cropland rent and labor costs, respectively. The fifth, sixth and seventh terms reflect the government subsidy on rice purchase, set-aside lands and for planting energy crops to represent the social welfare in a closed market. The eighth and ninth terms represent the area under the excess demand curve and the 10th term stands for the area under the excess supply curve. The 11th term is tariff revenue. GHG emission is modified in the last term to reflect that GHG emissions reduce social welfare. Equation (2) is the balance constraint for commodities. Equations (3) and (4) are the resource endowment constraints. Equation (3) controls cropland and Equation (4) is the other resource constraint. Equation (5) is further modified to reflect the greenhouse gas balance which shows emissions emitted of CO_2e (including emissions from bioenergy production, land use change and fertilizer use but CH4 emissions from animal manures) cannot be greater than total emissions.

TABLE 1: Variables.

Variable	Description of Variables
Q_i	Domestic demand of ith product
Q_i^G	Government purchases quantity for price supported ith product
Q_i^M	Import quantity of ith product
Q_i^X	Export quantity of ith product
$\psi(Q_i)$	Inverse demand function of ith product
P_i^G	Government purchase prince on ith product
C_{ik}	Purchased input cost in kth region for producing ith product
X_{ik}	Land used for ith commodities in kth region
L_k	Land supply in kth region
$\alpha_k(L_k)$	Land inverse supply in kth region
R_k	Labor supply in kth region
$\beta_k(R_k)$	Labor inverse supply in kth region
P^L	Set-aside subsidy
AL_k	Set-aside acreage in kth region
SUB_j	Subsity on planting jth energy crop
EC_{jk}	Planted acreage of jth energy crop in kth region
$ED(Q_i^M)$	Invese excess import demand curve for ith product
$ES(Q_i^X)$	Inverse excess export supply curve for ith product
TRQ_i	Import quantity exceeding the quota for ith product
$EXED(TRQ_i)$	Inverse excess demand curve of ith product that the import quantity is exceeding quota
tax_i	Import tariff for ith product
$outtax_i$	Out-of-quota tariff for ith product
Y_{ik}	Per hectare yield of ith commodity produced in kth region
E_{gik}	gth greenhouse gas emmision from ith produce in kth region
P_{GHG}	Price of GHG gas
GWP_g	Global warming potential of gth greenhouse gas
GHG_g	Net greenhouse gas emissions of gth gas
$Baseline_g$	Greenhouse gas emissions under the baseline of the gth gas
f_{ik}	Labor required per hectare of commodity i in region k

The data sources of agricultural commodities largely come from published government statistics and research reports, which include the Taiwan Agricultural Yearbook, Production Cost and Income of Farm Prod-ucts Statistics, Commodity Price Statistics Monthly, Taiwan Agricultural Prices and Costs Monthly, Taiwan Area Agricultural Products Wholesale Market Yearbook, Trade Statistics of the Inspectorate-General of Cus-toms, Forestry Statistics of Taiwan. Demand elasticities of agricultural products come from various sources and were gathered and sent by Chang and Chen.

7.4 STUDY SETUP

This study examines Taiwan's bioenergy production from ethanol, con-ventional bioelectricity and pyrolysis based electricity, and GHG emis-sions offset by utilizing current set-aside land with the consideration of the emissions from fertilizer use and land use change. Three gasoline prices (NT\$20, 30, 40 per liter), two coal prices (NT\$1.7, 3.45 per kg), six GHG prices (NT\$ 5, 10, 15, 20, 25, 30 per ton) plus estimated emissions from fertilizer use and land use change. The simulated gasoline and coal prices are selected based on the ranges of their market prices in 2012. Since Tai-wan has not established a GHG trading mechanism and GHG emission is currently of no value in Taiwan, the study examines several potential GHG prices based on the opinion of Professor Chi-Chung Chen, who is familiar with and engaged in Taiwanese agricultural and environmental policies.

The net mitigation of CO_2 from ethanol is estimated by [44]. They show that net CO_2 emissions are reduced by 0.107 ton per 1000 liters of ethanol. For conventional electricity, McCarl [45] shows that poplar can offset about 71.3% of carbon dioxide emissions relative to the fossil fuel and 75.1% for switchgrass. We calculate that the emissions reduction is 0.195 kg CO_2 for poplar and 0.246 kg CO_2 for switchgrass.

GHG emissions from land use change are estimated by Liu et al. [10], who calculate that annual mean GHG fluxes from soil of plantation and orchard are 4.70 and 14.72 Mg CO_2-C ha^{-1} yr^{-1}, −2.57 and −2.61 kg CH_4-C ha^{-1} yr^{-1} and 3.03 and 8.64 kg N_2O-N ha^{-1} yr^{-1}, respectively. Qin et al. [29] also indicated that the average N_2O flux is 1.8 kg N ha^{-1} and most of the simulation results are less than 5 kg N ha^{-1}. Because CO_2 and N_2O emissions are highly correlated with each other [19], we assume that the emission profile of CO_2 and N_2O are staying at the same level. In addition, Snyder et al. [12] show that fertilizer induced N_2O emissions from soil equates to a GWP of 4.65 kg CO_2 kg^{-1} of N applied. With these estimates, we arrive at the estimated emission level from fertilizer use and land use change (Table 2). Biochar also offers GHG emissions offset potential. This study also incorporates the GHG effect for different uses of biochar to see how it affects the GHG emissions reduction, based on Kung et al.'s estimates [5].

TABLE 2: Estimated emission level from fertilizer use and land use change.

	GHG	Units	Estimated emission level
GHG emissions from fertilizer and land use change	CO_2	Mg ha^{-1} yr^{-1}	4.7
	CH_4	kg ha^{-1} yr^{-1}	−2.57
	N_2O	kg ha^{-1} yr^{-1}	26.86
Net emissions	CO_2e	Mg ha^{-1} yr^{-1}	11.62

7.5 RESULTS, POLICY IMPLICATIONS

The simulation result indicates that when Taiwan tries to enhance its energy security by developing bioenergy, net GHG emissions are likely to increase, especially when GHG price is low (see Figure 2 and Figure 3). As indicated in Figure 2, emissions reduction from Taiwan's bioenergy production is lower than the emissions increased from fertilizer use and land use change. Only when GHG price is high and gasoline price is low, net emissions reduction may be achieved, and when the gasoline price keeps increasing, net emissions will increase (Figure 3).

The result shows that when energy security is the first priority of Taiwanese government, net GHG emissions will not be reduced in most cases. In other words, the study indicates that Taiwan gains energy security at a cost of emitting more. This is partly due to the fact that the emission offset ability of ethanol is lower than that of pyrolysis based electricity. When the gasoline price is low, feedstocks will be used in ethanol and electricity production but when the gasoline price increases, more feedstocks are converted into the ethanol production and the total amount of emissions offset is reduced. Market price also affects the net emissions from fertilizer use and land use change. Under low energy prices, bioenergy production is less profitable and less set-aside land is used for energy crop plantation. Fewer plantations require fewer fertilizer and therefore, emissions from fertilizer use and land use change will be smaller. When energy prices increase, more land is converted and brings higher emissions from land use change and fertilizer use. Although bioenergy is considered as a carbon sequestration technology, lifecycle analysis including fertilizer use and land use change indicates that ethanol does not bring GHG emissions reduction while pyrolysis is possible to offset emissions under certain conditions.

In this study, bioenergy comes from various sources including ethanol, conventional bioelectricity and pyrolysis based electricity. The result indicates that conventional bioelectricity is driven out by pyrolysis based electricity and electricity is solely produced via pyrolysis. In general, pyrolysis produces three outputs including biooil, biogas and biochar, all of which can be used to generate electricity. Because biochar is found to enhance crop yield and store carbon in a more stable form when used as a soil amendment, various uses of biochar are incorporated into the study. The result indicates that when biochar is used as a soil amendment, bioenergy production is relatively lower than when biochar is burned in the pyrolysis plant (Figure 4, Figure 5 and Appendix). However, if biochar is burned to provide electricity, it is unlikely to provide net GHG emissions offset. Using biochar as a soil amendment is possible to offset GHG emissions only when the GHG price is high. If the GHG price is low, ethanol production is high and fast pyrolysis that will generate more electricity will dominate slow pyrolysis that yields more biochar. Only when the GHG price increases to a certain level, slow pyrolysis becomes a dominant technology and ethanol production decreases.

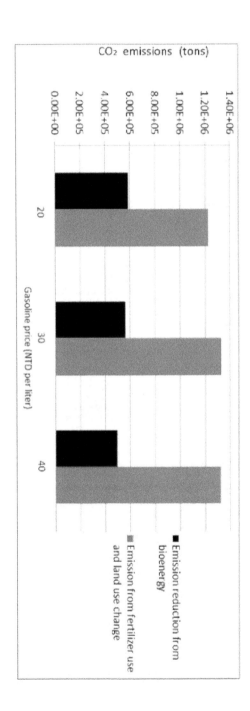

FIGURE 2: Estimated emissions from bioenergy production, fertilizer use and land use change at low greenhouse gas (GHG) price.

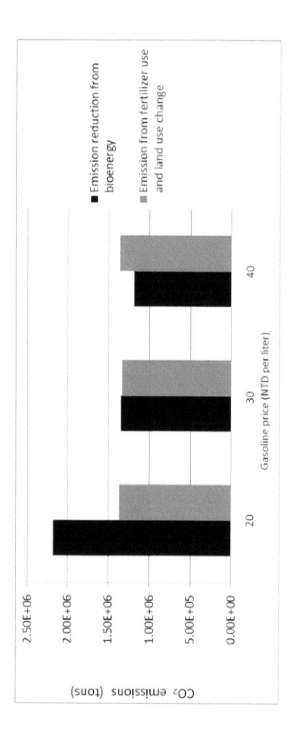

FIGURE 3: Estimated emissions from bioenergy production, fertilizer use and land use change at high GHG price.

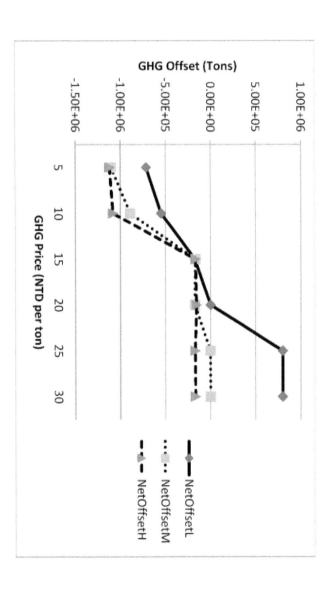

FIGURE 4: Net emissions from bioenergy production when biochar is used as a soil amendment. Net offset (L,M and H) represents the low, medium and high gasoline prices, respectively.

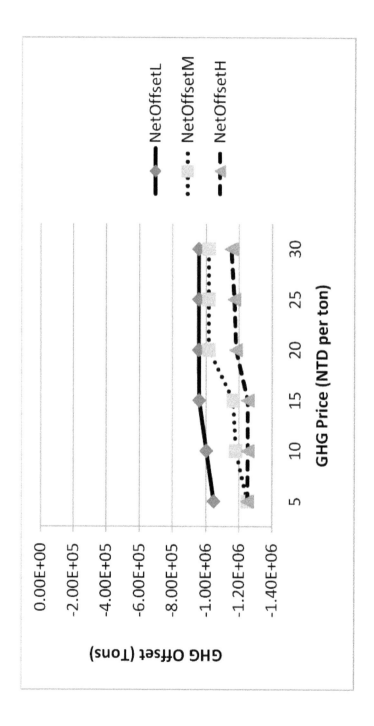

FIGURE 5: Net emissions from bioenergy production when biochar is used as an energy source. Net offset (L,M and H) represents the low, medium and high gasoline prices, respectively.

7.5.1 POLICY IMPLICATIONS

The study provides some insights about environmental effects on Taiwan's energy security concern via bioenergy production. The result shows that Taiwan's energy security can be enhanced by producing ethanol and electricity using currently set-aside land. However, bioenergy does not bring environmental benefits in most of cases when land GHG and fertilizer emissions are incorporated. In other words, bioenergy does not always offset GHG emissions in a broader lifecycle analysis. Some policy implications for Taiwan to gain economic and environmental benefits plus enhancement of energy security are provided including:

1. When Taiwan tries to develop a GHG emissions trading mechanism, effects of the trading system on domestic renewable energy production must be incorporated. As the study shows, bioenergy production is heavily impacted by GHG prices. Therefore, under a marketable GHG emissions trading system, effectiveness of energy security enhancement from bioenergy must be validated;

2. Development of the bioenergy industry requires long term planning. The simulation result indicates that Taiwan can enhance energy security from bioenergy production at a cost of higher emissions. However, under low energy prices, less set-aside land will be converted into the energy crop plantation and results in a net emissions offset. Bioenergy production will shrink under this situation. Therefore, in order to ensure energy security enhancement when the energy price is low, some government subsidies may be required for farmers to convert set-aside land into energy crop plantations;

3. This study shows that GHG emissions from fertilizer use and land use change are significant and have important impacts on both bioenergy production and net GHG emissions offset. Therefore, a proper estimation of these emission rates is required. The study examines the bioenergy production and GHG effects on Taiwan's set-aside land, located in the four major areas in Taiwan. Due to local soil and weather conditions, NOx emission rates from land use change and fertilizer application should not be the same in these

areas and future studies must be conducted in order to draw a more realistic picture;

4. Although energy security is the prior concern on Taiwan's bioenergy development, it may not always be so. As Taiwan is facing direct challenges from global climate shifts, GHG mitigation is another important issue that the Taiwanese must address. Bioenergy is one possible way to increase domestic energy supply, but it may not be an appropriate method for GHG emissions offset, especially for the significant effects from fertilizer and land use change emissions. As the result shows, Taiwan is not able to achieve the maximal bioenergy production and GHG emissions offset at the same time. The Taiwanese government must take this into account for future policy decisions;

5. Not all set-aside land can be used for bioenergy production. Joining the WTO releases some agricultural land but the Taiwanese government has been trying to utilize the idle land for other economic purposes including development of recreation sites and high economic value commodities. Therefore, using all set-aside land in bioenergy production may not be feasible. Further adjustments combining all existing and potential agricultural and associated policies may be required.

7.6 CONCLUSIONS

The study examines how much bioenergy can be produced and the consequent GHG emissions effect as Taiwan attempts to enhance its energy security. Simulation results indicate that while bioenergy indeed increases Taiwan's energy security, it is likely to increase net GHG emissions. This is somewhat contradictory to previous studies showing bioenergy provides both renewable energy and GHG emissions mitigation. Our result shows that emissions reduction by bioenergy is offset by the emissions of fertilizer use and land use change. Throughout 72 scenarios, only eight cases show net GHG emissions offset. GHG price is another important factor influencing the bioenergy production and GHG emissions offset. At a higher GHG price, ethanol production will shrink to a very low level and slow pyrolysis dominates all other bioenergy technologies. However, when Taiwan places energy security as its first priority, the impacts of GHG price on bioenergy production will be small.

7.6.1 Limitations The study has some limitations that must be addressed. First, potential uncertainties exist for many important factors. Depending on land types, GHG emissions from land use change, fertilizer use and soil carbon sequestration may differ, all of which lead to a different result. Second, Taiwan's GHG trading system has not been established and therefore, GHG prices used in this study are only based on the professional opinions rather than real market data. Further investigation is needed when the GHG emissions trading market is built. Third, the hauling distance of biofuel and biochar is estimated from McCarl et al.'s study [46], which assumes the pyrolysis plant is in the centre of a square surrounded by a grid layout of roads. This assumption may be released by combining GIS method to reflect a more accurate hauling distance and associated GHG emissions. Finally, CH_4 is another important GHG in agriculture, especially for rice paddies and animal manure. This study does not incorporate CH_4 emissions from specific agricultural commodities; instead, the study focuses on the CH_4 emissions from the land used for bioenergy crop plantation. CH_4 emissions must be incorporated into the analysis when rice straw and manure are used in bioenergy production (e.g., pyrolysis).

REFERENCES

6. Chen, C.C.; McCarl, B.A.; Chang, C.C.; Tso, C. Evaluation the potential economic impacts of Taiwanese biomass energy production. Biomass Bioenergy 2011, 35, 1693–1701.

7. Intergovernmental Panel on Climate Change (IPCC). Guidelines for National Greenhouse Gas Inventories; Cambridge University Press: Cambridge, UK, 2007.

8. Carbon Dioxide Information Analysis Center (CDIAC). List of Countries by Carbon Dioxide Emissions; Annual Report. United States Department of Energy: Washington, DC, USA, 2008.

9. Turner, D.P.; Lee, J.L.; Koerper, G.J.; Barker, J.R. The forest sector carbon budget of the United States: Carbon pools and flux under alternative policy options. Environ. Res. 1993, 8, 27–39.

10. Kung, C.C.; McCarl, B.A.; Cao, X.Y. Environmental impact and bioenergy production from pyrolysis in Taiwan. Energy Policy 2013, 60, 317–323.

11. Kung, C.C.; McCarl, B.A.; Cao, X.Y.; Xie, H.L. Bioenergy prospects in Taiwan using set—aside land-an economic evaluation. China Agr. Econ. Rev. 2013, 5, 489–511.

12. Wang, M.; Saricks, C.; Santini, D. Effects of Fuel Ethanol Use on Fuel-Cycle Energy and Greenhouse Gas Emissions; Report No. ANL/ESD-38; Center for Transportation Research, Argonne National Laboratory: Argonne, IL, USA, 1999.

13. Wang, M.; Han, J.; Dunn, J.B.; Cai, H.; Elgowainy, A. Well-to-wheels energy use and greenhouse gas emissions of ethanol from corn, sugarcane and cellulosic biomass for us use. Environ. Res. Lett. 2012, 7, 045905.

14. Searchinger, T.; Heimlich, R.; Houghton, R.A.; Dong, F.; Elobeid, A.; Fabiosa, J.; Tokgoz, S.; Hayes, D.; Yu, T.-H. Use of U.S. croplands for biofuels increases greenhouse gases through emissions from land-use change. Science 2008, 319, 1238–1240.

15. Liu, H.; Zhao, P.; Lu, P.; Wang, Y.S.; Lin, Y.B.; Rao, X.Q. Greenhouse gases fluxes from soils of different land-use types in a hilly area of south China. Agr. Ecosyst. Environ. 2007, 124, 125–135.

16. Grover, S.P.P.; Liverley, S.J.; Hutley, I.B.; Jamali, H.; Fest, B.; Beringer, J.; Butterbach-Bahl, K.; Arndt, S.K. Land use change and the impact on greenhouse gas exchange in north Australian savanna soils. Biogeosciences 2012, 9, 423–437.

17. Snyder, C.S.; Bruulsema, T.W.; Jensen, T.L.; Fixen, P.E. Review of greenhouse gas emissions from crop production systems and fertilizer management effects. Agr. Ecosyst. Environ. 2009, 133, 247–266.

18. Lehmann, J. A Handful of carbon. Nature 2007, 447, 143–144.

19. Lehmann, J.; Gaunt, J.; Rondon, M. Biochar sequestration in terrestrial ecosystems—a review. Mitig. Adapt. Strat. Gl. 2006, 11, 403–427.

20. Kung, C.C. Economics of Taiwanese Biochar Production, Utilization and GHG Offsets: A Case Study on Taiwanese Rice Fields. Ph.D. Thesis, Texas A&M University, College Station, Texas, TX, USA, 2010.

21. McCarl, B.A.; Peacocke, C.; Chrisman, R.; Kung, C.C.; Ronald, D. Economics of biochar production, utilization, and GHG offsets. In Biochar for Environmental Management: Science and Technology; Lehmann, J., Joseph, S., Eds.; Earthscan Publisher: London, UK, 2009; pp. 341–357.

22. Deluca, T.H.; MacKenzie, M.D.; Gundale, M.J. Biochar effects on soil nutrient transformations. In Biochar for Environmental Management: Science and Technology; Lehmann, J., Joseph, S., Eds.; Earthscan Publisher: London, UK, 2009; pp. 137–182.

23. Chan, K.Y.; Zwieten, L.; Meszaros, I.; Downie, A.; Joseph, S. Agronomic values of green waste biochar as a soil amendment. Aust. J. Soil. Res. 2007, 45, 629–634.

24. Wang, M. Well-to-wheels Energy and Greenhouse Gas Emission Results of Fuel Ethanol; Working paper. Argonne National Laboratory: Lemont, IL, USA, 2007.

25. Baggs, E.M.; Stevenson, M.; Pihlatie, M.; Regar, A.; Cook, H.; Cadisch, G. Nitrous oxide emissions following application of residues and fertilizer under zero and conventional tillage. Plant Soil 2003, 254, 361–370.

26. Baldos, U.L.C. A sensitivity analysis of the lifecycle and global land use change greenhouse gas emissions of U.S. corn ethanol fuel. Master thesis, Purdue University, West Lafayette, IN, USA, 2009.

27. Schaufler, G.; Kitzler, B.; Schindlbacher, A.; Skiba, U.; Sutton, M.A.; Zechmeister-Boltenstern, S. Greenhouse gas emissions from European soils under differ rent land use: effects of soil moisture and temperature. Eur. J. Soil. Sci. 2010, 61, 683–696.

28. Fargione, J.; Hill, J.; Tilman, D.; Polasky, S.; Hawthorne, P. Land clearing and the biofuel carbon debt. Science 2008, 319, 1235–1238.

29. Hill, J.; Polaskya, S.; Nelson, E.; Tilman, D.; Huo, H.; Ludwig, L.; Neumann, J.; Zheng, H.; Bonta, D. Climate change and health costs of air emissions from biofuels and gasoline. Pro. Nat. Acad. Sci. USA 2008, 106, 2077–2082.

30. Kim, H.; Kim, S.; Dale, B.E. Biofuels, land use change, and greenhouse gas emissions: some unexplored variables. Environ. Sci. Tech. 2009, 43, 961–967.

31. Timilsina, G.R.; Beghin, J.C.; Van Der Mensbrugghe, D.; Mevel, S. The impacts of biofuels targets on land-use change and food supply: A global CGE assessment. Agr. Econ. 2012, 43, 315–332.

32. Kwon, H.Y.; Mueller, S.; Dunn, J.B.; Wander, M.M. Modeling state-level soil carbon emission factors under various scenarios for direct land use change associated with united states biofuel feedstock production. Biomass Bioenergy 2013, 55, 299–310.

33. Farquharson, J.; Baldock, J. Concepts in modeling N2O emissions from land use. Plant Soil 2008, 309, 17–167.

34. Qin, Z.; Zhuang, Q.; Zhu, X. Carbon and nitrogen dynamics in bioenergy ecosystems: 1. Model development, validation and sensitivity analysis. GCB bioenergy 2013.

35. Qin, Z.; Zhuang, Q.; Zhu, X. Carbon and nitrogen dynamics in bioenergy ecosystems: 2. Potential greenhouse gas emissions and global warming intensity in the Conterminous United States. In GCB Bioenergy; 2013.

36. Samuelson, P.A. Spatial price equilibrium and linear programming. Am. Econ. Rev. 1950, 42, 283–303.

37. Takayama, T.; Judge, G.G. Spatial and Temporal Price Allocation Models; North-Holland Publishing Co.: Amsterdam, The Netherland, 1971.

38. McCarl, B.A.; Spreen, T.H. Price endogenous mathematical programming as a tool for sector analysis. Am. J. Agr. Econ. 1980, 62, 87–102.

39. Chang, C.C.; McCarl, B.A.; Mjedle, J.; and Richardson, J.W. Sectorial implications of farm program modifications. Am. J. Agr. Econ. 1992, 74, 38–49.

40. Adams, D.M.; Hamilton, S.A.; McCarl, B.A. The benefits of air pollution control: The case of the ozone and U.S. agriculture. Am. J. Agr. Econ. 1986, 68, 886–894.

41. McCarl, B.A.; Keplinger, K.O.; Dillon, C.R.; Williams, R.L. Limiting pumping from the edwards aquifer: An economic investigation of proposals, water markets and springflow guarantees. Water Resour. Res. 1999, 35, 1257–1268.

42. Reilly, J.M.; Tubiello, F.; McCarl, B.A.; Abler, D.; Darwin, R.; Fuglie, K.; Hollinger, S.;Izaurralde, C.; Jagtap, S.; Jones, J.; et al. U.S. Agriculture and climate change: New results. Clim. Change 2002, 57, 43–69.

43. McCarl, B.A.; Schneider, U.A. U.S. Agriculture's role in a greenhouse gas emission mitigation world: An economic perspective. Rev. Agr. Econ. 2000, 22, 134–159.

44. Chang, C.C.; Eddleman, B.R.; McCarl, B.A. Potential benefits of rice variety and water management improvements in the Texas gulf coast. J. Agr. Resource Econ. 1991, 16, 185–193.

45. Coble, K.H.; Chang, C.C.; McCarl, B.A.; Eddleman, B.R. Assessing economic im-
    plications of new technology: The case of cornstarch-based biodegradable plastics.
    Rev. Agr. Econ. 1992, 14, 33–43.
46. Chen, C.C.; Chang, C.C. The impact of weather on crop yield distribution in Taiwan:
    Some new evidence from panel data models and implications for crop insurance. J.
    Agr. Econ. 2005, 33, 503–511.
47. Burton, R.O.; Martin, M.A. Restrictions on herbicide use: An analysis of economic
    impacts on U.S. agriculture. N. Cent. J. Agr. Econ. 1987, 9, 181–194.
48. Chang, C.C. The potential impacts of climate change on Taiwan's agriculture. Agr.
    Econ. 2002, 27, 51–64.
49. Chen, C.C.; McCarl, B.A.; Chang, C.C.; Tso, C. Evaluation the potential economic
    impacts of Taiwanese biomass energy production. Biomass Bioenergy 2011, 35,
    1693–1701.
50. McCarl, B.A. Food, Biofuel, Global agriculture, and environment: Discussion. Rev.
    Agr. Econ. 2008, 30, 530–532.
51. McCarl, B.A.; Adams, D.M.; Alig, R.J.; Chmelik, J.T. Analysis of biomass fueled
    electrical power plants: Implications in the agricultural and forestry sectors. Ann.
    Oper. Res. 2000, 94, 37–55.

CHAPTER 8

Energy Potential and Greenhouse Gas Emissions from Bioenergy Cropping Systems on Marginally Productive Cropland

MARTY R. SCHMER, KENNETH P. VOGEL, GARY E. VARVEL, RONALD F. FOLLETT, ROBERT B. MITCHELL, AND VIRGINIA L. JIN

8.1 INTRODUCTION

Reduction in greenhouse gas (GHG) emissions from transportation fuels can result in near- and long-term climate benefits [1]. Biofuels are seen as a near-term solution to reduce GHG emissions, reduce U.S. petroleum import requirements, and diversify rural economies. Depending on feedstock source and management practices, greater reliance on biofuels may improve or worsen long-term sustainability of arable land. U.S. farmers have increased corn (*Zea mays* L.) production to meet growing biofuel demand through land expansion, improved management and genetics, increased corn plantings, or by increased continuous corn monocultures [2]–[4].

Energy Potential and Greenhouse Gas Emissions from Bioenergy Cropping Systems on Marginally Productive Cropland. © Schmer MR, Vogel KP, Varvel GE, Follett RF, Mitchell RB, and Jin VL. PLoS ONE **9**,3 (2014), http://journals.plos.org/plosone/article?id=10.1371/journal.pone.0089501. The work is made available under the Creative Commons CC0 public domain dedication, https://creativecommons.org/publicdomain/zero/1.0/.

Productive cropland is finite, and corn expansion on marginally-productive cropland may lead to increased land degradation, including losses in biodiversity and other desirable ecosystem functions [4]–[6]. We define marginal cropland as fields whose crop yields are 25% below the regional average. The use of improved corn hybrids and management practices have increased U.S. grain yields by 50% since the early 1980's [7] with an equivalent increase in non-grain biomass or stover yields. Corn stover availability and expected low feedstock costs make it a likely source for cellulosic biofuel. However, excessive corn stover removal can lead to increased soil erosion and decreased soil organic carbon (SOC) [8] which can negatively affect future grain yields and sustainability. Biofuels from cellulosic feedstocks (e.g. corn stover, dedicated perennial energy grasses) are expected to have lower GHG emissions than conventional gasoline or corn grain ethanol [9]–[13]. Furthermore, dedicated perennial bioenergy crop systems such as switchgrass (*Panicum virgatum* L.) have the ability to significantly increase SOC [14]–[16] while providing substantial biomass quantities for conversion into biofuels under proper management [17], [18].

Long-term evaluations of feedstock production systems and management practices are needed to validate current and projected GHG emissions and energy efficiencies from the transportation sector. In a replicated, multi-year field study located 50 km west of Omaha, NE, we evaluated the potential to produce ethanol on marginal cropland from continuously-grown no-tillage corn with or without corn residue removal (50% stover removal) and from switchgrass harvested at flowering (August) versus a post-killing frost harvest. Our objectives were to compare the effects of long-term management practices including harvest strategies and N fertilizer input intensity on continuous corn grain and switchgrass to determine ethanol production, potential petroleum offsets, and net energy yields. We also present measured SOC changes (0 to 1.5 m) over a nine year period from our biofuel cropping systems to determine how direct SOC changes impact net GHG emissions from biofuels. Furthermore, we evaluate the potential efficiency advantages of co-locating and integrating cellulosic conversion capacity with existing dry mill corn grain ethanol plants.

8.2 MATERIALS AND METHODS

This study is located on the University of Nebraska Agricultural Research and Development Center, Ithaca, Nebraska, USA on a marginal cropland field with Yutan silty clay loam (fine-silty, mixed, superactive, mesic Mollic Hapludalf) and a Tomek silt loam (fine, smectitic, mesic Pachic Argiudoll) soil. Switchgrass plots were established in 1998 and continuous corn plots were initiated in 1999. The study is a randomized complete block design (replications = 3) with split-split plot treatments. Main treatments are two cultivars of switchgrass, 'Trailblazer' and 'Cave-in-Rock', and a glyphosate tolerant corn hybrid. Main treatment plots are 0.3 ha which enables the use of commercial farm equipment. Switchgrass is managed as a bioenergy crop, and corn is managed under no-tillage conditions (no-till farming since 1999). Split-plot treatments are nitrogen (N) fertilizer levels and split-split plots are harvest treatments. Annual N fertilizer rates (2000–2007) were 0 kg N ha^{-1}, 60 kg N ha^{-1}, 120 kg N ha^{-1}, and 180 kg N ha^{-1} as NH_4NO_3, broadcast on the plots at the start of the growing season. The 0 kg N ha^{-1}, 60 kg N ha^{-1}, 120 kg N ha^{-1} fertilizer rates were used on switchgrass [19] while the 60 kg N ha^{-1}, 120 kg N ha^{-1}, and 180 kg N ha^{-1} fertilizer rates were used for corn. Switchgrass harvest treatments were initiated in 2000 and consist of a one-cut harvest either in early August or after a killing frost. Corn stover treatments were initiated in 2000 and are either no stover harvest or stover removal, where the amount of stover removed approximates 50% of the aboveground biomass after corn grain is harvested.

Baseline soil samples were taken in 1998 at the center of each subplot and re-sampled in 2007 at increments of 0–5, 5–10, 10–30, 30–60, 60–90, 90–120, and 120–150 cm depths [15]. Average changes in total SOC (0–1.5 m) from 1998–2007 were used to estimate direct soil C changes. Further management practices and detailed soil property values from this study have been previously reported [15], [20]. Summary of petroleum offsets (GJ ha^{-1}), ethanol production (L ha^{-1}), greenhouse gas (GHG) emissions (g CO_2e MJ^{-1}), net GHG emissions (Mg CO_2e ha^{-1}), and GHG reductions (%) for corn grain, corn grain with stover removal, and switchgrass are presented in Table S1 in File S1.

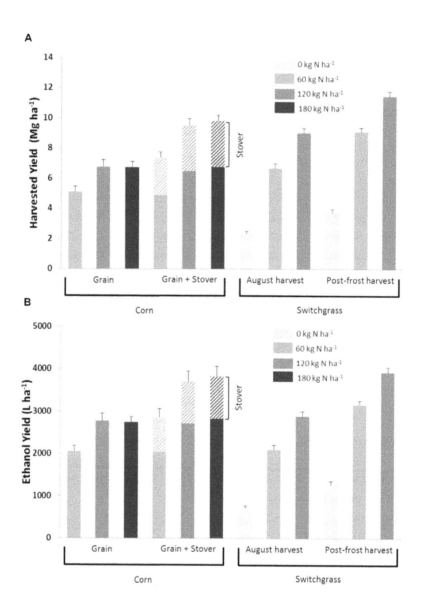

FIGURE 1: Harvested mean annual yield ± standard error (A) and ethanol energy ± SEM (B) for no-till continuous corn (grain-only harvest or grain and stover harvest) and switchgrass (August harvest or Post-frost harvest) under variable nitrogen rates on marginally-productive rainfed cropland for 2000–2007 (n = 3 replicate corn system plots and 6 replicate switchgrass plots).

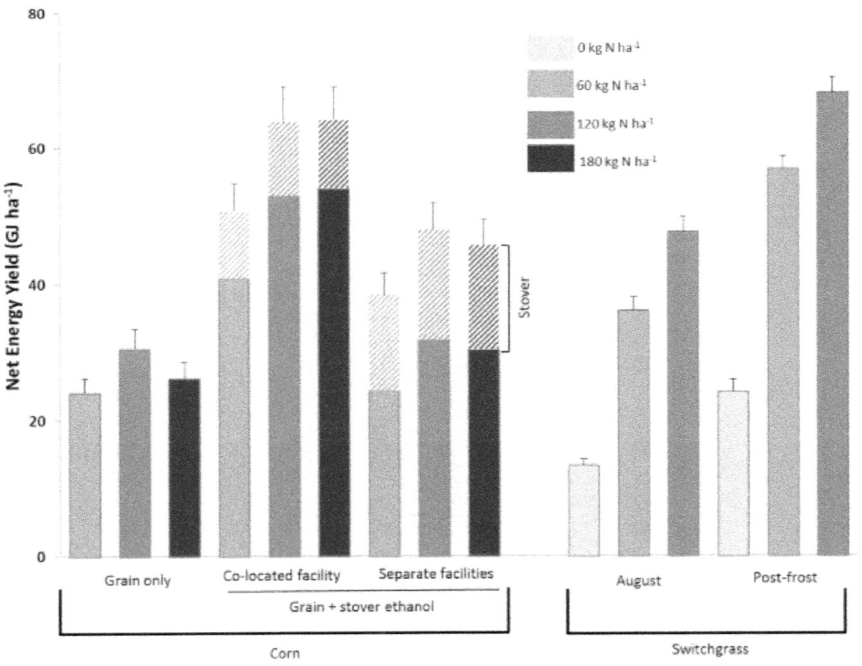

FIGURE 2: Net energy yield ± standard error for no-till continuous corn (grain-only or grain and stover harvest) and switchgrass (August harvest or post-frost harvest) under variable nitrogen rates on marginally-productive cropland (n = 3 replicate corn system plots and 6 replicate switchgrass plots). Conversion processes evaluated include corn grain-only harvest at a natural gas (NG) dry mill, corn grain with stover harvest at a co-located facility (lignin portion of stover used as primary energy source for grain and cellulose conversion), corn grain with stover harvest at separate ethanol facilities (NG dry mill and a cellulosic ethanol plant), and switchgrass (cellulosic ethanol plant).

8.2.1 STATISTICAL ANALYSES

Yield data analyzed were from 2000 to 2007, where 2000 was the initiation of harvest treatments for continuous corn and switchgrass and 2007 was the last year that SOC was measured for this study. Data from switchgrass cultivars were pooled together based on their similar aboveground biomass yields over years and similar changes in SOC [15]. Data were analyzed using a linear mixed model approach with replications considered a random effect. Mean separation tests were conducted using the Tukey-Kramer method. Significance was set at $P \leq 0.05$.

8.2.2 LIFE-CYCLE ASSESSMENT

For energy requirements in the production, conversion, and distribution of corn grain ethanol and cellulosic ethanol, values from the Greenhouse Gases, Regulated Emissions, and Energy Use in Transportation (GREET v. 1.8) [21], Energy and Resources Group Biofuel Analysis Meta-Model (EBAMM) [22], and Biofuel Energy Systems Simulator (BESS) [23] life cycle assessment models were used as well as previous agricultural energy estimates for switchgrass [12]. Energy use in the agricultural phase consisted of agricultural inputs (seed, herbicides, fertilizers, packaging), machinery energy use requirements, material transport, and diesel requirements used in this study. Stover energy requirements from the production phase were from the diesel requirements to bale, load, and stack corn stover and the embodied energy of the farm machinery used. A proportion of the N fertilizer and herbicide requirements were allocated to the amount of stover harvested.

Multiple biorefinery configurations are presented to evaluate different conversion scenarios and how this affects GHG emissions, petroleum offset credits, and net energy yield (NEY) values. Biorefinery scenarios evaluated in this study are: (i) a natural gas (NG) dry mill corn grain ethanol plant with dry distillers grain (DDGS) as a co-product for the corn grain-only harvests [23]–[25], (ii) a co-located dry mill corn grain and cellulosic ethanol plant with combined heat and power (CHP) and DDGS co-product, where corn stover is primarily used to displace dry mill ethanol plant natural gas requirements [25], [26], (iii) and a standalone cellulosic (switchgrass or corn stover) ethanol plant (sequential hydrolysis and fermentation) with CHP capability and electricity export [22], [27]–[29]. Chemical and enzyme production costs and related GHG emissions for corn grain and cellulosic conversion to ethanol were also incorporated [28]. Ethanol recovery for corn grain was estimated to be 0.419 L kg^{-1} [23]. Ethanol recovery for corn stover and switchgrass were based on cell wall composition from harvested biomass samples. Ground aboveground switchgrass samples were scanned using a near-infrared spectrometer to predict cell wall and soluble carbohydrate biomass composition [30]. Ground corn stover samples were analyzed using a near-infrared spectrometer-based calibration equation developed by the National Renewable Energy Laboratory to predict corn stover cell wall composition [31]. Switchgrass and corn stover cell wall conversion to ethanol was based on composition components of glucan, xylose and arabinose [30], [31]. Glucan to ethanol conversion was assumed to be 85.5%, and xylose and arabinose was estimated to have 85% ethanol recovery efficiency [29]. Estimated ethanol recovery for corn stover was 327 L Mg^{-1} which was similar to other findings [29]. For switchgrass, ethanol recovery based on glucan, xylose, and arabinose concentrations was estimated to be 311 L Mg^{-1} and 344 L Mg^{-1} for an August harvest and a post-frost harvest, respectively.

Ethanol plant size capacity was estimated to be 189 million L yr^{-1} for the corn grain-only and cellulosic-only scenarios. For the co-located facility, total plant size was assumed to be 378 million L yr^{-1} capacity. Fossil fuel energy requirements for the conventional corn grain ethanol plant is assumed to be 7.69 MJ L^{-1} for natural gas to power the plant and to dry DGS, 0.59 MJ L^{-1} for corn grain transportation from farm to ethanol plant, 0.67 MJ L^{-1} for electricity purposes, 0.13 MJ L^{-1} to capital depreciation costs, and 0.58 MJ L^{-1} for wastewater processing and effluent restoration [10], [22]. Fossil fuel requirements for the corn grain/cellulosic ethanol plant are feedstock transportation 0.63 MJ L^{-1} for corn stover, 0.59 MJ L^{-1} for corn grain transportation from farm to ethanol plant, 0.44 MJ L^{-1} to capital depreciation costs, and 0.58 MJ L^{-1} for wastewater treatment and processing (Table S2 in File S1). Cellulosic ethanol plant fossil fuel requirements are 0.63 MJ L^{-1} for switchgrass transportation from field to ethanol plant, 0.06 MJ L^{-1} diesel requirements for biomass transport within the ethanol plant grounds, 0.44 MJ L^{-1} to capital depreciation costs, and 0.58 MJ L^{-1} for wastewater processing, effluent restoration, and recovery (Table S2 in File S1).

For the co-located corn grain and cellulosic facility, we assumed (i) power and electrical utilities were shared [26]; (ii) power requirements were supplied mainly from the lignin portion of stover with combined ethanol purification from the starch and cellulosic ethanol conversion pathways [26]; and (iii) extra stover biomass would be required in addition to the lignin to meet steam requirements. A co-location facility would require additional unprocessed bales to be used in addition to lignin which lowered the amount of ethanol being generated from stover at a co-located facility compared to a standalone cellulosic facility that uses stover as their primary feedstock (Table S1 in File S1). Electricity would be imported from the grid in this scenario and DDGS exported as the only co-product. Recent analysis [29] of converting cellulose to ethanol has estimated a higher internal electrical demand than previously assumed [26]; suggesting electricity export under this configuration would be unlikely. The value of DDGS as animal feed would likely preclude its use in meeting power requirement in a co-located facility. We based our total biomass energy requirement on the lignin concentration in stover and the expected biomass energy use requirements to power a co-located ethanol plant [25]. Estimated biomass requirements were 11 MJ L^{-1} ethanol and embodied energy value of 16.5 MJ kg^{-1} (low heating value) for stover biomass.

Net energy yield (NEY) values (renewable output energy – fossil fuel input energy) were calculated for each feedstock and conversion scenario. Output energy was calculated from ethanol output plus co-product credits. Co-product credit for DDGS is 4.13 MJ L^{-1} for the corn grain-only ethanol plant and the co-located corn grain/cellulosic ethanol plant [32]. Electricity co-product credit for standalone cellulosic ethanol was estimated to be 1.68 MJ L^{-1} [29]. Petroleum offsets (GJ ha^{-1}) were calculated in a similar fashion as NEY with total ethanol production (MJ ha^{-1}) along with petroleum displacement from co-products minus petroleum inputs consumed in the production, conversion, and distribution phase (Tables S1 and S3 in File S1). Petroleum offsets were calculated as the difference between ethanol output and petroleum inputs from the agricultural, conversion, and distribution phase (Table S1 in File S1). Petroleum requirements for each cropping system were calculated from input requirements from this study and derived values from the EBAMM model [22]. For input requirements without defined petroleum usage, we used the default parameter in EBAMM that estimates U.S. average petroleum consumption at 40% for input source. Petroleum offset credits associated with corn grain ethanol co-products were estimated to be 0.71 MJ L^{-1} while credits for corn stover and switchgrass cellulosic ethanol co-products (standalone facility) were 0.12 MJ L^{-1} (Table S3 in File S1). Petroleum offset credits were calculated from GREET (v 1.8).

8.2.3 GREENHOUSE GAS EMISSIONS

Greenhouse gas offsets associated with the production of corn grain and cellulosic ethanol were modeled from the EBAMM and BESS models [22], [23]. Agricultural GHG emissions were based on fuel use, fertilizer use, herbicide use, farm machinery requirements, and changes in SOC. Direct land use change by treatment plot can either be a GHG source or a GHG sink depending on SOC changes from this study [15]. Co-product GHG credits for DDGS or electricity export were derived from the BESS [23] and GREET (v. 1.8) models [21]. Co-product GHG credits for DDGS was −347 g CO_2e L^{-1} ethanol and −304 g CO_2e L^{-1} ethanol for cellulosic electricity export (Table S4 in File S1). Indirect land use changes for corn grain ethanol or switchgrass were not estimated in this analysis. GHG offsets were calculated on both an energy and areal basis (Table S1 in File S1).

Greenhouse gas emissions from N fertilizer were evaluated from the embodied energy requirements and subsequent nitrous oxide (N_2O) emissions (Table S4 in File S1). Direct and indirect nitrous oxide emissions were calculated in this study using Tier 1 Intergovernmental Panel on Climate Change calculations. Greenhouse gas emission values for the agricultural phase are included in Table S4 in File S1 and for the conversion and distribution phase in Table S5 in File S1. For the agricultural phase, total GHG emissions were calculated from the production of fertilizers, herbicides, diesel requirements, drying costs for corn grain, and the embodied energy in farm machinery minus direct soil C changes occurring for the study period (Table S4 in File S1). GHG emissions were reported on an energy basis, areal basis, and the difference between ethanol and conventional gasoline (Table S1 in File S1). For net GHG emissions (Mg CO_2e ha^{-1}), calculations were based on GHG intensity values (g CO_2e MJ^{-1}) multiplied by biofuel production (MJ ha^{-1}) for each cropping system. GHG reductions (Table S1 in File S1) were calculated as the percent difference from conventional gasoline as reported by the California Air Resource Board (99.1 g CO_2 MJ^{-1}) [33].

8.3 RESULTS AND DISCUSSION

Harvest and N fertilizer management treatments affected grain and biomass yields in both crops over eight growing seasons (Fig. 1A). Switchgrass harvested after a killing frost had 27% to 60% greater biomass yields compared with an August harvest under similar fertilization rates. Highest harvested biomass yields (mean = 11.5 Mg ha^{-1} yr^{-1}) were from fertilized (120 kg N ha^{-1}) switchgrass harvested after a killing frost while continuous corn showed similar grain and stover yields [factorial analysis of variance (ANOVA), P = 0.72] under the highest N fertilizer levels (180 kg N ha^{-1}) (Fig. 1A).

Potential ethanol yields varied from 2050 to 2774 L ha^{-1} yr^{-1} for corn grain-only harvests while those for corn grain with stover removal ranged from 2862 to 3826 L ethanol ha^{-1} yr^{-1} (Fig. 1B). Ethanol contribution from corn stover ranged from 820 to 998 L ha^{-1} yr^{-1} when stover is converted at a standalone cellulosic plant (Fig. 1B). Separate ethanol facilities showed slightly higher potential ethanol yields (L ha^{-1}) than at a co-located facility (Table S1 in File S1) because a larger portion of corn stover biomass was required to meet thermal power requirements at a co-located facility (SI text in File S1). Unfertilized switchgrass had potential ethanol yield values similar to corn stover. Switchgrass under optimal management practices had 17% higher biomass yields than the highest yielding corn with stover removal treatment. Potential ethanol yield for switchgrass, however, was similar (factorial ANOVA, P>0.05) to corn with stover removal (Fig. 1B) due to lower cellulosic ethanol recovery efficiency than exists for corn grain ethanol conversion efficiency. Switchgrass ethanol conversion efficiency from this study was based on updated biochemical conversion processes [29] using known cell wall characteristics [30] that result in lower conversion rates than previous estimates [12], [18].

Net energy yield (NEY) (renewable output energy minus fossil fuel input energy) and GHG emission intensity (grams of CO_2 equivalents per megajoule of fuel, or g CO_2e MJ^{-1}) are considered the two most important metrics in estimating fossil fuel replacement and GHG mitigation for bio-fuels [34]. Switchgrass harvested after a killing frost (120 kg N ha^{-1}) and the co-located grain and stover conversion pathway (120 kg N ha^{-1} and 180 kg N ha^{-1} treatments) had the highest overall NEY values (Fig. 2). Net energy yields for continuous corn were higher at a co-located facility because stover biomass and lignin replaced natural gas for thermal energy (Fig. 2). Ethanol conversion of corn grain and stover at separate facili-ties was intermediate in NEY while traditional corn grain-only natural gas (NG) dry mill ethanol plants had the lowest NEY values for the continu-ous corn systems. Delaying switchgrass harvest from late summer to after a killing frost resulted in significant improvement in NEY and potential ethanol output under similar N rates. Unfertilized switchgrass had similar NEY values compared with corn grain processed at a NG dry mill ethanol plant (factorial ANOVA, P = 0.12) while fertilized switchgrass harvested after a killing frost had higher NEY values (factorial ANOVA, P<0.0001) than NG dry mill corn grain ethanol plants (Fig. 2).

Both the continuous corn and switchgrass systems showed significant petroleum offset (ethanol output minus petroleum inputs) capability, with the intensified bioenergy cropping systems having the highest petroleum offsets (Fig. 3). Petroleum use varied by cropping system in the agricul-tural phase with continuous corn systems having higher overall petroleum requirements than switchgrass. Petroleum requirements (mainly diesel fuel) to harvest corn stover are small relative to corn grain harvest as a result of low harvested stover yields. Lowest petroleum offsets for con-tinuous corn systems were from stover harvests at a separate dedicated cellulosic facility (Table S1 in File S1). Corn grain-only harvests offset less petroleum compared with grain and stover at separate ethanol facili-ties under similar fertilizer rates (factorial ANOVA, P<0.01). Management practices in switchgrass resulted in the largest variation in petroleum off-set credits (Fig. 3B). Petroleum offsets (GJ ha^{-1}) were positively associ-ated with NEY values [−1.81+0.84 (Petroleum offset); (P<0.0001); (R^2 =

0.76)], indicating that bioenergy cropping systems with large NEY values will likely result in higher petroleum displacement.

All bioenergy cropping systems evaluated in our study had SOC sequestration rates exceeding 7.3 Mg CO_2 yr^{-1} (Table S4 in File S1), with over 50% of SOC sequestration occurring below the 0.3 m soil depth [15]. Soil organic C increased even with corn stover removal, indicating that removal rates were sustainable in terms of SOC and grain yield for this time period. No-tillage continuous corn systems have lower stover retention requirements to maintain SOC than continuous corn with tillage or corn-soybean (*Glycine max* (L.) Merr.) rotations [8]. Consequently, all conversion pathways had negative GHG emission values as a result of SOC sequestration offsetting GHG emissions from the production, harvest, conversion and distribution phases for corn grain ethanol and cellulosic ethanol. For switchgrass, SOC storage values were similar to other findings within the same ecoregion [16] and a long-term Conservation Reserve Program grassland [35]. Measured SOC storage from the continuous corn systems (Table S4 in File S1) were significantly higher than modeled SOC storage estimates from this region [36]. Corn grain grown with low N rates (60 kg ha^{-1}) had GHG intensity values similar to continuous corn under optimum N rates (120 kg ha^{-1}) but resulted in lower ethanol yields and lower petroleum offset potential (Fig. 3A). Lowest GHG emission intensity values on an energy basis (g CO_2e MJ^{-1}) were from unfertilized switchgrass (Table S1 in File S1) due to lower ethanol yields, lower agricultural energy emissions, and similar SOC storage compared with the other biofuel cropping systems. For switchgrass, management practices that resulted in the lowest GHG emission on an energy basis resulted in the lowest petroleum offset potential (Fig. 3B). Direct N_2O emissions (Table S4 in File S1) were estimated using Intergovernmental Panel on Climate Change methodology and are in agreement with study site N2O flux measurements from a later time series which indicated N rate as the major contributor to N_2O emissions [37]. When evaluating GHG emissions on a per unit area basis (g CO_2e ha^{-1}), unfertilized switchgrass and corn grain-only systems showed similar results with the more intensified cropping systems (Table S1 in File S1).

Both switchgrass and continuous corn with stover removal produced similar ethanol potential, NEY values, petroleum offsets, and GHG emissions but overall values and metric efficiencies were dependent on management practices and downstream conversion scenarios. Dedicated perennial grass systems used for bioenergy will need to have similar or greater yield potential than existing annual crops for widespread adoption to meet renewable energy demands and provide similar economic returns to producers. We have previously shown that switchgrass ethanol yields were comparable with regional corn grain ethanol yields [12]. Here we demonstrate that when switchgrass is optimally managed, ethanol potential is similar to a continuous corn cropping system with stover removal and exceeds ethanol yield for corn grain-only systems on marginally-productive cropland. Furthermore, breeding improvements for bioenergy specific switchgrass cultivars have shown higher yield potential than cultivars evaluated here [38].

Coupling sustainable agricultural residue harvests with dedicated energy crops improves land-use efficiency and reduces biomass constraints for a mature cellulosic biofuel industry. Recent analysis has shown that sufficient land exists in the U.S. Corn Belt to support a cellulosic ethanol industry without impacting productive cropland [18], [39], [40]. The effect of dedicated energy crops and corn grain on indirect land use change varies significantly based on the assumptions and models used [13], [41], [42] but bioenergy crops grown on marginally-productive cropland will have less impact on indirect land use change than bioenergy crops grown on more productive cropland. Likewise, model assumptions underlying direct SOC sequestration will impact system evaluations of GHG emissions and mitigation. Measured SOC sequestration values presented here were based on production years evaluated and were not extrapolated beyond this time-frame. Extrapolating SOC values from this time-frame to a 30-yr time horizon or 100-yr time horizon is still larger than current life cycle assessment assumptions on SOC sequestration potential of switchgrass or no-till corn [12], [42], [43]. This highlights the importance of accounting for direct SOC changes at depth to accurately estimate GHG emissions for biofuels under both marginal and productive cropland. Further long term evaluation of management practices (e.g. tillage, stover removal) on SOC sequestration potential for corn grain systems under irrigated conditions on productive cropland is warranted [44].

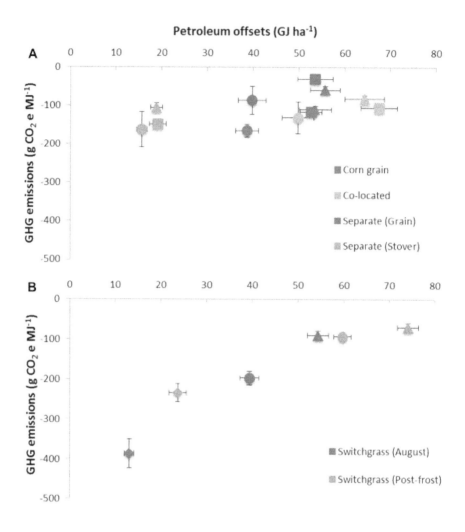

FIGURE 3: Petroleum offsets compared with GHG emissions (g CO_2e MJ^{-1} ethanol) for continuous corn and switchgrass grown on marginally-productive cropland (n = 3 replicate corn system plots and 6 replicate switchgrass plots). (A) Continuous corn values represent harvest method (stover harvested or retained) and ethanol conversion pathway (co-located facility or at a separate ethanol facilities). (B) Switchgrass values are based on harvest date and N fertilizer rate. Fertilizer rates are 0 kg N ha^{-1} (♦), 60 kg N ha^{-1} (•), 120 kg N ha^{-1}, and 180 kg N ha^{-1} (•). Error bars indicate standard errors of the mean.

A multi-feedstock, landscape approach minimizes economic and environmental risks in meeting feedstock demands for cellulosic ethanol production by providing sufficient feedstock availability while maintaining ecosystem services. A co-located cellulosic biorefinery is expected to have economic advantages by reducing capital costs requirements for cellulosic conversion and through sharing of infrastructure costs. In this study, we used corn stover as the feedstock for the co-located cellulosic biorefinery but the benefits will apply to other cellulosic feedstocks. A co-located facility can increase NEY values by decreasing natural gas use for thermal energy, but current and forecasted U.S. natural gas prices [45] may affect large scale adoption of co-location unless there are incentives for displacing fossil energy in existing NG dry mill ethanol plants [46]. Integrating cellulosic refining capacity with existing corn grain ethanol plants can improve the sustainability of first generation biofuels and enable the implementation of cellulosic biofuels into the U.S. transportation sector.

REFERENCES

1. Unger N, Bond TC, Wang JS, Koch DM, Menon S, et al. (2010) Attribution of climate forcing to economic sectors. Proc Natl Acad Sci USA 107(8): 3382–3387. doi: 10.1073/pnas.0906548107
2. Wallander S, Claassen R, Nickerson C (2011) The ethanol decade: An expansion of U.S. corn production, 2000–09. USDA-ERS Economic Information Bulletin No. (EIB-79).
3. Claassen R, Carriazo F, Cooper J, Hellerstein D, Ueda K (2011) Grassland to cropland conversion in the northern plains: The role of crop insurance, commodity, and disaster programs Economic Research Report No. (ERR-120).
4. Wright CK, Wimberly MC (2013) Recent land use change in the western corn belt threatens grasslands and wetlands. Proc Natl Acad Sci USA 110(10): 4134–4139. doi: 10.1073/pnas.1215404110
5. Wiens J, Fargione J, Hill J (2011) Biofuels and biodiversity. Ecol Appl 21(4): 1085–1095. doi: 10.1890/09-0673.1
6. Donner SD, Kucharik CJ (2008) Corn-based ethanol production compromises goal of reducing nitrogen export by the Mississippi river. Proc Natl Acad Sci USA 105(11): 4513–4518. doi: 10.1073/pnas.0708300105
7. USDA-NASS. 2013 U.S. Corn Yields. Available: http://www.nass.usda.gov/Charts_and_Maps/Field_Crops/cornyld.asp.
8. Wilhelm WW, Johnson JMF, Karlen DL, Lightle DT (2007) Corn stover to sustain soil organic carbon further constrains biomass supply. Agron J 99(6): 1665–1667. doi: 10.2134/agronj2007.0150

9. Adler PR, Del Grosso SJ, Parton WJ (2007) Life-cycle assessment of net green-house-gas flux for bioenergy cropping systems. Ecol Appl 17(3): 675–691. doi: 10.1890/05-2018

10. Farrell AE, Plevin RJ, Turner BT, Jones AD, O'Hare M, et al. (2006) Ethanol can contribute to energy and environmental goals. Science 312(5781): 506–508. doi: 10.1126/science.1121416

11. Sheehan J, Aden A, Paustian K, Killian K, Brenner J, et al. (2003) Energy and environmental aspects of using corn stover for fuel ethanol. J Indust Ecol 7(3–4): 117–146. doi: 10.1162/108819803323059433

12. Schmer MR, Vogel KP, Mitchell RB, Perrin RK (2008) Net energy of cellulosic ethanol from switchgrass. Proc Natl Acad Sci USA 105(2): 464–469. doi: 10.1073/pnas.0704767105

13. Wang M, Han J, Dunn JB, Cai H, Elgowainy A (2012) Well-to-wheels energy use and greenhouse gas emissions of ethanol from corn, sugarcane and cellulosic biomass for US use. Environ Res Letters 7(4): 045905. doi: 10.1088/1748-9326/7/4/045905

14. Frank AB, Berdahl JD, Hanson JD, Liebig MA, Johnson HA (2004) Biomass and carbon partitioning in switchgrass. Crop Sci 44: 1391–1396. doi: 10.2135/crop-sci2004.1391

15. Follett RF, Vogel KP, Varvel GE, Mitchell RB, Kimble J (2012) Soil carbon sequestration by switchgrass and no-till maize grown for bioenergy. BioEnerg Res 5: 866–875. doi: 10.1007/s12155-012-9198-y

16. Liebig MA, Schmer MR, Vogel KP, Mitchell RB (2008) Soil carbon storage by switchgrass grown for bioenergy. BioEnerg Res 1(3–4): 215–222. doi: 10.1007/s12155-008-9019-5

17. DOE US (2011) U.S. billion-ton update: Biomass supply for bioenergy and bioproducts industry. ORNL/TM-2011/224.

18. Gelfand I, Sahajpal R, Zhang X, Izaurralde R, Gross KL, et al. (2013) Sustainable bioenergy production from marginal lands in the US Midwest. Nature 493: 514–517. doi: 10.1038/nature11811

19. Vogel KP, Brejda JJ, Walters DT, Buxton DR (2002) Switchgrass biomass production in the Midwest USA: Harvest and nitrogen management. Agron J 94: 413–420. doi: 10.2134/agronj2002.0413

20. Varvel GE, Vogel KP, Mitchell RB, Follett RF, Kimble J (2008) Comparison of corn and switchgrass on marginal soils for bioenergy. Biomass Bioenergy 32(1): 18–21. doi: 10.1016/j.biombioe.2007.07.003

21. Greenhouse Gases, Regulated Emissions, and Energy Use in Transportation (GREET) (2012) Argonne National Laboratory. Available: http://greet.es.anl.gov/.

22. Renewable and Applicable Energy Laboratory (2007) Energy and resources group biofuel analysis meta-model. Available: http://rael.berkeley.edu/sites/default/files/EBAMM/.

23. Liska AJ, Yang HS, Bremer VR, Erickson G, Klopfenstein T, et al.. (2008) BESS: Biofuel energy systems simulator; life-cycle energy and emissions analysis model for corn-ethanol biofuel (v. 2008.3.0).

24. Liska AJ, Yang HS, Bremer VR, Klopfenstein TJ, Walters DT, et al. (2009) Improvements in life cycle energy efficiency and greenhouse gas emissions of corn-ethanol. J Ind Ecol 13(1): 58–74. doi: 10.1111/j.1530-9290.2008.00105.x

25. Wang M, Wu M, Huo H (2007) Life-cycle energy and greenhouse gas emission impacts of different corn ethanol plant types. Environ Res Letters 2(2): 024001. doi: 10.1088/1748-9326/2/2/024001

26. Wallace R, Ibsen K, McAloon A, Yee W (2005) Feasibility study for co-locating and integrating ethanol production plants from corn starch and lignocellulosic feedstocks. USDOE Rep No. NREP/TP-510-37092.

27. Spatari S, Bagley DM, MacLean HL (2010) Life cycle evaluation of emerging lignocellulosic ethanol conversion technologies. Bioresource Technol 101(2): 654–667. doi: 10.1016/j.biortech.2009.08.067

28. MacLean HL, Spatari S (2009) The contribution of enzymes and process chemicals to the life cycle of ethanol. Environ Res Letters 4(1): 014001. doi: 10.1088/1748-9326/4/1/014001

29. Humbird D, Davis R, Tao L, Kinchin C, Hsu D, et al.. (2011) Process Design and Economics for Biochemical Conversion of Lignocellulosic Biomass to Ethanol. USDOE (Department of Energy NREL/TP-5100-47764, Golden, CO), pp 136.

30. Vogel K, Dien BS, Jung HG, Casler MD, Masterson SD, et al. (2011) Quantifying actual and theoretical ethanol yields for switchgrass strains using NIRS analyses. BioEnerg Res 4(2): 96–110. doi: 10.1007/s12155-010-9104-4

31. Templeton DW, Sluiter AD, Hayward TK, Hames BR, Thomas SR (2009) Assessing corn stover composition and sources of variability via NIRS. Cellulose 16: 621–639. doi: 10.1007/s10570-009-9325-x

32. Graboski MS (2002) Fossil energy use in the manufacture of corn ethanol. National Corn Growers Association

33. California Air Resources Board (2013) Low carbon fuel standard. Available: http://www.arb.ca.gov/Fuels/Lcfs/Lcfs.htm.

34. Liska AJ, Cassman KG (2008) Towards standardization of life-cycle metrics for biofuels: Greenhouse gas emissions mitigation and net energy yield. J Biobased Materials and Bioenergy 2(3): 187–203. doi: 10.1166/jbmb.2008.402

35. Gelfand I, Zenone T, Jasrotia P, Chen J, Hamilton SK, et al. (2011) Carbon debt of conservation reserve program (CRP) grasslands converted to bioenergy production. Proc Natl Acad Sci USA 108(33): 13864–13869. doi: 10.1073/pnas.1017277108

36. Davis SC, Parton WJ, Del Grosso SJ, Keough C, Marx E, et al. (2012) Impact of second-generation biofuel agriculture on greenhouse-gas emissions in the corn-growing regions of the US. Front Ecol Environ 10: 69–74. doi: 10.1890/110003

37. Jin VL, Varvel GE, Wienhold BJ, Schmer MR, Mitchell RB, et al.. (2011) Field emissions of greenhouse gases from contrasting biofuel feedstock production systems under different N fertilization rates. ASA-CSSA-SSSA International Annual Meetings. Oct 16–19, San Antonio TX.

38. Vogel KP, Mitchell RB (2008) Heterosis in switchgrass: Biomass yield in swards. Crop Sci 48(6): 2159–2164. doi: 10.2135/cropsci2008.02.0117

39. Mitchell RB, Vogel KP, Uden DR (2012) The feasibility of switchgrass for biofuel production. Biofuels 3(1): 47–59. doi: 10.4155/bfs.11.153

40. Uden DR, Mitchell RB, Allen CR, Guan Q, McCoy T (2013) The feasibility of producing adequate feedstock for year-round cellulosic ethanol production in an intensive agricultural fuelshed. BioEnerg Res 6(3): 930–938. doi: 10.1007/s12155-013-9311-x

41. Searchinger T, Heimlich R, Houghton RA, Dong F, Elobeid A, et al. (2008) Use of U.S. croplands for biofuels increases greenhouse gases through emissions from land-use change. Science 319(5867): 1238–1240. doi: 10.1126/science.1151861

42. Dunn J, Mueller S, Kwon H, Wang M (2013) Land-use change and greenhouse gas emissions from corn and cellulosic ethanol. Biotechnology for Biofuels 6(1): 51. doi: 10.1186/1754-6834-6-51

43. Fargione J, Hill J, Tilman D, Polasky S, Hawthorne P (2008) Land clearing and the biofuel carbon debt. Science 319(5867): 1235–1238. doi: 10.1126/science.1152747

44. Follett RF, Jantalia CP, Halvorson AD (2013) Soil carbon dynamics for irrigated corn under two tillage systems. Soil Sci Soc Am J 77: 951–963. doi: 10.2136/sssaj2012.0413

45. Energy Information Agency (2013) Annual energy outlook for 2013 with projections to 2040. DOE/EIA-0383. doi: 10.1787/9789264179233-en

46. Plevin RJ, Mueller S (2008) The effect of CO2 regulations on the cost of corn ethanol production. Environ Res Letters 3(2): 024003. doi: 10.1088/1748-9326/3/2/024003

There are several supplemental files that are not available in this version of the article. To view this additional information, please use the citation on the first page of this chapter.

CHAPTER 9

Streamflow Impacts of Biofuel Policy-Driven Landscape Change

SAMI KHANAL, ROBERT P. ANEX, CHRISTOPHER J. ANDERSON, AND DARYL E. HERZMANN

9.1 INTRODUCTION

As demand for renewable fuels grows, biofuels from lignocellulosic feedstock are considered a promising alternative to corn-based ethanol [1]–[2]. Cellulosic biofuels are expected to be both environmentally and energetically superior to grain-based biofuels [3]–[6]. The mandate set by the Renewable Fuel Standard [7] to use 16 billion gallons of cellulosic biofuel per year by 2022 is projected to have significant impact on agricultural land use in the U.S. as lands are converted for the production of bioenergy crops [8]. Prior studies [6], [9] have investigated yields, land use, economics and greenhouse gas emissions of bioenergy crops, but one key factor often overlooked is the hydrologic balance associated with bioenergy crop production.

*Streamflow Impacts of Biofuel Policy-Driven Landscape Change. Khanal S, Anex RP, Anderson CJ, and Herzmann DE. PLoS ONE **9**,10 (2014), http://journals.plos.org/plosone/article?id=10.1371/journal.pone.0109129. The work is made available under the Creative Commons Attribution 4.0 International License, http://creativecommons.org/licenses/by/4.0/.*

There is strong coupling between the land surface and atmosphere that is heavily influenced by the vegetative land cover [10]–[13]. Change in land cover thus has the potential to impact local and regional climate through alteration of the energy and moisture balances of the land surface [14]–. The longer growing season and greener vegetative cover of biofuel crops result in higher water loss to the atmosphere through evapotranspiration (ET), decline in soil water depth [17], [19] and reduced surface runoff [20] relative to annual cropping systems. Changes in soil moisture and runoff determine streamflow, groundwater recharge and influence water quality.

Bioenergy crops, e.g., switchgrass and miscanthus, can transpire as much as 38% more than corn over a growing season [20]. Replacing traditional annual cropping systems with switchgrass in the Midwest and High Plains may cause additional stress to water resources because the agricultural crop production in large portions of these areas (e.g., Kansas and Nebraska) is dependent upon irrigation water from already stressed local resources [21]. Streamflow volume (Q) is responsive to changes in both climate and land cover [22]–[23], and changes in Q have important biological and socioeconomic implications [24]–[26]. Anthropogenic alteration of Q has been shown to impair aquatic communities and ecosystems, and the likelihood of impairment rises rapidly with increasing severity of reduced Q [25].

Water may be a significant limiting factor for biofuel crop production in many agricultural regions. It is important that we develop projections of future water use for agricultural crop production under climate change induced by land use change and to account for the impact of that water use on critical water resources. Prior studies have examined the potential for biofuel crops to affect regional climate [15]. The climate feedback of biofuel crops examined in these studies, however, is based on hypothetical scenarios that do not account for the socio-economic responses of land managers and thus do not represent plausible land use patterns that might result from current biofuel policies. To our knowledge, no prior studies have explored the changes in streamflow volume in response to climate change induced by land use/land cover (LULC) change; certainly none have examined this under the constraints of enacted legislation. Future projections of climate and climate-driven streamflow under plausible landscape scenarios will aid state and federal agencies in assessing the local cost of adaptation, increase public awareness, and guide the development of new mitigation programs related to water resources.

In this study, we examine changes in hydrologic processes including precipitation (P), ET, PET, runoff and Q that result from modification of local/regional climate driven by switchgrass cropping systems predicted to replace current cropping systems in the High Plains (hereafter referred to as the "biofuel scenario"). A regional climate model coupled to a land surface model is used to capture feedback between changes in the vegetation canopy due to switchgrass planting and regional climate processes. The change in Q to climate change under the biofuel scenario compared to the current cropping system scenario (hereafter referred to as the "baseline scenario") is estimated based on widely used non-parametric approaches [22]–[23], [27]. These non-parametric approaches utilize the concept of elasticity of Q that is usually derived using the historic relationship between Q, P and PET. Following the similar approach, we first derived the elasticity of Q to climate, and later used it in combination with projected changes in P and PET to derive changes in Q under the biofuel scenario relative to baseline across the conterminous U.S.

9.2 MATERIALS AND METHODS

9.2.1 REGIONAL CLIMATE MODELING FRAMEWORK

The Weather Research Forecast (WRF) model version 3.1.1 [28]–[29] and NOAH land surface model (NOAH LSM) are used for regional climate simulations. The simulation domain covers the continental U.S. at a resolution of 0.25 degree (i.e., 24 km). The NOAH LSM coupled to the WRF model is used to represent the interaction of soil and vegetation with the atmosphere [30]. Regional climate simulations were produced for the period of 1979–2004 under the two land use scenarios. The choice of the simulation period is constrained by 1) availability of data describing the baseline scenario; 2) the accuracy of extrapolating land use categories derived from 1991–1995 satellite data further into the future; and 3) the large computational burden associated with longer simulations. The baseline scenario represents land use categories and monthly phenology based on satellite derived data from

1991–1995, and the biofuel scenario represents projected alternative (i.e., switchgrass) land use categories (Figure 1) with identical atmospheric forcing data. Following the recent studies [16], [31]–[32] that have used 2-years as a minimum length for spin-up, we discarded the first two years (i.e., 1979 and 1980) of each simulation to allow for adjustment of the land surface with the atmosphere. Details of the model configuration are provided in the Anderson et al [19], and thus are not described here.

9.2.2 LAND USE SCENARIOS

The baseline scenario uses the NOAH LSM default settings of land use and vegetation parameters, including 24 vegetation classes, a vegetation parameter table and satellite-based (1991–1995) monthly vegetation fraction from which leaf area index (LAI) and albedo are derived (Table S1). The projection of LULC change produced by the Policy Analysis System (POLYSYS) model [33]–[34] in support of the DOE study report "U.S. Billion-Ton Update" [7] is used to create the biofuel land use scenario. The Billion Ton Update study examines the feasibility of attaining annual production of one billion dry tons of biomass feedstock by 2030 based on projections of future biomass demand, inventory, production capacity, availability, and technology. In the 2011 update, the greatest conversion of traditional cropping system to switchgrass occurs in the Great Plains, with 20–30% in Kansas and 30–45% in northern and northwest Oklahoma [35]. The POLYSYS simulation contained county level switchgrass and crop production for 2022, the first year Renewable Fuel Standard (RFS) goals reach maximum levels, at a $60/dry ton farm gate switchgrass price with the Billion-Ton Study baseline assumptions, including an extension of the USDA 10-year yield forecast for major food and forage crops to 2022. An area weighted method was used to resample county-level POLYSIS estimates of switchgrass conversion to the WRF grid [19]. The NOAH LSM uses a single vegetation category for each grid cell, and vegetation parameters are homogenous within each grid cell. Thus, a grid cell that contains a mixture of vegetation types does not explicitly account for each type, but represents an average of vegetation parameters over all vegetation types present in the grid cell.

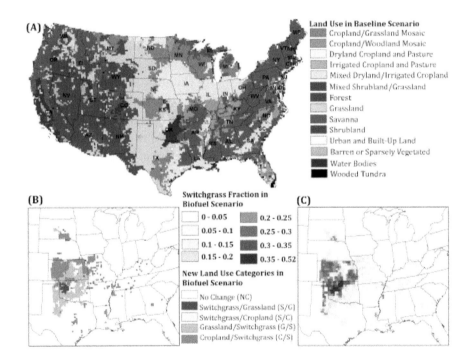

FIGURE 1: Default land use categories in the WRF model (A); new land use categories defined for the biofuel scenario (B); and fraction of land use that is switchgrass in the biofuel scenario (C).

For the biofuel scenario, four new vegetation classes, their related vegetation parameters and monthly vegetation fraction are introduced in the WRF model based on default land use categories (Figure 1A). The new vegetation classes are modified to represent a mixture of the default land use categories and switchgrass (Figure 1B). This reflects that a mixture of biofuel and conventional crops are expected in regions where biofuel crops are adopted rather than complete replacement of conventional crops with switchgrass. Two of the new classes (Switchgrass/Grassland Mosaic and Switchgrass/Cropland Mosaic) are used for grid cells in which the switchgrass fraction exceeds 30%, and two (Grassland/Switchgrass Mosaic and Cropland/Switchgrass Mosaic) are used for grid cells in which the switchgrass fraction is below 30% [36] (Figures 1B and 1C). We prevented land use change to switchgrass in regions beyond that projected by the Billion ton update by only changing the parameters based on latitude/latitude. For example, land cover changes were mostly in Oklahoma and Kansas (Figure 1B); we did not change parameters in California.

As the new land use categories under the biofuel scenario reflect a mixture of switchgrass and conventional crops rather than complete replacement of switchgrass, the phenology for new land use categories is characterized in our simulations by adjusting the satellite-based monthly vegetation fraction based on discussion with scientists working on field trials of switchgrass. The monthly greenness fraction is the basis for LAI and albedo calculations in NOAH LSM, and both LAI and albedo increase with vegetation fraction. During the growing season, spectrally weighted albedo (which is required in WRF) increases as a crop is greening up. To represent change in phenology consistent with managed plots of switchgrass stands, vegetation fraction is increased during February–October to simulate earlier greening, denser foliage at peak LAI, and later senesce of switchgrass (a perennial grass) compared to annual crops and rangeland. Increase of vegetation fraction ranges 10–20%, except north central Oklahoma where it is increased by 90% to offset low LAI due to winter wheat harvest [19].

The maximum and minimum values of LAI and albedo are adjusted to reflect changes in vegetation under the biofuel scenario (Table S1). And, the same monthly phenology is imposed for each simulated year. When the vegetation fraction is at its peak, the LAI and albedo are as well. Maximum LAI is based upon observations of field stands of managed perennial grass that grows a denser canopy than prairie grass. Although LAI>6 as measured in field trials of managed switchgrass is used in simulations of switchgrass production [37]–[38], maximum LAI<6 is set to reflect a regional vegetation mixture. This approach is consistent in simulations with Van Loocke et al [17].

9.2.3 CLIMATE ELASTICITY OF STREAMFLOW AND CHANGES IN STREAMFLOW

In this version of WRF model, surface runoff is computed as the excess of precipitation that does not infiltrate into the soil [39]. Although NOAH-LSM describes the canopy and root zone in detail, the interactions between groundwater, the root zone, and surface water were not yet included at the time our project was undertaken and completed. This version parameterizes surface runoff with a simple infiltration-excess scheme rather than terrain slope channel routing, and it treats baseflow as a linear function of bottom soil-layer drainage [40]. Thus, runoff estimates from WRF are not representative of the changes in streamflow. For the purpose, we used non-parametric approaches that are demonstrated as or more robust than complex and detailed hydrologic models for evaluating the sensitivity of streamflow to climate [22], [27], [41]. These non-parametric approaches use the concept of climate elasticity of streamflow (ε_x), computed based on historic P, PET and Q data. The climate elasticity of streamflow is defined by the proportional change in Q to the change in a climate variable (x), such as P or PET [42]. It is an index commonly used to quantify the sensitivity of Q to changes in climate. Often this index (i.e., ε_x) is derived from the historic climate and hydrologic data (i.e., P, PET and Q) [23], [42]. Streamflow in unimpaired watersheds (i.e., watersheds in which streamflow are not subject to regulation or diversion, and defined as reference watershed in this study) can be modeled as a function of P and PET [23]. The changes in Q due to changes in P and PET can be approximated as:

$$\Delta Q / Q = \varepsilon_p \Delta P / P + \varepsilon_{pet} \Delta PET / PET \qquad (1)$$

In equation 1, ΔQ, ΔP and ΔPET are changes in Q, P and PET, respectively; ε_p and ε_{pet} are the elasticities of streamflow with respect to P and PET. Prior studies [23], [27], [43] have proposed non-parametric approaches to estimate εx from observed climatic data. Of the various approaches (Table S2, Figure S2) found in the literature, we have no reason to favor one over the others, and thus use the average of ε_p estimated from all available non-parametric approaches to predict average changes in Q under the alternative LULC scenario. Details about the non-parametric approaches used in this study and hydro-climatology of the conterminous U.S are discussed in Figure S3.

To compute elasticity estimates, we used historical annual streamflow, precipitation and PET information for 1,845 reference watersheds across the conterminous United States (see Figure S1). To compute ε_x estimates from the historic climate record, we used PET instead of ET similar to prior studies [22]–[23] due to data limitations associated with the estimation of actual ET from 1950–2009. Further, as ε_{pet} was computed using historic PET estimates, we used PET instead of ET from the regional climate model simulations to estimate changes in Q to maintain consistency in the methodology. These climate elasticity values were combined with the differences in mean annual P and PET between the biofuel and baseline scenarios, expressed as a percentage of the baseline, to compute the relative change in Q across the nation under the biofuel scenario (equation 1).

9.2.4 SENSITIVITY OF STREAMFLOW CHANGE

Uncertainty in the estimated change in Q under the biofuel scenario is evaluated based on: 1) difference in elasticity estimates computed from various non-parametric approaches (discussed in the supporting material), and 2) year to year changes in simulated P and PET for the period 1981–2004. To evaluate the sensitivity of changes in streamflow to changes in both climate and elasticity estimates, we calculated the standard deviation (std) of percent annual change in P and PET between the biofuel and baseline scenarios, and the std of elasticity estimates from the seven different non-parametric methods as shown in equations 2 and 3.

$$\Delta Q = (P \pm std\ P) \times \varepsilon_p + (PET \pm std\ PET) \times \varepsilon_{pet} \qquad (2)$$

$$\Delta Q = P \times (\varepsilon_p \pm std\ \varepsilon_p) + PET \times (\varepsilon_{pet} \pm std\ \varepsilon_{pet}) \qquad (3)$$

$$\Delta Q = (P \pm std\ P) \times \varepsilon_p + PET \times \varepsilon_{pet} \qquad (4)$$

In equations 2, 3 and 4, P and PET indicate the percent change in precipitation and evapotranspiration under the biofuel scenario relative to the baseline. ε_p and ε_{pet} indicate the mean values of ε_p and ε_{pet} from the seven empirical methods.

The sensitivity of estimated change in Q is evaluated by varying the percent change in mean annual P and PET under the biofuel scenario (equation 2; Figure S6A and S6B), and the mean estimates of ε_{pet} and ε_p (equation 3; Figures S6C and S6D) by ± their standard deviation. We estimated the sensitivity of predicted change in Q by varying the percent change in mean annual P and PET simultaneously because P observed in many regions including High Plains are correlated with PET [44] and ET [45]. Sensitivity measures are also computed varying only P (equation 4; Figure S7) because the impacts of varying P and PET simultaneously will tend to cancel each other in the regions where impacts are inversely correlated, thus not reflecting the full contribution of P or PET to Q (equation 4).

9.3 RESULTS

9.3.1 CLIMATE ELASTICITY OF STREAMFLOW

Precipitation elasticity of streamflow is estimated in the range of 1–3.6 with a mean of 2.2 for watersheds across the U.S., implying that a 1% change in P will result in more than a 1% change in Q. The relationship between P and Q is generally non-linear, and this non-linearity is influenced by catchment properties including storage processes, ET and vegetation properties, and these factors are implicitly factored into elasticity estimates. Approximately 46% of all watersheds examined have εp higher than 2, and these watersheds are clustered in the Southwest, Midwest and Southeastern parts of the nation (Figure 2A) where PET usually exceeds P. Only a few basins in the Northwest have ε_p less than 1.5.

Evapotranspiration elasticity of streamflow for watersheds across the U.S. is estimated to be in the range of negative 3.6 to negative 1, with a mean of negative 1.9, indicating that a 1% reduction in PET would result in about a 1.9% increase in Q. The geographical distribution of ε_{pet} (Figure 2B) in the conterminous U.S. is similar to the distribution of ε_p with lower values ε_{pet} in the arid and semiarid regions of the Southwest and Midwest.

9.3.2 VARIABILITY IN CLIMATE ELASTICITY OF STREAMFLOW

The variability in elasticity estimates computed using seven different non-parametric approaches (expressed as a standard deviation) is high in the arid and semi-arid regions of the Midwest and Southwest, and lower in the humid and semi-humid regions (Figures S4A and S4B). Also, the standard deviation of εpet estimates is higher than the standard deviation of ε_p estimates. This is due to differences among the seven different approaches; five of seven approaches [46]–[51] depend upon aridity index (PET/P) to estimate elasticity estimates, while the other two depend on Q and P or PET [23], [27] (see Table S2 for details).

9.3.3 PROJECTED CLIMATE CHANGE UNDER BIOFUEL SCENARIO

9.3.3.1 CHANGE IN ANNUAL PRECIPITATION AND EVAPOTRANSPIRATION.

Under the biofuel scenario, the mean annual change in P relative to the baseline is projected to be in the range of negative 10% to positive 10% (Figure 3A). About 84% of the conterminous U.S. is predicted to experience changes in P in the range of negative 5% to positive 5% under the biofuel scenario; and a general decrease in P is predicted over 52% of the area. The magnitude of projected change in mean annual P in the main bio-fuel crop producing region (i.e., Kansas and Oklahoma) is between 2.5% to 5% relative to the baseline scenario. Under the biofuel scenario, western Kansas and Oklahoma show 5 to 15 mm higher annual P than under the baseline scenario.

The projected change in mean annual PET under the biofuel scenario is estimated to be in the range of negative 6% to positive 16% relative to the baseline (Figure 3C). Higher PET in the switchgrass planted region is consistent with lower mean temperatures due to earlier green-up and the higher LAI of switchgrass compared to current vegetation, and higher net radiation under the biofuel scenario.

In addition to the observed changes in climate in switchgrass plant-ed region, we observed changes in climate patterns in areas away from the switchgrass concentrated area. Under the biofuel scenario, southern regions including parts of Arizona, New Mexico and Texas, the High Plains including western Kansas and Oklahoma, the Midwest includ-ing eastern part of Nebraska and Iowa, and a large region of the eastern states show an increase in annual P of between 2.5% to 10% relative to the baseline scenario. A large decline in annual P (i.e., between 2.5% to 10% relative to the baseline) is predicted across much of the northern (i.e., northern Minnesota, South Dakota, Wisconsin), Midwestern (i.e., Wyoming, Idaho, northern Colorado) and the High Plains (i.e., Missouri) regions. Due to the internal non-linear climate dynamics, a single simu-lation is insufficient to conclude that they are systematically caused by land use change in the Great Plaines. They could, in fact, be an artifact of the initial atmospheric conditions.

PET is usually a good representation of actual ET when there is no plant water stress, and is thus commonly used in precipitation-runoff modeling applications [41]. We observed similar trends in PET and ET in the switchgrass perturbed region (Figures 4B and 4C) although they differed in magnitude. Increases in ET increase low-level humidity and the potential for more P [19]. Change in mean annual P, PET, ET and runoff when examined by land use (i.e., switchgrass altered and unal-tered) categories in the High Plains, suggests that the difference in ET, PET and runoff represent a change induced by land cover perturbation (Figures 4 and S5).

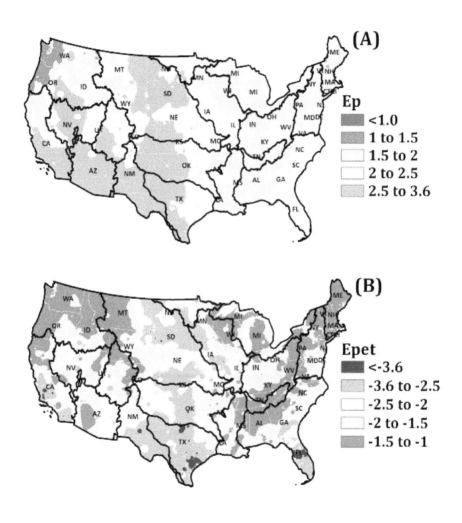

FIGURE 2: Hydro-climatology of the conterminous US; (A) Precipitation elasticity of streamflow (ε_p) and (B) Evapotranspiration elasticity of streamflow (ε_{pet}).

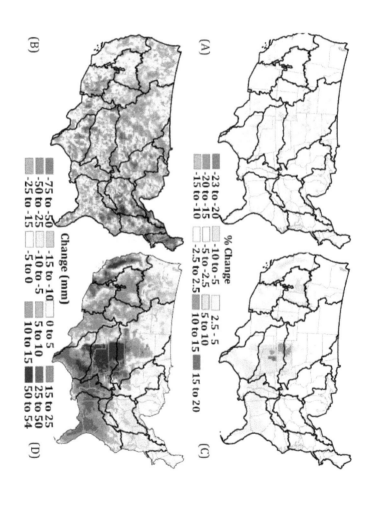

FIGURE 3: Change in mean annual precipitation and potential evapotranspiration for 1981–2004 expressed in A) percentage change in mean annual precipitation; B) change in mean annual precipitation (millimeters); C) percentage change in mean annual PET; and D) change in PET (millimeters) under the biofuel scenario. Percent change under the biofuel scenario relative to the baseline scenario for a given location is estimated as the mean of: (Baseline$_{xi}$ – Biofuel$_{xi}$) / Baseline$_{xi}$ × 100 where x is either P or PET and i is year between 1981 and 2004.

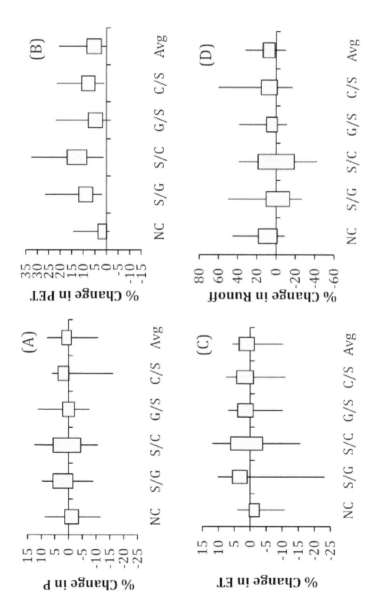

FIGURE 4: Percent difference in (A) annual precipitation; (B) PET averaged by land use categories in switchgrass altered regions in Kansas and Oklahoma as shown in Figure 1. Box top and bottom edges are the interquartile range of percent difference for each year, and whiskers are maximum and minimum annual values. X-axis labels are land use categories: No Change (NC), Switchgrass/Grassland (S/G), Switchgrass/Cropland (S/C), Grassland/Switchgrass (G/S), Cropland/Switchgrass (C/S), and average over all categories (Avg).

Differences in climate between the baseline and biofuel projections vary annually in both magnitude and direction between 1981 and 2004. The magnitude of mean annual change in PET and ET is higher than the year to year changes in PET and ET in switchgrass planted regions under the biofuel scenario. Also, the mean annual runoff in the switchgrass dominated region is lower than other regions in Kansas and Oklahoma. Compared to land cover with a lower fraction of switchgrass (i.e., cropland/switchgrass and grassland/switchgrass), the land cover with a large fraction (i.e., >30%) of switchgrass (i.e., switchgrass/grassland and switchgrass/cropland) demonstrated a higher magnitude of change in PET, ET and runoff in the biofuel scenario (Figure 4). This indicates that the observed changes in PET, ET and runoff associated with biofuel feedstock production are large and significant. Decrease in runoff results from lower soil moisture levels due to higher evapotranspiration of switchgrass during the growing season [19]. Conversely, the inter-annual variability of change in P is as large as or larger than the magnitude of mean annual change in P (Figure 4). Thus it is hard to conclude that P is changed under the biofuel scenario.

9.3.3.2 STREAMFLOW RESPONSE TO PROJECTED CLIMATE CHANGE.

Mean annual change in Q in response to changes in P and PET under the biofuel scenario is shown in Figure 5. Across the conterminous U.S., the change in mean Q under the biofuel scenario is estimated to be in the range of negative 56% to positive 20% relative to the baseline scenario. An increase in Q with magnitude greater than 5% is predicted over 12% of the area. Lower PET but higher P in New Mexico and Arizona are estimated to increase Q under the biofuel scenario. However, a decrease in Q with magnitude greater than 5% is predicted over 30% of the area. The increase in P is smaller than the increase in PET under the biofuel scenario, and this causes a net decline in Q in the High Plains. In the High Plains, Q is predicted to be about 20% lower than the baseline. Streamflow in the biofuel crop region within the High Plains is 18% lower relative to the baseline (Figure 5). The switchgrass areas in the biofuel crop region show decreases in streamflow that are twice as large as the decrease in the unperturbed area.

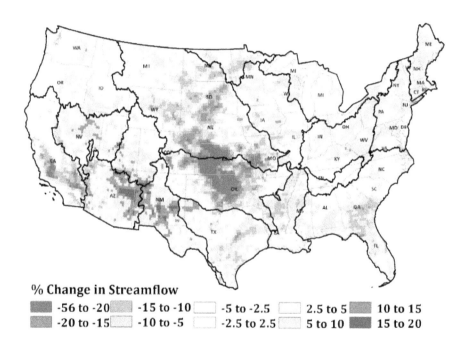

FIGURE 5: Percent change in mean annual streamflow as a function of change in annual precipitation and potential evapotranspiration under the biofuel scenario.

9.3.3.3 SENSITIVITY OF STREAMFLOW.

The change in Q under the biofuel scenario is observed to be highly sensitive to mean annual P and PET for large parts of the conterminous U.S., however, it is less sensitive in the switchgrass planted region where change in PET is larger than P (Figure S6). As parameters associated with precipitation and evapotranspiration are very likely to be correlated with each other, PET and P were changed independently in the sensitivity test to examine the relative significance of changes in P and PET on streamflow estimates (see the section 'Materials and Method' for details). Under these sensitivity analyses, Q always decreases under the biofuel scenario, even in the case when PET is held constant and P is increased by one standard deviation (Figure S7). Sensitivity analysis also showed that predicted changes in streamflow are robust to differences in ε_p and ε_{pet} stimation approaches. Change in Q across the range of ε_p and ε_{pet} estimation approach (Figure S6C and S6D) is less than the change in Q resulting from changes in P and PET equal to their inter-annual variability (Figure S6A and S6B).

9.4 DISCUSSION

Climate change is a likely result of a policy that encourages expansion of energy crop production in the High Plains and Midwest U.S [19]. This climate change is caused by alteration of the surface energy and moisture balances induced by changes in land cover when current cropping systems are replaced by energy crops. Notable changes include lower temperature, higher PET, lower runoff and a decline in Q in the region where switchgrass is predicted to replace current vegetation. Our analyses show changes in the mean annual PET, ET and streamflow to be a stronger climate change impact of biofuel crop production than changes in mean annual P.

The main goal of this study is to use a simple and robust model to communicate to policy makers and analysts the potential implications of biofuels policy-induced climate change on streamflow to inform the search for a sustainable renewable fuel production system. While adopting the simple but robust model for estimating streamflow changes, we made several assumptions, and these assumptions are likely to introduce uncertainties in our streamflow estimates. Despite some of the limitations (as discussed below), our conclusion that increase cellulosic feedstock production are likely to reduce water yield is found to be in agreement with other studies conducted [52], [20] in other parts of the nation.

Weather conditions simulated with regional and global climate models over short periods are sensitive to their initial conditions. A change in the vegetation type at the initial condition will result in a different sequence of weather conditions. Thus, differences between simulations may be a combination of a transient response from unpredictable nonlinear dynamics acting upon a different initial state as well as a systematic response from a structural change in forcing of local climate, in this case land use change [53]–[54]. The average change for a transient response is expected to be zero given either a long data record or multiple simulations from alternative initial conditions. Changes in Q estimated in this study thus include uncertainties inherent in P and PET estimated over a relatively short time series (in this case 24 years). Anderson et al [19], using one-way ANOVA analysis, found statistically significant change of monthly values for ET and P. However, the small change in annual ET resulted from offsetting statistically significant monthly changes. To examine further whether the ET and P responses were possibly transient, they performed a second simulation of a single year of the control design beginning from a different initial condition. The difference of ET in the two control simulations was much smaller than the difference between the control and LULC scenario. Precipitation, however, showed substantial sensitivity to the initial condition. This sensitivity of P to initial conditions indicates a larger number of simulations would be needed to identify a forced response in P, if one exists at all. Therefore, we are unable to state that a forced change in P exists anywhere in our simulation domain. Since P, ET and PET are correlated, we conclude that outside of the switchgrass planted region, change in PET and Q should not be considered a forced response. Within the switchgrass planted region, however, change of PET and Q is a forced response. Another source of uncertainty in our estimates can be driven by the use of temperature based approach to estimate PET instead of approach like Penman-Monteith (see the supporting material). As temperature-based PET estimates are likely to overestimate the impacts of changes in temperature, our streamflow estimates might have been overestimated to some extent.

Another possible area of limitation includes elasticity estimates. In the study, we interpolated the elasticity estimates based on reference watershed for the conterminous US. This means, our elasticity estimates may not accurately reflect the connection of climate to Q in non-reference watersheds where streamflow is heavily influenced by land use practices and ground water pumping. Predicted change in Q in non-reference watersheds thus might be over or under estimated in the biofuel scenario. Despite this limitation, this study is useful in providing the direction of change in streamflow under the biofuel scenario without requiring the use of detailed hydrologic models that are computationally complex and often provide ambiguous results when compared [27].In the study we show a potential change in annual streamflow volume as an outcome of a landscape influenced climatic system. Our analyses also suggest that under the biofuel scenario, there is a change in seasonal P and ET. In the growing seasons (i.e., April–June), P decreases and ET increases. Evapotranspiration increases until soil moisture nears wilting point, eliminating transpiration and inhibiting further decline in soil moisture [19]. Decrease in P and increase in ET suggests the possibility for a higher magnitude of change in Q seasonally than annually, and we recommend that this possibility should be explored further with detailed hydrologic models.

This study examines landscape induced climate change and ignores projected climate change due to atmospheric concentration of greenhouse gases (GHGs). There is a strong coupling among landscape processes, atmospheric GHGs and climate, since landscape acts as a sink or source to GHGs like CO_2 which affect the distribution of heat over land and in the atmosphere and feedback to the climate [55]. However, these processes are often uncoupled when we make future projections, and this is likely to introduce biases in the projection of future changes in climate. Thus, we recommend the direct coupling of landscape and other climate forcing factors (e.g., GHGs) to better predict the range of future climate change and its impacts on environmental resources [19]. In this study, we examine landscape induced climate change and its impact on Q on an annual basis. Changing landscape and climate alter not only Q, but also stream quality due to changes in the nutrient and sediment content of runoff, particularly from agricultural fields. Evaluation of stream quality under landscape driven climate change is thus recommended as a topic for future research.

In this study, we find interaction between land use change and climate change in Q that is not considered in previous studies. Although processes related to landscape characteristics, such as soil moisture, infiltration, and surface roughness could affect Q, but are not considered in estimation in Q in this study, we assume that the effect on streamflow of changing climate is larger than that resulting from change in landscape characteristics [56]–[58]. For example, Tu [56] examined change in streamflow in eastern Massachusetts under three different climate change scenarios under IPCC and land use scenarios relative to current conditions, individually and in combination. Tu [56] found that the change in streamflow under both climate and land use scenarios is similar to the streamflow changes examined under a climate change scenario only. Although the process-based hydrologic models (e.g., Soil and Water Assessment Tool (SWAT) and the Variable Infiltration Capacity (VIC)) have the ability to integrate the effects of land use change and climate change on Q and streamflow quality, they are computationally expensive given that they operate at the field- or sub-field scale and need to be calibrated with large amounts of empirical data. Using the method developed here, areas of concern can be identified and then detailed hydrologic models can be used to examine in more detail the specific trade-offs between land use and management options and streamflow.

The results here suggest that changes induced in the climate system by biofuel crop production may increase stress on the water resources of the High Plains. Under the biofuel scenario, increased ET reduces soil moisture [19], and lower soil moisture during the growing season can cause plant water stress and reduce crop yield. Irrigation can reduce water stress but additional irrigation will increase pressure on already strained water resources in arid agricultural regions such as the High Plains. The predicted 20% decline in Q in the biofuel crop producing region under the biofuel scenario would exacerbate on-going conflicts over water allocation between agriculture and other uses. As we develop and implement policies to pursue more sustainable cellulosic biofuel production, we should carefully consider potential water limitations and other impacts to the hydrologic cycle.

REFERENCES

1. Hill J (2007) Environmental costs and benefits of transportation biofuel production from food-and lignocellulose-based energy crops. A review. Agronomy for Sustainable Development 27: 1–12. doi: 10.1051/agro:2007006
2. Hess JR, Foust TD, Hoskinson R, Thompson D (2003) Roadmap for agriculture biomass feedstock supply in the United States. DOE/NE-ID-11129, US Department of Energy, Washington DC, USA.
3. Costello C, Griffin WM, Landis AE, Matthews HS (2009) Impact of biofuel crop production on the formation of hypoxia in the Gulf of Mexico. Environ Sci Technol 43: 7985–7991. doi: 10.1021/es9011433
4. Donner SD (2007) Surf or turf: A shift from feed to food cultivation could reduce nutrient flux to the Gulf of Mexico. Global Environ Change 17: 105–113. doi: 10.1016/j.gloenvcha.2006.04.005
5. Schnoor J, Doering III OC, Entekhabi D, Hiler EA, Hullar TL, et al.. (2008) Water implications of biofuels production in the United States. National Academy of Sciences, Washington DC, USA.
6. Tilman D, Hill J, Lehman C (2006) Carbon-negative biofuels from low-input high-diversity grassland biomass. Science 314: 1598–1600. doi: 10.1126/science.1133306
7. U.S. Department of Energy (2011) U.S. Billion-Ton Update: Biomass Supply for a Bioenergy and Bioproducts Industry. Perlack, R.D., Stokes, B.J.; Leads, ORNL/TM-2011/224. Oak Ridge National Laboratory, Oak Ridge, TN. 227.
8. Marshall E, Caswell M, Malcom S, Motamed M, Hrubovcak J, et al.. (2011) Measuring the indirect land-use change associated with increased biofuel feedstock production. A review of modeling efforts. US Department of Agriculture-ERS, AP054. Washington DC, USA.
9. Schmer MR, Vogel KP, Mitchell RB, Perrin RK (2008) Net energy of cellulosic ethanol from switchgrass. PNAS 105: 464–469. doi: 10.1073/pnas.0704767105
10. Diffenbaugh N (2009) Influence of modern land cover on the climate of the United States. Climate Dynamics 33: 945–958. doi: 10.1007/s00382-009-0566-z
11. Pielke Sr RA, Adegoke J, Beltran-Przekurat A, Hiemstra CA, Lin J, et al. (2007) An overview of regional land-use and land-cover impacts on precipitation. Tellus B 59: 587–601. doi: 10.3402/tellusb.v59i3.17038
12. Twine TE, Kucharik CJ, Foley JA (2004) Effects of land cover change on the energy and water balance of the mississippi river basin. J Hydrometeor 5: 640–655. doi: 10.1175/1525-7541(2004)005<0640:eolcco>2.0.co;2
13. Bonan GB (2001) Observational evidence for reduction of daily maximum temperature by croplands in the Midwest United States. J Clim 14: 2430–2442. doi: 10.1175/1520-0442(2001)014<2430:oefrod>2.0.co;2
14. Halgreen W, Schlosser CA, Monier E, Kicklighter D, Sokolov A, et al. (2013) Climate impacts of a large-scale biofuels expansion. Geophys Res Lett 40: 1624–1630. doi: 10.1002/grl.50352

15. Georgescu M, Lobell D, Field C (2011) Direct climate effects of perennial bio-energy crops in the United States. PNAS 108(11): 4307–4312. doi: 10.1073/pnas.1008779108

16. Mishra V, Cherkauer KA, Niyogi D, Lei M, Pijanowski BC, et al. (2010) A regional scale assessment of land use/land cover and climatic changes on water and energy cycle in the upper Midwest United States. Int. J. of Climatology 30(13): 2025–2044. doi: 10.1002/joc.2095

17. Vanloocke A, Bernacchi CJ. Twine TE (2010) The impacts of Miscanthus× giganteus production on the Midwest US hydrologic cycle. GCB Bioenergy, 2, 180–191. Skamarock WC, Klemp JB (2008) A time-split nonhydrostatic atmospheric model for weather research and forecasting applications. Journal of Computational Physics 227: 3465–3485.

18. Pielke Sr RA (2005) Land use and climate change. Science 310: 1625–1626. doi: 10.1126/science.1120529

19. Anderson CJ, Anex RP, Arritt RW, Gelder BK, Khanal S, et al. (2013) Regional climate impact of a biofuels policy projection. Geophys Res Lett 40: 1217–1222 doi:10.1002/grl.50179.

20. Le PVV, Kumar P, Drewry DT (2011) Implications for the hydrologic cycle under climate change due to the expansion of bioenergy crops in the Midwestern United States. PNAS 108: 15085–15090. doi: 10.1073/pnas.1107177108

21. Stone K, Hunt P, Cantrell K, Ro K (2010) The potential impacts of biomass feedstock production on water resource availability. Bioresource Technology 101: 2014–2025.

22. Chiew F, Teng J, Vaze J, Kirono D (2009) Influence of global climate model selection on runoff impact assessment. Journal of Hydrology 379: 172–180. doi: 10.1016/j.jhydrol.2009.10.004

23. Zheng H, Zhang L, Zhu R, Liu C, Sato Y, et al.. (2009) Responses of streamflow to climate and land surface change in the headwaters of the Yellow River Basin. Water Resources Research 45 W00A19.

24. Anderson CJ (2012) Local adaption to changing flood vulnerability in the Midwest. Climate change in the Midwest; Pryor, S.C., Eds.; Indiana University Press, ISBN: 978-0-253-00682-0.

25. Carlisle DM, Wolock DM, Meador MR (2011) Alteration of streamflow magnitudes and potential ecological consequences: a multiregional assessment. Front Ecol Environ 9(5): 264–270. doi: 10.1890/100053

26. Vogel RM, Wilson I, Daly C (1999) Regional regression models of annual streamflow for the United States. Journal of Irrigation and Drainage Engineering 125: 148–157. doi: 10.1061/(asce)0733-9437(1999)125:3(148)

27. Sankarasubramanian A, Vogel RM, Limbrunner JF (2001) Climate elasticity of streamflow in the United States. Water Resources Research 37: 1771–1781. doi: 10.1029/2000wr900330

28. Skamarock W, Klemp JB (2008) A time-split nonhydrostatic atmospheric model for weather research and forecasting applications. J. of Computational Physics. 227(7): 3465–3485. doi: 10.1016/j.jcp.2007.01.037

29. Skamarock W, Klemp JB, Dudhia J, Gill DO, Barker DM, et al.. (2008) A description of the Advanced Research WRF version 3. NCAR Technical Note NCAR/TN-475 STR.

30. Ek M, Mitchell K, Lin Y, Rogers E, Grunmann P, et al.. (2003) Implementation of NOAH land surface model advances in the National Centers for Environmental Prediction operational mesoscale Eta model. J Geophys Res 108: doi:10.1029/2002JD003296.

31. Steiner AL, Pal JS, Rauscher SA, Bell JL, Diffenbaugh NS, et al. (2009) Land surface coupling in regional climate simulations of the West African monsoon, Climate Dynamics. 33(6): 869–892. doi: 10.1007/s00382-009-0543-6

32. Elía R, Caya D, Côté H, Frigon A, Biner S, et al. (2008) Evaluation of uncertainties in the CRCM-simulated North American climate. Climate Dynamics 30: 113–132. doi: 10.1007/s00382-007-0288-z

33. Ugarte DG De La Torre, Ray DE (2000) Biomass and bioenergy applications of the POLYSYS modeling framework. Biomass Bioenergy 18: 291–308. doi: 10.1016/s0961-9534(99)00095-1

34. Ray DE, Moriak TF (1976) POLYSIM: A National Agricultural Policy Simulator. Agricultural Sector Models for the United States 28: 14–21.

35. Dicks MR, Campiche J, Ugarte DDLT, Hellwinckel C, Bryant HL, et al.. (2009) Land use implications of expanding biofuel demand. J. of Agric. And App. Econ. 41(2): 435–453.

36. Khanal S, Anex RP, Anderson CJ, Herzmann DE, Jha MK (2013) Implications of biofuel policy-driven land cover change for rainfall erosivity and soil erosion in the United States. GCB Bioenergy: doi:10.1111/gcbb.12050.

37. Miguez F, Maughan M, Bolero GA, Long SL (2012) Modeling spatial and dynamic variation in growth, yield, and yield stability of the bioenergy crops Mixcanthus x giganteus and Panicum virgatum across the conterminous United States, Global Change Biology-Bioenergy doi:10.1111/j.1757-1707.2011.01150.x.

38. Mitchell R, Schmer MR (2012) Switchgrass harvest and storage. In A. Monti (ed.) Switchgrass: A valuable biomass crop for energy (Green Energy and Technology), pp. 113–127.

39. Chen F, Dudhia J (2001) Coupling an Advanced Land Surface-Hydrology Model with the Pen State-NCAR MM5 Modeling System. Part I: Model Implementation and Sensitivity. Monthly Weather Review1 29: 569–585. doi: 10.1175/1520-0493(2001)129<0569:caalsh>2.0.co;2

40. Rosero E, Gulden LE, Yang Z, Goncalves L, Nui G, et al. (2011) Ensemble Evaluation of Hydrologically Enhanced Noah-LSM: Partitioning of the Water Balance in High-Resolution Simulations over the Little Washita River Experimental Watershed. J. Hydrometeor 12: 45–64. doi: 10.1175/2010jhm1228.1

41. Chiew FHS (2006) Estimation of rainfall elasticity of streamflow in Australia. Hydrological Sciences 51: 613–625. doi: 10.1623/hysj.51.4.613

42. Schaake JC (1990) From climate to flow. In: Climate change and US water resources, Waggoner, P.E., Eds.; J. Wiley and Sons, pp. 177–206.

43. Arora VK (2002) The use of the aridity index to assess climate change effect on annual runoff. Journal of Hydrology 265: 164–177. doi: 10.1016/s0022-1694(02)00101-4

44. Zhang L, Dawes W, Walker G (2001) Response of mean annual evapotranspiration to vegetation changes at catchment scale. Water Resources Research 37: 701–708. doi: 10.1029/2000wr900325

45. Koster RD, Dirmeyer PA, Guo Z, Bonan G, Chan E, et al. (2004) Regions of strong coupling between soil moisture and precipitation. Science 305: 1138–1140. doi: 10.1126/science.1100217

46. Zhang L, Dawes W, Walker G (2001) Response of mean annual evapotranspiration to vegetation changes at catchment scale. Water Resources Research 37: 701–708. doi: 10.1029/2000wr900325

47. Pike J (1964) The estimation of annual run-off from meteorological data in a tropical climate. Journal of Hydrology 2: 116–123. doi: 10.1016/0022-1694(64)90022-8

48. Budyko MI (1963) Evaporation under natural conditions, Israel Program for Scientific Translations; [available from the Office of Technical Services, US Dept. of Commerce, Washington].

49. Turc L (1954) The water balance of soils. Relation between precipitation evaporation and flow. Ann Agron. 5: 491–569.

50. Ol'Dekop E (1911) On evaporation from the surface of river basins. Trans. Met. Obs. lurevskogo, Univ. Tartu, 4.

51. Schreiber P (1904) Über die Beziehungen zwischen dem Niederschlag und der Wasserführung der Flüsse in Mitteleuropa. Meteorologische Zeitschrift 21: 441–452.

52. Wu Y, Liu S (2012) Impacts of biofuels production alternatives on water quantity and quality in the Iowa River Basin. Biomass and Bioenergy 36: 182–191. doi: 10.1016/j.biombioe.2011.10.030

53. Braun M, Caya D, Frigon A, Slivitzky M (2011) Internal variability of Canadian RCM's hydrological variables at the basin scale in Quebec and Labrador. Journal of Hydrometeorology 13: 443–462. doi: 10.1175/jhm-d-11-051.1

54. Solomon S, Qin D, Manning M, Chen Z, Marquis M, et al.. (2007) Climate Change 2007: The Physical Science Basis, IPCC Fourth Assessment Report, Miller, H.L., Eds.; Cambridge University Press, United Kingdom and New York, NY, USA, pp. 996.

55. Marland G, Pielke RA Sr, Apps M, Avissar R, Betts RA, et al. (2003) The climatic impacts of land surface change and carbon management, and the implications for climate-change mitigation policy. Climate Policy 3: 149–157. doi: 10.3763/cpol.2003.0318

56. Tu J (2009) Combined impact of climate and land use changes on streamflow and water quality in eastern Massachusetts, USA. Journal of Hydrology 379: 268–283. doi: 10.1016/j.jhydrol.2009.10.009

57. Guo G, Hu Q, Jiang T (2007) Annual and seasonal streamflow responses to climate and land-cover changes in the Poyang Lake Basin China. Journal of Hydrology 355: 106–122. doi: 10.1016/j.jhydrol.2008.03.020

58. Hu Q, Willson GD, Chen X, Akyuz A (2004) Effects of climate and landcover change on stream discharge in the Ozark highlands, USA. Environ. Model Assess 10: 9–19. doi: 10.1007/s10666-004-4266-0

There are several supplemental files that are not available in this version of the article. To view this additional information, please use the citation on the first page of this chapter.

PART III

BIOMASS CHALLENGES

CHAPTER 10

Trading Biomass or GHG Emission Credits?

JOBIEN LAURIJSSEN AND ANDRÉ P. C. FAAIJ

10.1 INTRODUCTION

All countries that ratified The Kyoto Protocol, committed themselves to specific targets to reduce GHG levels. Participating countries can reduce emissions within their country borders. However, apart from this so-called 'domestic action' also some flexible mechanisms were designed that enable participating countries to cooperate with other countries, to enhance cost-effectiveness of climate change mitigation e.g. Joint Implementation (JI) and Clean Development Mechanism (CDM).

Biomass can play an important and dual role in greenhouse gas mitigation: it can be used as an energy source to substitute for fossil fuels, or as a carbon reservoir. Several studies have analyzed the potential contribution of bio-energy to the future world's energy supply (e.g. Smeets et al. 2007; Hoogwijk et al. 2003; Berndes et al. 2002; Hoogwijk et al. 2005). Substantial variation in the estimates of the biomass production potentials exists

Trading Biomass or GHG Emission Credits?. © Laurijssen J and Faaij APC. Climatic Change **94,3–4** *(2009). DOI 10.1007/s10584-008-9517-7. Reprinted with permission from the authors.*

within and between different studies; ranges are mostly in between 200 and 1200 EJ/year in 2050, of which the forestry biomass potential is now assessed at 12–74 EJ/year (IPCC 2007). Despite the variation in potential, considering the current global energy use of around 400 EJ/year, these studies indicate that the contribution of bio-energy to the future world's energy supply could be very significant. Worldwide biomass potentials are, however, unequally distributed over the world; a large potential biomass production capacity can be found in developing countries and regions such as Latin America, Sub-Saharan Africa and Eastern Europe (Smeets et al. 2007). Countries in these areas can use the biomass within their borders to substitute for fossil fuels; they can, however, also become suppliers of biomass to other countries. This creates important future opportunities for such regions as the export of biomass can generate considerable income sources for the relatively poor regions of the world with large biomass potentials. Countries that have an emission reduction obligation under the Kyoto Protocol can import the biomass/bio-fuels to substitute for fossil fuels (see also a study by Damen and Faaij 2006). In this case, the importing country reduces GHGemissions by avoiding the use of fossil fuels. The GHG mitigation potential is thus affected by the carbon intensity of the energy system in the importing country (the reference system).

Another option is to convert the biomass locally; by investing in renewable energy projects in these countries, Annex I countries can acquire GHG-emissions credits, which help them reach their targets. The host countries can also benefit from these projects, as the introduction of modern biomass conversion technologies can enable communities to be self-sufficient and reduce the dependency of fossil fuel imports (Hall and House 1994).

Physical trade of biomass/bio-fuels and the trade of emission-credits derived from bio-energy projects (JI and CDM) are thus two options to reduce CO_2-eq levels by using biomass. Which of those options is optimal from a land-use, cost and a GHG mitigation perspective will depend on various crucial criteria and will probably differ under different circumstances. In their paper "Should we trade biomass, electricity, renewable certificates or CO_2-credits?" members of IEA Task Groups 38 and 40 already briefly discussed different trading options and the dependency upon the specific situations of the "exporting" and "importing" country (Schlamadinger et al. 2004). In this study, we will compare bio-transportation-fuel (biofuel) production in different countries to reveal some regional differences in the production of bio-mass and its further treatment. Biofuels are also interesting because to date no CDM methodology has been approved for transportation fuels, except for a methodology on the production of biodiesel based on waste oils and/or waste fats from biogenic origin (UNFCCC 2007). The rules and regulations for accounting carbon balances of JI projects, CDM projects and trading of bio-fuels are different and we will investigate the impact of these differ-ences. Induced land-use changes (leakage) are a central and crucial point in the current debate around the net GHG impact of bio-fuels (Fargione et al. 2008; Searchinger et al. 2008). Moreover, other issues like diversifica-tion and flexibility may influence the decision for a trading system in both exporting and importing country. Summarizing, the aim of this study is as follows: given the large global bio-energy production potential, how should we make optimal use of this renewable source, regarding, amongst others, land-use, costs and avoided emissions?

In Section 2, for both trading systems a methodological framework is composed, where all factors, accounting methods and legislation influenc-ing the amount and costs of CO_2 reduction are represented. In Section 3, two case studies are presented: Mozambique and Brazil. In both case studies, physical trading and emission credit trading are compared for the amount of credits generated, the amount of land-use and the associated costs. Section 4 provides a discussion of the key findings, followed by conclusions and recommendations in Section 5.

10.2 METHODOLOGY

For both trading systems, all factors influencing the costs of CO_2-reduction need to be recognized and are represented in a methodological framework that will be the basis of study.

10.2.1 PHYSICAL BIOMASS TRADE

For physical biomass trade (Fig. 1), the chain starts with biomass produc-tion and harvesting in the exporting country.Depending on the form in which it is transported, different pre-treatment options are needed. Storage and transportation might occur in both exporting and importing country and international transport occurs between those countries. Finally, the biomass is converted in the importing country. The total amount of GHG-emissions and costs for this chain can be calculated by adding them for all steps in the logistic chain. To determine the costs and amount of GHGemission reduction, the chain needs to be compared with the total costs and GHGemissions of the reference energy system of the importing country. In the case of residual biomass, emission and costs related to biomass production and harvesting need to be allocated for residues, moreover, they need to be compared with the residue reference system (the use of residues if not used for energy generation). Since the conversion efficiencies of biomass and fossil fuels might be different, the systems need to be compared based on the same amount of final energy produced. In the case of dedicated crops, there might also be a sequestration component to account for. If the exporting country is an Annex I country, they can include this sequestration in their NationalGHGInventory (net-net or gross-net accounting approach, depending on the type of activity IPCC 2003). (The sequestration benefit lies, however, with the exporter and it would probably (in reality) not affect the costs of GHG-emission for the importing country.) If the exporting country is a non-Annex I country, sequestration is not accounted for since there are no National Inventory and Kyoto targets to comply with. More worrying is if the biomass production leads to a net carbon loss, e.g. due to unsustainable harvesting in existing forests, there is, so far, no mechanism to account for this loss.

Currently, in e.g. the Netherlands, the avoided GHG-emissions by using bio energy are calculated by assuming that biomass has, by default, zero emissions. Upstream emissions for both biomass and fossil fuels chains are ignored (or assumed to be equivalent). The amount of emission reduction is calculated only by determining the amount of avoided emissions during conversion of the reference fuel. Thus, also the emissions (or reductions) related to reference land-use and the alternative fates of residues are ignored.

For the purpose of this study, two options to determine the costs of GHG-emission reduction for physical biomass trade will be explored:

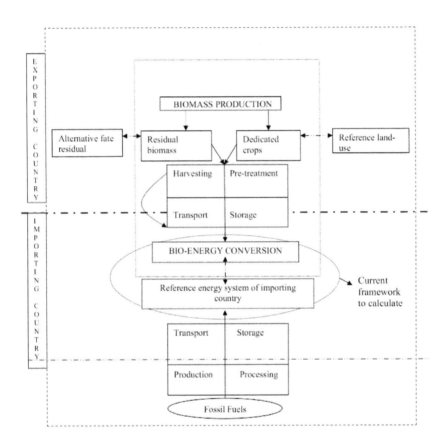

FIGURE 1: Methodological framework for physical biomass trade

The amount of emission reduction is determined by calculating the amount of GHG-emission during conversion of the avoided fossil fuels only. The associated costs, are the differences in the costs of the biomass chain and the fossil fuel chain.

$$C_{avoidance} = \frac{C_B - C_R}{E_R}$$

(1)

where:

- $C_{avoidance}$ = Costs per avoided GHG emission (€/CO_2-eq)
- E_R = GHGemissions for reference (fossil fuel) chain, excluding up-stream emissions (CO_2-eq)
- C_B = Costs for biomass chain (€)
- C_R = Costs for reference (fossil fuels) chain (€)

To determine the "real" emission reductions in the physical biomass trading system, in the second scenario, all factors as represented in Fig. 6 will be included. In this scenario, the costs of the avoided emissions for physical biomass trading can be calculated as follows:

$$C_{avoidance} = \frac{C_B - C_R \pm C_A}{E_R - E_B \pm E_A \pm E_L \pm E_K}$$

(2)

where:

- $C_{avoidance}$ = Costs per avoided GHG emission (€/CO_2-eq)
- C_B = Costs for biomass chain (€)
- C_R = Costs for reference chain (€)
- C_A = Costs/benefits related to alternative use of residues (€)
- E_R = GHG emissions for reference (fossil fuel) chain (CO_2-eq)
- E_B = GHG emissions for biomass chain (CO_2-eq)
- E_A = Emissions (benefits) related to alternative fate of residues (CO_2-eq)
- E_L = Carbon gains or losses compared to reference land-use (CO_2-eq) (Note: the costs or benefits (€) that occur due to carbon losses or carbon sequestration are not directly included in the model as they will be reflected in the final costs of CO_2 avoidance. If there is for example a net sequestra-

tion effect, the amount of avoided carbon will increase, resulting in lower costs of CO_2 avoidance for the whole chain
- E_K = Emissions (benefits) due to leakage (CO_2-eq)

EMISSION CREDIT TRADE

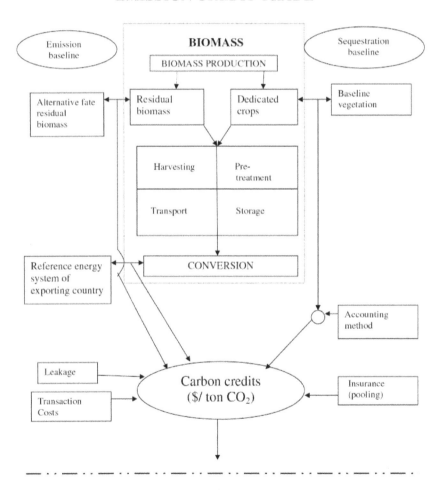

FIGURE 2: Methodological framework for emission credit trade

10.2.2 EMISSION CREDIT TRADE

For emission credit trade, the core part of the methodological framework (Fig. 2) is the same as for physical biomass trade, only the whole chain occurs in the exporting country. Two types of baselines can emerge for bio-energy projects:

1. The GHG emission baseline. In general this baseline represents: "the anthropogenic emissions by sources that would occur in the absence of the proposed project activity".
2. The sequestration baseline: to consider carbon stock changes associated with biomass production.

The biomass project might thus be split up in two parts: a fuel-switch project (A) and a sequestration project (B) (the contribution of (B) can be very significant to the total amount of carbon mitigated (Schlamadinger et al. 2001)). For sequestration projects, a distinction should be made between JI and CDMprojects. Firstly, because the range of eligible land-use, land-use change and forestry (LULUCF) activities under theCDMis limited to afforestation and reforestation, while for JI all LULUCF activities are eligible. Secondly, because of the non-permanence risk (i.e. fire, diseases, harvesting etc.) related to sequestration projects. The threat associated with this non-permanence issue is that credits are issued for carbon sequestration while the carbon is lost and the credits cannot be taken back. The issue of permanence in LULUCF projects does not arise for JI projects because they are implemented in Annex I countries with established national GHG inventories. For LULUCF sector reporting by Annex I countries, it has been agreed under the Kyoto Protocol that the "Stock Change" (SC) approach will be used. (Guidelines on this approach are described in the Good Practice Guidance for LULUCF (IPCC2003)). For CDM projects, on the other hand, the non-permanence issue plays a large role since there is no mechanism to compensate for reductions in carbon stocks. This complex and highly debated issue in the negotiations towards the implementation of the Kyoto Protocol, leaded to various proposals of accounting methods to address this issue. The accounting method to be applied for A/R projects in the CDM for the first commitment period (Temporary Crediting) was ultimately decided upon at COP 9 in Milan. In this study, four accountingmethods (Stock Change, Average Storage, Ton Year and Temporary Crediting) will be modelled in order to determine the effect of different approaches on the costs of GHG-emission reduction.

When the project is implemented and monitored, credits, resulting from the project activity, can be calculated and should be adjusted for leakage (UNFCCC 2001). Apart from the direct investment and operation & maintenance costs that come with the implementation of bio energy projects, CDM and JI projects lead to extra transaction cost. Transaction costs for CDM projects can be divided in market, pre-implementation and implementation transaction costs that arise from the CDMproject cycle (Krey 2004). For JI projects, transaction costs vary depending on which "track" is applied. In the Track 1 procedure, the project design, the project performance and emission reduction calculations do not need to be validated, monitored and verified by an entity officially designated by the UNFCCC or the COP; this will reduce the transaction costs. The Track 2 procedure is very similar to the CDM project cycle.

With the implementation of projects, there is always the risk of failure and consequently, the inability to generate the foreseen credits. There are various ways to reduce this risk, one often seen example is the bundling of many (smaller) projects into a pool of projects.Although this will generate extra costs, it can also be beneficial in terms of transaction costs. Transaction costs are especially high for small-scale projects, and by bundling several smaller project into a pool, the costs of some (fixed) components of the transaction can be divided among different projects. (The Executive Board of the CDM, also recognized the large transaction cost burden of small-scale CDM projects, and adopted the simplified CDM rules for small-scale CDM projects (UNFCCC 2002). Another way of reducing the risk of non-delivery on a project-by-project base is that sellers reserve a certain percentage (approx. 20%) of the credits from each year's production into a non-delivery buffer. Whatever type of insurance is selected, it will always bring along extra cost, which should be included when calculating the cost of emission reduction in the emission credit trade system.

The costs of GHG-reduction for emission credit trade can now be calculated as follows:

$$C_{avoidance} = \frac{C_B + C_T - C_R}{E_R - E_B \pm E_L \pm E_K - E_I}$$

(3)

where:

- $C_{avoidance}$ = Costs per avoided GHG emission (€/CO_2-eq)
- C_B = Costs for biomass project (I en O&M) (€)
- C_T = Transaction costs (€)
- C_R = Costs for (avoided) reference energy (€)
- E_R = GHG emissions for reference system (=baseline emissions) (CO_2-eq)
- E_B = GHG emissions for biomass chain (=project emissions) (CO_2-eq)
- E_L = Carbon gains or losses compared to reference land-use (CO_2-eq)
- E_K = Emissions (benefits) due to leakage (CO_2-eq)
- E_I = Reduction of carbon credits due to insurance (buffer) (CO_2-eq) (We choose here to express the insurance costs in carbon credits, because that is the method applied with buffering. However, it could also be expressed in monetary units.)

10.3 CASE STUDIES

Two case study countries are selected: Mozambique and Brazil. For both countries, two fictional trading systems are analyzed: physical trading of biomass fuels from the country of origin to the Netherlands and the trade of emission credits derived from biofuel projects in the country of biomass origin. Transportation fuels are chosen in both cases since the transportation sector is a large contributor to global GHGemissions, transportation fuels are easy to trade and there is a large potential for biobased transportation fuels (Faaij 2006). Data collection is based on existing studies in both cases.

10.3.1 MOZAMBIQUE

10.3.1.1 CASE DESCRIPTION

In Fig. 3, the considered chains, including the reference systems, are summarised. System 1 represents the physical trading case. This system includes the production and harvesting of eucalyptus in Mozambique. After harvest, the biomass is transported by trucks to a local gathering point where it is converted to pyrolysis oil.

The pyrolysis oil is transported by trucks to the harbour for international shipping. In the Rotterdam Harbour, conversion into Fischer-Tropsch diesel via Entrained Flow gasification ($1000MW_{th}$) takes place. Finally, the FT-diesel is distributed to the fuel stations where cars are filled, after which the final conversion occurs in the car. The reference fuel is assumed to be (fossil fuel) diesel. Fischer-Tropsch diesel can be used in common Internal Combustion Engines (ICEs) without adaptation and is thus com-

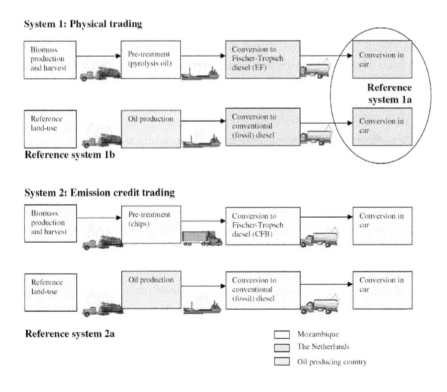

FIGURE 3: Mozambique: biomass trading systems with reference systems

pletely interchangeable with fossil fuel diesel. Reference system 1a corresponds to the current calculation method, where emissions and land use changes of both system 1 and the reference system are ignored. Reference system 1b indicates the complete chain.

System 2 represents the emission credit trade case. The system denotes a fictive CDM project in Mozambique where a eucalyptus plantation is established. The harvested wood is transported by trucks to a local gathering point where chips are produced. The chips are transported by trucks to a conversion facility where Fisher- Tropsch diesel is produced via CFB gasification (387 MW_{th}). The FT-diesel is finally transported to the fuel stations after which it is converted in the car. Reference system 2a corresponds to the baseline situation in Mozambique.

Land-use. The debate over 'carbon debt' created by land use changes has currently expanded to include the issue of competition between food and fuel, as the production of biofuels will put additional pressure on land. Recent studies have justly debated that including the greenhouse gas emissions from direct or indirect land clearance could drastically worsen or even revert the greenhouse gas emission balance of growing plants for bio energy (Fargione et al. 2008; Searchinger et al. 2008). Other studies, however, have indicated that it is well possible to avoid leakage due to bio-energy production if, at the same time, the productivity of agriculture and cattle-breeding is improved (Batidzirai et al. 2006; Smeets and Faaij 2008). Presently, the productivity of agriculture in Mozambique is very low compared to what is technically feasible. Batidzirai et al. (2006) showed that around 25 million hectares could be made available for bio-energy production in 2015 when the level of agricultural technology is increased from low to intermediate. In this case study we assume that the land for bio-energy production of dedicated crops is limited to surplus land not required for food production. However, as the leakage effect of bio-energy production is such an important issue it will be taken up further in the discussion.

10.3.1.2 DATA

Data on costs and emissions related to the production, logistics and conversion of the fuels (fuel switch) are given in Table 1. Data on carbon sequestration or loss related to the switch in land use (land use change) are shown in Table 2. For the CDM case (system 2) data on transaction costs and risks are also provided.

TABLE 1: Overview of data for emissions and costs related to fuel switch

	System 1	System 1a	System 1b
	FT-diesel	Diesel	Diesel
Heating value (MJ_{HHV}/l)	36.9[a]	37.9[a]	37.9[a]
Fuel economy NL (km/l)	14.5	14.5	14.5
Fuel costs Rotterdam (€/GJ)	6.5[b]	9.5[c]	9.5[c]
Distribution costs NL (€/GJ)[d]	1.33	1.33	1.33
Total cost of fuel chain (€/km)[e]	$C_b = 0.020$	$C_r = 0.028$	$C_r = 0.028$
WTT emissions (gCO_{2eq}/km)	9.4[f]	n.o.r.	26[g]
TTW emissions (gCO_{2eq}/km)	0h	138[gi]	138[gi]
Total emissions (gCO_{2eq}/km)	$E_b = 9.4$	$E_r = 138$	$E_r = 164$
	System 2	System 2a	
	FT-diesel	Diesel	
Heating value (MJ_{HHV}/l)	36.9[a]	37.9[a]	
Fuel economy Mozambique (km/l)	11.8[j]	11.8[j]	
Fuel costs Rotterdam (€/GJ)	9.1[b]	n.o.r	
Shipping costs (€/GJ)	0.22[b]	n.o.r	
Fuel costs Mozambique (€/GJ)	8.88	9.5[c]	
Distribution costs Africa (€/GJ)[k]	0.70	0.70	
Total cost of fuel chain (€/km)[e]	$C_b = 0.030$	$C_r = 0.033$	
WTT emissions Rotterdam ($kgCO_2/GJ_{HHV}$)	2.5[b]	n.o.r.	
Emissions due to shipping ($kgCO_2/GJ_{HHV}$)	0.6[b]	n.o.r.	
WTT emissions Mozambique (gCO_{2eq}/km)	6[l]	26[f]	
TTW emissions (gCO_{2eq}/km)	0[h]	180[m]	
Total emissions (gCO_{2eq}/km)	$E_b = 6$	$E_r = 206$	

n.o.r. Not of relevance [a]Beer et al. (2001) [b]Batidzirai et al. (2006) [c]Average Rotterdam Harbour diesel costs (including oil price) over the year 2005 is 70$/bbl or 0.44 $/l (159 l/ bbl). Dollar exchange rate = 1.22 $/€ →0.36 €/l. Energy content diesel = 37.9MJ_{HHV}/l .The price of fossil diesel can than be calculated: 9.5 €/GJ_{HHV}. Prices are assumed to be equal worldwide [d]Hamelinck (2004) [e]Calculated by adding fuel (sys 1: 6.5 €/GJ; sys 2: 8.88 €/ GJ) and distribution costs (sys 1: 1.33 €/GJ; sys 2: 0.70 €/GJ) and convert to €/km (using *heating values and fuel economies)* [f]*Batidzirai et al. (2006) found GHG-emissions of 3.7 kgCO₂/GJ_{HHV} for FT-diesel produced via EF gasification in Rotterdam (from pyrolysis oil produced from eucalyptus in Mozambique). With assumed FT-diesel heating value of 36.9 MJ/l and a NL fuel economy of 14.5 km/l, we calculated WTT emissions of 9.4 gCO₂/km* [g]*Edwards et al. (2005) (based on average European 5-seater sedan)* [h]*IPCC default* [i]*This number is relatively low compared to other studies (182 gCO₂/km (Little 1999) or ±152 gCO₂/km (Van den Broek et al. 2003)). However, especially the former study is much older and technology has improved since then. Besides, the European Automobile Manufacturers Association has, by means of a voluntary environmental agreement with the European Commission, set a target of 140 gCO₂/km for average vehicle emissions from new vehicles sold in Europe by 2008 (Commission of the European Communities 1999). Taking into account that a larger share of the European car fleet is petrol fuelled and petrol-fuelled cars have higher CO₂ emissions, we assume that the previously described number of 138 gCO₂/km is a realistic number for diesel-fuelled cars in The Netherlands* [j]*IEA/WBCSD (2004)* [k]*www.shell.com (distribution costs in composition of retail price for diesel in Southern Africa)* [l]*Calculated by subtracting the emissions due to shipping from the WTT emissions for the fuel chain to Rotterdam (converted to CO₂eq/km)* [m]*Little (1999)*

Fuel-switch. For the physical trading system, the fuel costs are composed of the costs to deliver the diesel in Rotterdam (including production, logistic and conversion costs) and costs to deliver the fuel at the fuel stations (distribution costs). The conversion of FT fuels is assumed to be by a new facility that uses the Entrained Flow (EF) gasification technique (1000MW_{th}) and is fed with pyrolysis oil. Emissions can be divided in upstream or well-to-tank (WTT) emissions that occur during the production and transportation of the fuel and tailpipe or tank-to-wheel (TTW) emissions that occur during conversion in the car.

For the emission trading system, we estimate fuel costs and emissions based on the study by Batidzirai et al. (2006). The conversion of FT fuels is assumed to be by a new build plant that uses the Circulating Fluidised Bed (CFB) gasification technique and is fed with chips. The capacity of the conversion facility is assumed to be 387 MWth on an input basis. Batidzirai et al. (2006) calculated fuel costs of 9.1 €/GJ_{HHV} and WWT emissions of 2.5 kg CO₂/GJ_{HHV} for FT-diesel, produced in Mozambique and delivered in the Rotterdam Harbour. Since we assume that the FT-diesel is not transported to the Netherlands but is used in Mozambique itself, the emissions and costs of the international ship transport are subtracted to derive fuel costs and emissions. Reference system 2a represents the production and use of conventional diesel in Mozambique. Most cars in Mozambique are older than current European cars; therefore, the fuel economy of African diesel cars is lower than the fuel economy of current European cars. In the reference case, we assume a direct emission (TTW) of 180 g/km for diesel cars in Mozambique. The assumption is based on several (older) studies (i.e. Little 1999) for cars in Europe. Specific data for Africa could not be found. For the WTT part we assume a value of 26 g CO_2eq/km, which was found for the upstream processes of conventional diesel in Europe (Edwards et al. 2005).

TABLE 2: Overview of data on carbon content related to land use changes

	Eucalyptus plantation		Pasture	Cropland
Total area (ha)	50,000		50,000	50,000
Rotation (years)	7		n.o.r.	n.o.r.
Total years (years)	21		n.o.r.	n.o.r.
Carbon fraction (tC/tdm)	0.5		0.5	0.5
Productivity (tdm/ha)	12[a]		3[b]	0[b]
Root/shoot ratio	0.45[b]		2.8[b]	–
	After pasture	After cropland		
Soil carbon content (tC/ha)	45[c]	35.4[d]	50	30

n.o.r. Not of relevance [a]Batidzirai et al. (2006) [b]Values taken from IPCC Good Practice Guidance for LULUCF (IPCC 2003), the productivity value (0) of cropland follows from the (IPCC) assumption that all crops (including roots) are harvested before land-use switch [c]Assuming a 10% decrease in soil carbon content with a land use change from pasture to plantation (Guo and Gifford 2002) [d]Assuming a 18% increase in soil carbon content with a land use change from cropland to plantation (Guo and Gifford 2002)

Land-use change. We assume, for both systems 1 and 2, that a eucalyptus plantation of 50,000 ha (Based on the capacity of the "fictive" CFB conversion facility in Mozambique, The assumed base scale of the CFB conversion facility is 387 MWth on an input basis (Batidzirai et al. 2006). Assuming a plant that operates 8000 h/year, results in an 11.15 PJ demand of energy input. The energy content of the wood is 19.4 GJ/tdm, this means an input of 575×10^3 tdm/year. The production of eucalyptus is assumed to be 12 tdm/ha × year, therefore an area of 47,876 ha is needed. In order to account for losses, we assume here an area of 50,000 ha to provide the plant with biomass.) is planted with a rotation length of 7 years. The assumed lifetime of the projects is 21 years. This period is rather long concerning techno-economic projects, but rather low for forestry projects. The influence of the chosen timescale will be analysed later. We consider two different reference land-use types in this study: Pasture, since almost 50%of the land in Mozambique is in use as permanent pasture (FAO. FAOSTAT Database. http://faostat.fao.org. Cited 3 Nov 2005), and cropland. We postulate that land for bio-energy production is only available after increasing the level of agricultural technology for food production (Smeets et al. 2007; Batidzirai et al. 2006). Leakage effects (Fargione et al. 2008; Searchinger et al. 2008) are thus not studied here, but are considered very important and will, as mentioned before, be discussed later.

To determine the carbon gains and losses related to the change in land-use, changes in carbon content in the different carbon pools need to be estimated. Five major types of carbon pools can be distinguished according to the IPCC Good Practice Guidance for LULUCF (IPCC 2003). In this study, we take into account only aboveground and belowground biomass and soil organic carbon because the other carbon stocks (litter and dead wood) are generally small and can often be ignored. For cropland, it is assumed that the aboveground and belowground biomass is 0 tdm/ha, since all crops are assumed to be harvested before the land is switched to a biomass plantation (IPCC 2003). The other belowground biomass values are estimated using root-to-shoot ratios as provided by the IPCC (2003). Data on soil carbon content are based on a study by Guo and Gifford (2002). They concluded that a shift from pasture to plantation will generally result in a decrease of soil carbon content with 10%, a shift from cropland to plantation will lead to a mean increase of 18%. Initial soil carbon content of grasslands and croplands were estimated to be 50 tdm/ha and 30 tdm/ha respectively (confirmed by Cowie, pers.comm) based on indicative values for native vegetation and cropland as presented in the IPCC Good Practice Guidance for LULUCF; considering a tropical dry to moist climate for Mozambique (Table 3) and relative stock change factors for the cropland management activities as presented in Table 4.

Transaction costs and risks. Michaelowa et al. (2003) analysed the relation between project size and transaction costs for CDM projects. They found transaction costs of 0.1–1.0 €/ton CO_2 for large (20,000 ton CO_2/year–200,000 ton CO_2/year) to very large (>200,000 ton CO_2/year) projects. To reduce risks of CDM projects, often a so-called non-delivery buffer is used. The percentage of this buffer can vary, and is mostly in between 10–30% (UNDP 2003; IFC 2006). Here we use a value of 20%.

TABLE 3 Default reference (under native vegetation) soil organic C stocks

Region	HAC soils	LAC soils	Sandy soils	Spodic soils	Volcanic soils	Wetlands soils
Boreal	68	NA	10#	117	20	146
Cold temperate, dry	50	33	34	NA	20	87
Cold temperate, moist	95	85	71	115	130	87
Warm temperate, dry	38	24	19	NA	70	88
Warm temperate, moist	88	63	34	NA	80	88
Tropical, dry	38	35	31	NA	50	86
Tropical, moist	65	47	39	NA	70	86
Tropical, wet	44	60	66	NA	130	86

Source: IPCC (2003)

TABLE 4: Relative stock change factors for cropland

	Land use factor	Tillage factor	Input factor	
Estimated SOC native (tC/ha)	Long term cultivated, tropical, dry	Full tillage, tropical	Low, tropical, dry	Estimated SOC cropland (tC/ha)
50	0.69	1.0	0.92	31.7a

ᵃCalculated by multiplying the reference native SOC value with all three stock change factors (50 × 0.69 × 1.0 × 0.92 = 31.7). The value should, however, be seen as a rough assumption

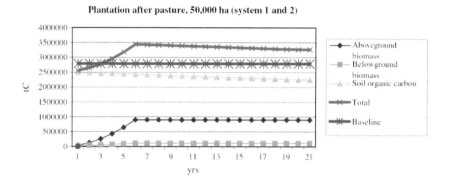

FIGURE 4: Cumulative carbon storage in the Eucalyptus plantation compared to the carbon storage in the baseline (pasture)

10.3.1.3 RESULTS

Figures 4 and 5 show the cumulative carbon storage over the lifetime of the plantation compared to the different baselines. From Fig. 4 it can be seen that the establishment of a eucalyptus plantation on pasture land firstly leads to a decrease in carbon. Later, the carbon accumulation in the plantation is above the baseline. Establishing a plantation on former cropland has a positive effect on carbon stocks from the beginning. The exact amount of carbon sequestration that can be attributed to both scenarios depends on the accounting rule to be applied and the chosen timeframe. The total amount of CO_2 sequestration for the two scenarios, calculated with the 4 different accounting methods (Stock Change (SC), Average Storage (AS), Ton Year (TY) and Temporary Crediting (TC)) is given in Table 5. (The results are converted to CO_2/km to be compatible with formula (2); the total amount of harvest over 21 years (tdm) is converted to GJ, assuming eucalyptus has a higher heating value of 19.4 GJ/tdm. The conversion efficiency from biomass via pyrolysis oil to FT-fuel is 48%. (conversion efficiency from biomass to pyrolysis oil is 67% and to FT is 71% (Batidzirai et al. 2006)). To convert to km driven, the heating value of FT-diesel (36.9 MJ/l) and the fuel economy of 14.5 km/l are used.) For all methods, a discount rate of 10% is used. Since the fuel efficiency in trading systems 1 and 2 is different, differences exist between calculated values for both systems. Table 5 shows that the establishment of a plantation on former pasture land, generates far lower carbon credits than on former cropland, independent of the accounting rule applied.

Since the carbon benefits occur mainly in the first 20 years, the carbon benefits per km would be considerably lower if they would be attributed to a longer project period; indicating the impact of project lifetime on the results.

The total emission reductions, expressed in kilometers driven in order to be able to compare the outputs of systems with different efficiencies, are shown in Fig. 6. The first four scenarios represent the pasture (P) baseline situation, whereas cropland (C) is the baseline vegetation in the last four scenarios. For both baselines, four different accounting rules (Stock Change (SC), Average Storage (AS), Ton Year (TY) and Temporary Crediting (TC)) are applied. The total emission reductions do not vary much between the physical and emission credit systems, if considering the complete chains (1b and 2). Larger differences are however found between reference systems 1a and 1b and between the cropland an pasture scenarios. The effect of the different carbon accounting methods is largest in the cropland reference scenarios, as already could be seen in Tables 5 and 6.

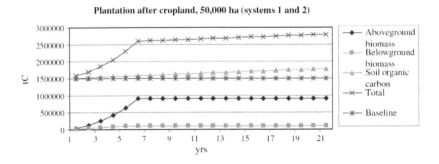

FIGURE 5: Cumulative carbon storage in the Eucalyptus plantation compared to the carbon storage in the baseline (cropland)

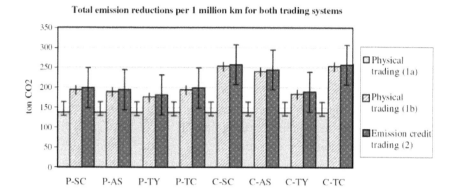

FIGURE 6: Total emission reduction per 1 Mkm driven for the two trading systems (3 reference situations) with two different baseline vegetations and four accounting rules (timescale 21 years)

TABLE 5: CO_2 benefits (gCO_2/km) for 2 baseline scenarios; based on 4 accounting rules and a lifetime of 21 years

Accounting method	EL Plantation after pasture (1)		EL Plantation after cropland (2)	
	System 1	System 2	System 1	System 2
Stock change	40 gCO_2/km	49 gCO_2/km	99 gCO_2/km	122 gCO_2/km
Average storage	35 gCO_2/km	42 gCO_2/km	86 gCO_2/km	105 gCO_2/km
Ton-year	22 gCO_2/km	27 gCO_2/km	30 gCO_2/km	36 gCO_2/km
Temporary credits	40 gCO_2/km	49 gCO_2/km	99 gCO_2/km	122 gCO_2/km

TABLE 6: CO_2 benefits (gCO_2/km) for 2 baseline scenarios; based on 4 accounting rules and a lifetime of 60 years

Accounting method	EL Plantation after pasture (1)		EL Plantation after cropland (2)	
	System 1	System 2	System 1	System 2
Stock change	11 gCO_2/km	14 gCO_2/km	27 gCO_2/km	34 gCO_2/km
Average storage	10 gCO_2/km	12 gCO_2/km	24 gCO_2/km	29 gCO_2/km
Ton-year	6 gCO_2/km	8 gCO_2/km	8 gCO_2/km	10 gCO_2/km
Temporary credits	11 gCO_2/km	14 gCO_2/km	27 gCO_2/km	34 gCO_2/km

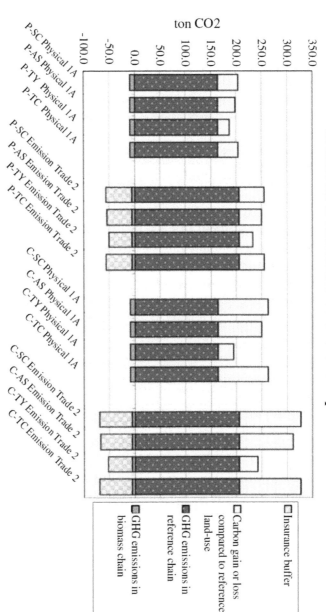

FIGURE 7: Avoided CO$_{2eq}$-emissions per 1,000,000 km driven for 2 trading systems, 2 baseline vegetations and 4 different accounting methods

The breakdowns of carbon gains in Fig. 7 show that emissions in the reference energy system have the largest impact on the total emission reductions. Land use changes have a considerable contribution, especially in the cropland baseline scenarios. Due to the higher emission in the reference system in Mozambique (206 gCO_2/km), compared to the Netherlands (164 gCO_2/km), the effect of fuel switch is largest in the emission credit system. The total emission reductions are, however, almost equal for both systems (Fig. 6) due to the considerable amount of carbon credits reserved for insurance in system 2.

The costs of CO_2-avoidance are shown in Fig. 8. The results show that for all scenarios the physical trading systems deliver carbon credits with the highest benefits, although the uncertainties are high. Due to the largely fluctuating oil prices, economic benefits can turn into costs as soon as diesel prices start to drop. In Figs. 9, 10 and 11 the effect of parameter variation, within uncertainty ranges, on the cost of CO_2 avoidance for the different trading systems is shown for one scenario: pasture as baseline vegetation and temporary crediting as accounting method. The results show that the respective prices of fossil diesel and FT-diesel have by far the largest influence on the results. Fuel prices thus dominate the economic results.

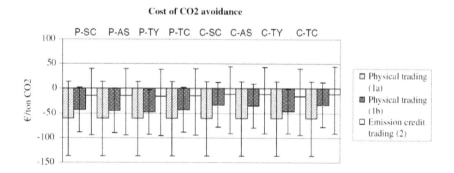

FIGURE 8: Cost of CO_2 avoidance (€/ton CO_2) for the two trading systems (3 reference situation) with two different baseline vegetations and four accounting rules (lifetime 21 years)

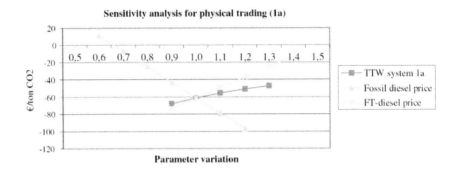

FIGURE 9: Effect of parameter variation on cost of CO_2 avoidance (1a)

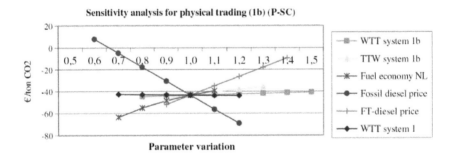

FIGURE 10: Effect of parameter variation on cost of CO_2 avoidance (1b) (P-TC)

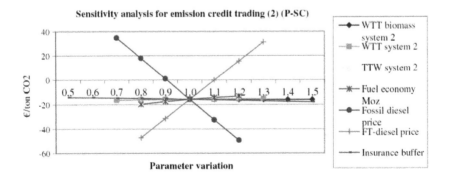

FIGURE 11: Effect of parameter variation on cost of CO_2 avoidance (2) (P-TC)

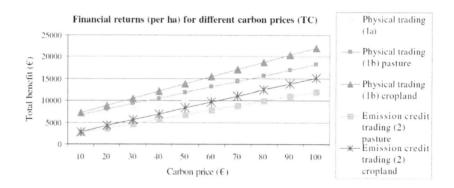

FIGURE 12: Financial returns per hectare at different carbon prices

From the emission reduction related parameters, the tailpipe emissions (TTW) in the reference energy systems and the fuel economy have the largest influence on the results.

Although the benefits per ton of carbon avoided can be a good indicator for the performance of the respective trading systems, other indicators can also be helpful in analyzing the trade-off between both systems. Therefore, we also analyze the financial returns per hectare at different carbon prices (Fig. 12).

Results are shown for the two baseline vegetation scenarios, and Temporary crediting accounting method. The results show that with current carbon prices of 20–30 €/ton CO_2, physical trading is more beneficial. When carbon prices increase to over 85 €/ton CO_2, emission credit trading becomes more attractive than physical trading (1a) in one scenario (plantation based on former cropland).

10.3.1.4 DISCUSSION AND CONCLUSION

The results have shown that the carbon benefit of physical trading depends largely on the reference system used (1a or 1b). Although the effect of ignoring or including upstream emissions was relatively small here, the benefits related to the switch in land-use (excluding leakage) were large and almost equal to the benefits related to the fuel switch. Due to limited available information on SOC values, however, these results should be considered with care. A comparison of the two trading options (system 1 and system 2) showed that there is only a very small difference in carbon gains, if considering the whole chain (ref 1b and 2). Emission credit trading delivers slightly more credits than physical trading.

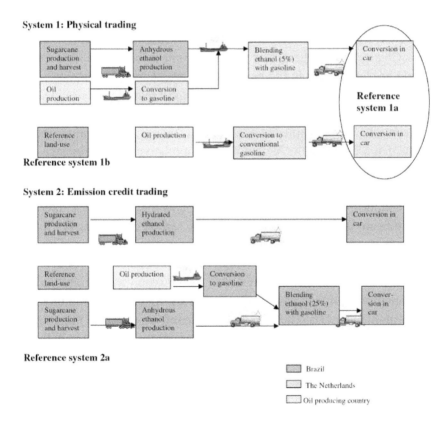

FIGURE 13: Brazil: biomass trading systems with reference systems

The economic performance of the emission credit trading system is, however, much lower than the performance of the physical trading system. This difference can not be attributed to the extra transaction costs involved in system 2; these have a negligible effect. The main difference in economic benefits is related to the relatively more expensive production costs of FT-diesel in Mozambique. Even though in the latter system no costs for overseas transport are made, the production costs in Mozambique are still larger than the production costs in Rotterdam. Only when carbon prices triple compared to the price today, emission credits trading can become financially more attractive, in this case study.

It turned out that the total amount of credits, derived for the land use change part of the projects, vary greatly per accounting method, previous land use and timeframe. Between the different accounting methods, a large difference also exists in the moment when the credits are generated. The Stock Change method generates credits as soon as the carbon stocks in the project are larger than in the baseline, but credits should also be "paid", as soon as carbon stocks are lower in the project than in the baseline, as is the case in the first period of the plantation after pasture. Because of the discount rate, credits earned at a later phase have a smaller value, and the depths that are paid at the beginning have a relatively larger impact. This effect is not apparent in the Average Storage approach, since the average carbon storage is calculated and no credits have to be "paid" at the start.

In conclusion, the case study results showed that in this case, there are no significant differences in the amount of carbon credits per km for the two trading options if the total chains are compared. However, since carbon benefits related to land use change can at this moment not be accounted for in the physical trading systems, the emission trading system will generate more carbon credits than the physical trading system although it goes with higher costs. The emissions in the reference energy systems had the largest influence on the amount of carbon credits available in both systems, whereas the differences in fuel prices dominate the financial results. Due to the large scale of the fictive CDM project, transaction costs had a negligible influence on the total costs. Both different baseline vegetations and different accounting rules had a considerable influence on the results. With current carbon prices, physical trading system is most beneficial in economic sense.

10.3.2 BRAZIL

The second case study country is Brazil. Brazil has a unique position in the bio-fuel market since it is the world's largest ethanol fuel producer as well as consumer. It gained valuable experience in the various aspects of sugarcane ethanol production during the 30 years of the Brazilian Alcohol Program (PROALCOOL). Currently, the ethanol blending requirement in Brazil is 25%.

10.3.2.1 CASE DESCRIPTION

Physical trading is represented in system 1 (Fig. 13) Here, ethanol, produced from sugarcane in Brazil, is transported overseas to the Rotterdam harbour where it is mixed with conventional gasoline to form a blend fuel (5%, since this is the current maximum according to the fuel quality directive of the EU (98/70/EC). Directive 98/70/EC of the European Parliament and of the Council of 13 October 1998 relating to the quality of petrol and diesel fuels and amending Council Directive 93/12/EEC.). The fuel is transported to fuel stations where cars are filled and final conversion occurs in the vehicle. Reference system 1a corresponds to the tailpipe emissions of conventional gasoline cars in the Netherlands, whereas in reference system 1b also upstream emissions of gasoline production and reference land use are taken into account.

Since alcohol production for the blend of up to 25% is not a candidate for CDM projects in Brazil (it corresponds to a baseline before the base year for the Kyoto Protocol) (Coelho et al. 2005), we consider as the emission credit trading scenario a project where ethanol vehicles (100% hydrated) are subsidized. The scenario is based upon an agreement between the German and Brazilian governments, where Germany plans to invest a total amount of 40 million dollars, subsidizing 100,000 alcohol vehicles, to purchase carbon credits as part of its Kyoto Protocol commitments. System 2, therefore, represents the production of hydrated ethanol (100%) from sugarcane in Brazil and the conversion in the alcohol vehicle. Reference system 2a indicates the baseline situation in Brazil: the production of anhydrous ethanol and gasoline that are mixed in a blend of 25% and converted in the car.

Land use. Recent studies have indicated that it is possible to avoid leakage due to bio-energy production if, at the same time, the productivity of agriculture and cattle-breeding is improved (Smeets and Faaij 2008). According to Coelho et al. (2005) and Macedo et al. (2004) competition between bio-fuel crops and food crops has, until now, been avoided in Brazil; the great rise in productivity resulting from technological developments allowed the growth of sugarcane production without excessive land-use expansion and the expansion of agriculture over the past 40 years took place mostly in degraded pasture areas. In this case study, we also assume no leakage effects for land-use. However, as mentioned before, the current debate on carbon penalties related to displacements is a serious issue of concern and will be discussed later.

10.3.2.2 DATA

Fuel switch. In the physical trading system (system 1), fuel costs are the costs of ethanol delivered in the Rotterdam Harbour (production plus transportation costs) added to these costs are the costs of mixing the blend and distributing it to the fuel stations. Total costs are expressed in €/km$_{\text{(EToH)}}$ (Table 7). In the reference case, fuel costs are composed of the gasoline costs in the Rotterdam harbour, plus the distribution costs for delivery at the fuel stations. CO_2-eq emissions in system 1 consist of the emissions related to the production, conversion, transportation of ethanol (WTT). Emissions during conversion in the car (TTW) are assumed to be zero for bio-fuels (IPCC default approach). Emissions in reference system 1a relate to the tailpipe (TTW) emissions of conventional gasoline cars in the Netherlands. In reference system 1b, well-to-wheel emissions for gasoline in average Dutch gasoline cars are considered.

In the CDMscenario (system 2) we assume that the cost of the project is 40 million dollars. 100,000 Ethanol vehicles will be subsidised with this money. We assume that the ethanol vehicles will be intensively used, driving at least 25,000 km/year.With this consumption pattern, the cars are assumed to have a lifetime of 10 years. It should be noted that a methodology for this CDM project has to date neither been approved by, nor submitted to, the CDM Executive Board.

Land-use change. Data on above- and belowground carbon contents have been estimated using data from the IPCC Good Practice Guidance for LULUCF (IPCC 2003). Data on soil carbon content are taken from Silveira et al. (2000). They found a decrease in SOC of 24% (over 20 years) when forest is turned into pasture land in Brazil, followed by a decrease of 22% over 20 years when a sugarcane plantation is established on the pasture land. Initial (forest) SOC was found to be 61.5 tC/ha (Silveira et al. 2000). We assume in this study that sugarcane SOC is equal to cropland SOC, this will be discussed later. An overview of the data on carbon content related to land use changes for system 1 and 2 is provided in Table 8

TABLE 7: Overview of data for emissions and costs related to fuel switch (Brazil)

	System 1	System 1a	System 1b
	Gasohol(E5)	Gasoline	Gasoline
Energy content (GJ_{HHV}/tonnedry)	29.8[a]	47.3[a]	47.3[a]
Density (kg/m³)	791[a]	745[a]	745[a]
Energy content (MJ_{HHV}/l)	23.6	35.2	35.2
Fuel efficiency (km/GJfuel)	440[b]	430[b]	430[b]
Fuel costs ethanol ($/l)	0.23[c]		
Fuel costs gasoline ($/bbl)		65d	65d
Mixing costs (€/l)	0.05[e]		
Distribution costs NL (€/l)	0.1[f]	0.1[f]	0.1[f]
Total costs of fuel chain (€/GJ)	14.36	12.35	12.35
Total cost of fuel chain (€/km)	$C_b = 32.6 \times 10^{-3}$	$C_r = 28.7 \times 10^{-3}$	$C_r = 28.7 \times 10^{-3}$
WTT emissions (gCO_{2eq}/km)	28[g]	n.o.r.	28[g]
TTW emissions (gCO_{2eq}/km)	0h	168[g]	168[g]
Total emissions (gCO_{2eq}/km)	$E_b = 28$	$E_r = 28$	$E_r = 28$
	System 2	System 2a	
	Ethanol (E100)	Gasohol (E25)	
Energy content (GJ_{HHV}/tonnedry)	29.8[a]	42.7[a]	
Density (kg/m³)	791[a]	756.5[a]	
Energy content (MJ_{HHV}/l)	23.6	32.3	
Fuel efficiency (km/GJfuel)	540[b]	440[b]	
Average (km/year)	25,000		
Lifetime car (years)	10		

TABLE 7: *Cont.*

	System 2	System 2a
	Ethanol (E100)	Gasohol (E25)
Total project costs (Me)	40	
WTW gasoline brazil (kgCO$_2$/l)		2.82[i]
WTW anhydr ethanol Brazil (kgCO$_2$/l)		0.4[j]
WTW hydr. ethanol Brazil (kgCO$_2$/l)	0.39[k]	
WTW total (kgCO$_2$/l)	0.39	2.212
WTW total (gCO$_2$/MJ)	16.5	68.6
Total emissions (gCO$_{2eq}$/km)	$E_b = 31$	$E_r = 156$

[a]*Hamelinck (2004) Appendix, Table A.1. For system 2a (Gasohol E25) energy content and density values have been calculated based on a mixture of 25%vol ethanol and 75%vol gasoline* [b]*Hamelinck (2004). Table 8, Page 39. (For gasohol E5 and E25 the values from E10 are taken)* [c]*Based on data from Coelho et al. (2005), average export price in the period 2001–2003* [d]*Rotterdam Harbour price of gasoline at the beginning of year 2006 (www.iea.org)* [e]*Elam (2000)* [f]*Van den Broek et al. (2003)* [g]*EUCAR et al. (2005)* [h]*Biofuels assumed to have zero CO$_2$ emissions (TTW)* [i]*Macedo et al. (2004)* [j]*Macedo et al. (2004). Ethanol life-cycle emissions 34.5 kgCO$_2$/ton of sugarcane. 86,0 l anhydrous ethanol per ton of sugarcane* [k]*Macedo et al. (2004). Ethanol life-cycle emissions 34.5 kgCO$_2$/ton of sugarcane. 88.6 l hydrous ethanol per ton of sugarcane.*

TABLE 8: Overview of data on carbon content related to land use changes for systems 1 and 2

	Sugarcane plantation		Pasture	Cropland
Total area (ha)	45,000[a]		45,000[a]	45,000[a]
Productivity (tdm/ha)	65		–	–
Carbon fraction (tC/tdm)	0.5		0.5	0.5
Aboveground biomass (tdm/ha)	0[b]		6.2[b]	0[b]
Root/shoot ratio (tdm/tdm)	0[b]		1.6[b]	0[b]
	After pasture	After cropland		
Soil carbon content (tC/ha)	36.5[c]	36.5[c]	47.0[c]	36.5[c]

[a]*Amount of extra hectares needed to have 100,000 ethanol vehicles driving 25,000 km per year with a fuel economy of 8 l/km. 88.6 l of hydrous ethanol can be produced from 1 ton of sugarcane and 65 tons of cane are produced from 1 ha. The baseline is 10,000 ha which is needed to have E25 vehicles driving 2.5 × 10^9 km/year* [b]*IPCC Good Practice Guidance*

for LULUCF (IPCC 2003) ͨSilveira et al. (2000) found an initial SOC of 61.5 tC/ha in Sao Paulo forest. Conversion to pasture land lead to a SOC decrease of 24%, resulting in a SOC of 47 tC/ha for pasture land. Final conversion into sugarcane lead to a further decrease in SOC of 22%, meaning a SOC of 36.5 tC/ha on sugarcane plantations

Transaction costs and risks. Transaction costs are assumed to be already included in the costs of the CDM project (system 2). Risks are assumed to be insured by using an insurance buffer of 20%.

10.3.2.3 RESULTS

Since it is assumed that cropland substitution (sugarcane production on former cropland) has no significant influence on carbon storage, only carbon changes related to the replacement of pasture land by sugarcane plantations are reflected. From Fig. 14, it can be seen that the total carbon storage decreases when sugarcane crops are grown on previous pasture lands (approximately 0.5 Mton C on 45,000 ha over 20 years). This carbon loss can almost completely be attributed to decreases in soil carbon content resulting from the land use change.

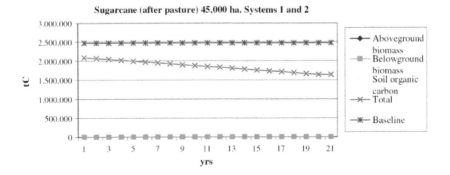

FIGURE 14: Cumulative carbon storage in the sugarcane plantation compared to the carbon storage in the baseline (pasture)

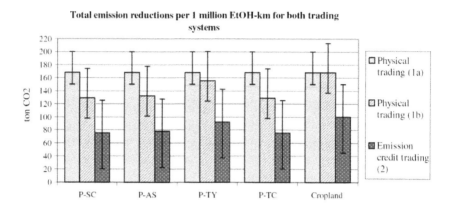

FIGURE 15: Total emission reductions per 1,000,000 EtOH km driven for system 1 and 2, for two vegetation baselines and four accounting rules

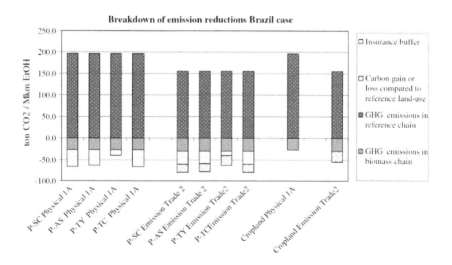

FIGURE 16: Avoided CO_{2eq} emissions per 1,000,000 EtOH km driven for 2 trading systems, 2 baseline vegetations and 4 different accounting methods

Figure 15 shows that including the carbon effects of upstream emissions and landuse changes (1b) results in less carbon credits compared to the same scenario where these steps are excluded (1a). This is mainly due to land-use changes since this difference does not appear in the cropland scenario. Emission credit trading delivers the fewest emission credits in all scenarios. The reason for this can be found when looking at the breakdowns in Fig. 16. Applying different accounting rules has a considerable influence on the results, as we have already seen in the Mozambique case study.

Figure 16 shows that (avoided) emissions in the reference system account for the largest part of the emission reductions. Emission in the biomass chain (WTT) and emissions related to land-use change (mainly SOC) account for a decrease in the avoided emissions. The emissions in the reference energy system in emission trading (E25 Brazil) are considerably lower than the emissions in the physical trading reference system. The reason is that in the baseline (E25) already 25% of gasoline is replaced by ethanol. GHG emissions in the biomass chain and carbon losses due to land-use change are similar to the physical trading scenario. However, additional losses result from the extra emission credits reserved in an insurance buffer.

In Fig. 17, the costs of CO_2 avoidance are displayed. The results show that in contradiction to the Mozambique case study, here, for all scenarios costs are associated with CO_2 avoidance. Costs of CO_2 avoidance in physical trading are mainly given by relative differences in oil and ethanol prices (taxes and import tariffs were not taken into account). With current gasoline and ethanol prices, the physical trading systems produce carbon credits with a price of 20–30 euros, but due to fluctuating oil prices, uncertainties are high. With increasing oil prices, the costs of physical trading will be reduced. The costs of carbon credits in the emission credit trading scenario are largely given by the investment of 40,000,000 dollars that is foreseen to facilitate the purchase of a new fleet of 100,000 ethanol vehicles. In the physical trading scenarios, no such investments are necessary for the cars, since the fuel can be used in the currently available vehicles. Uncertainties in the emission credit trading system (2) are related to the uncertainties in the amount of CO_2 avoided. That is, amongst others, determined by the amount of kilometres that the fleet will drive, the lifetime of the cars and the fuel efficiency of the fleet.

Sensitivity analyses have shown that fuel prices (both ethanol and gasoline) have the largest influence on the costs of CO_2 reduction in the physical trading scenarios. The WTW emissions in the gasoline fraction of the reference energy system, the assumed amount of kilometres driven per year, the lifetime of the cars and the assumed project lifetime have a considerable influence on the results in the emission credit trading system.

Figure 18 shows the financial returns per hectare, at different carbon prices, for both trading systems with pasture as baseline vegetation scenario. Emission credit trading is favourable at low carbon prices; with increasing carbon prices physical trading becomes more financially attractive. The switching point, with current oil and ethanol prices, is around 30 € per ton of carbon for system 1a, and around 70 € per ton of carbon for system 1b (including upstream emissions and land-use changes).

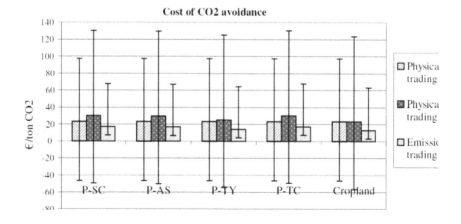

FIGURE 17: Costs of CO_2 avoidance (e/ton CO2) for physical trading (1a and 1b) and the emission credit trading system with two different baseline vegetations and four accounting rules

10.3.2.4 DISCUSSION AND CONCLUSION

Due to the strict rules for registration of CDM or JI projects, all carbon fluxes related to the project must be taken into account in emission credit trading. In physical trading, however, current practice is to only consider the avoided emissions related to the conversion of the substituted fossil fuel. Especially when this change in land-use leads to carbon losses, this current approach can be very precarious.

From Fig. 15, it could be seen that the carbon losses related to land-use change occur primarily during the first 20 years after the sugarcane plantation is established. In this study, we attributed these carbon losses completely to the first 20 years of ethanol production. Consequently, if ethanol production from the established sugarcane plantation is continued after 20 years, the carbon losses per km driven would be lower. This indicates again the large influence of the chosen time scale on carbon balances. Fuel prices are, also in this case study, one of the most important variables determining the cost of CO_2 avoidance. Given the large uncertainties and fluctuations concerning fuel prices in the future, it might be hard to predict the financial gains or losses related to the switch of fossil fuels to bio-fuels. However, if we expect fossil fuel prices to increase over the years, an

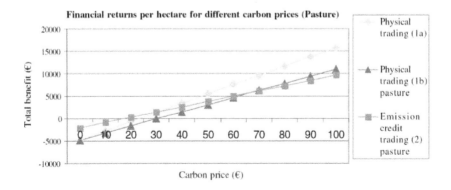

FIGURE 18: Financial returns at different carbon prices

increase in bio-fuel trading can be expected. It is then important to ensure that this increase will not lead to carbon losses related to land-use changes.

10.4 DISCUSSION

- In the physical trading scenarios, we did not take into account an insurance buffer since few is known about such practices in physical trading, in contradiction to CDM projects. However, some insurance might be desired to ensure a reliable supply of resources. This would increase the costs of this trading system.
- In the Mozambique case study we assumed a larger conversion facility in the Rotterdam Harbour (1000 MWth) than in Mozambique (387 MWth). Although Rotterdam Harbour is one of the largest harbours in the world, where, in theory, large amounts of biomass could be imported and converted, these fictive examples had considerable impacts on the results.
- We assumed sugarcane as cropland and the IPCC Guidelines (2003) prescribe no significant changes in carbon content if cropland remains cropland. Sugarcane is, however cultivated in a ratoon system and only every four or five years, the complete plants are removed and replaced by new plants. Sugarcane is therefore closer to wood species then 'normal' crops. The assumption of no carbon change when cropland is replaced by sugarcane is therefore on the conservative side, there is probably an increase in SOC, but the exact numbers are not known.
- Throughout this study, it appeared very difficult to find reliable sources for SOC values. The values we used in this report are based on IPCC default assumptions (IPCC 2003), on a meta-analysis by Guo and Gifford (2002) that was based on several studies and for Brazil specific data for Sao Paulo could be found. Furthermore, our assumptions were confirmed by Cowie (pers. comm.). However, one should be careful with conclusions based on non-measured soil carbon content data. Especially since the SOC values had a considerable impact on the results.
- Four accounting rules were explored in the case studies of this report. Although currently a decision for the Temporary Crediting method has been made, we still consider it valuable to analyse the effect that the choice for a certain method can have on the results. Besides, the described accounting rules may be reconsidered in a new accounting period.

The case study results showed that direct land use changes can have such a large influence (both positive and negative) on total carbon balances of the trading systems, mainly due to changes in soil carbon, that it would be dubious to ignore them (as currently done in physical trading). With the implementation of a certification system [This certification system would be needed for biomass from non-Annex I countries only, since carbon stocks in Annex I countries are accounted for in National GHG Inventories anyhow.] it could be ensured that no carbon losses occur during production of the biomass. Although carbon changes from land use changes can be taken into account in CDM and JI projects, the chosen timeframe is rather arbitrary and has a large influence on the results as shown in this study.

In this study, carbon leakage effects from indirect land-use change (ILUC), which could have large consequences on the GHG balance of a biofuel (Fargione et al. 2008; Searchinger et al. 2008), are excluded. We acknowledge the importance and relevance of studying these effects and to including them into carbon accounting of both trading systems. Considering the results of this study, however, we refer to recent studies that have indicated that it is well possible to avoid leakage due to bio-energy production if, at the same time, the productivity of agriculture and cattlebreeding is improved (e.g. Smeets and Faaij 2008). Furthermore, the exclusion of leakage effects has been done for both trading systems and for both case studies which makes the results at least comparable.

This study has further shown that transportation and transaction costs have a negligible influence on the financial results in large scale trading systems as explored in this study. Since oil prices, although fluctuating, are almost the same anywhere in the world (thus similar for importing and exporting country), the cheapest trading system to reduce GHG-emission depends mainly on 1) the amount of GHG-emission that can be avoided in the exporting respectively importing country, where the balance between the emissions in the baseline and the emissions in the bio-fuel system turns out to be most important and 2) the associated cost of the bio-fuel system in either of those countries. If the conversion of biomass is cheaper in the importing country, physical trading would be beneficial. If, on the other hand, conversion is cheaper in the country of biomass origin, both emission credit trading and physical trading are good alternatives since transportation costs are found to be negligible at large trading scales. It then depends on other choices, which trading option is favourable. Besides, subsidies, taxes, policies and legislation might affect both bio-fuel trading and domestic production opportunities.

10.5 CONCLUSIONS AND RECOMMENDATIONS

The results of both case studies are summarized in Table 9. The total emission reductions (tCO_2/km; tCO_2/ha) in the Brazilian case study are mostly higher than in Mozambique. The costs of CO_2 avoidance (€/tCO_2) are, however,much higher in the Brazilian case study as compared to Mozambique, where there are benefits instead of costs. This effect also becomes apparent in the financial returns. The optimization method to be used depends on individual preferences (see also Schlamadinger et al. 2005). Moreover, for companies, taxes and import tariffs, that were not included in this case study, play a great role, whereas governments can influence these factors to optimize their strategies. Besides, other issues than CO_2 avoidance and costs may play a role in defining the best trading option.

Given the large global bio-energy production potential, how should we make optimal use of this renewable source, regarding, amongst others, land-use, costs and avoided emissions? To answer this question, various factors that have been discussed before should be included: a) which option delivers the most GHG-emission credits b) which option is cheaper or delivers the highest economic benefits, but also c) other factors might be important in decision making.

TABLE 9: Overview of main results from both case studies

Optimization methods↓	Baseline	Pasture			Cropland		
	Trading system	1a	1b	2	1a	1b	2
Total emission reductions (tCO_2/Mkm)	Moz	138	194	199	138	254	257
	Bra	168	129	76	168	168	100
Total emission reductions (tCO_2/ha)	Moz	90	127	106	90	165	137
	Bra	205	157	116	205	205	154
Cost of CO_2 avoidance (e/tCO_2)	Moz	−61	−43	−14	−61	−33	−11
	Bra	23	30	17	23	23	13
Financial returns at a carbon price of 30 euros (ke/ha)	Moz	8.2	9.3	4.7	8.2	10.5	5.6
	Bra	1.4	−0.1	1.5	1.4	1.4	2.6

Two factors turned out to play a major role in question a): 1) sequestration or carbon losses as a result of land use change and 2) the difference in the emissions in the reference energy systems and the biomass systems. The first factor is dependent on the trading system used. Carbon changes related to land-use change can currently only be included in emission credit trading. The second factor, can vary per country, per energy system and thus per individual case. Therefore, no hard conclusions can be drawn here.

b) We found that the cheapest trading system to reduce GHG-emission depends mainly on the amount of GHG-emission that can be avoided in the exporting or importing country since both bio- and fossil transportation fuel prices are world market prices if transportation costs are low.

Among other factors c) that play a role in decision making are, diversification of energy sources, logistical capacities, domestic market protection, enhancing sustainable development, job creation and policies and regulations that are already in place. All other things being equal, physical trading could be the preferred trading alternative, because with the same amount of money and emission credits, one can also benefit from the energy. Further advantages are the diversification in energy sources and possible job creation. Besides, current policies aim at increasing the share of renewables and biomass in particular. The Netherlands, with Rotterdam as one of largest harbours of the world, would be an optimal suited country to invest in large-scale conversion facilities for biomass, thereby becoming a major player in physical biomass trading. Physical trading could also, in addition to CDM and JI, contribute to enhance development in developing countries. Poorer countries could benefit from the establishment of a global biomass market that can provide consistent demands and generate considerable income sources for these countries. We conclude that physical trading could be the most desirable trading option for biomass, unless emission reductions can be much higher in the country of biomass origin (as a result of higher emissions in the reference energy system or lower emissions in the bio-fuel project). A solution has to be found, however, to account for the carbon effects of direct and indirect land-use changes (see also Lewandowski and Faaij 2006; Fargione et al. 2008; Searchinger et al. 2008). Starting points for addressing this key issue lay in developing integrated land-use policies and management strategies that address total land-use in a region. The most important component of such a strategy is that development of bioenergy crop production is done in balance with improvements of management of agriculture and livestock management. This could be an important element for future certification systems currently widely discussed for biofuel production

The case studies all concern transportation fuels. This is a very critical category for CDM and it will probably become extremely important in the short term. We hope that this paper can contribute to the development of a methodology to determine GHG-impacts of bio-transportation fuel projects, related to ET, CDM and the certification of traded bio fuels.

REFERENCES

1. Batidzirai B, Faaij APC, Smeets E (2006) Biomass and bioenergy supply from Mozambique. Energy for Sustainable Development X(1):54–81
2. Beer T, Grant T, Morgan G, Lapszewicz J, Anyon P, Edwards J, Nelson P, Watson H, Williams D (2001) Comparison of transport fuels: life-cycle emissions analysis of alternative fuels for heavy vehicles. CSIRO for Australian Greenhouse Office
3. Berndes G, Hoogwijk MM, Van den Broek R (2002) The contribution of biomass in the future global energy system: a review of 17 systems. Biomass Bioenergy 25(1):1–28
4. Coelho ST, Goldemberg J, Lucon O, Guardabassi P (2005) Brazilian sugarcane ethanol: lessons learned. Energy for Sustainable Development X(2):26–39
5. Commission of the European Communities (1999) Commission recommendation of 5 February 1999 on the reduction of CO2 emissions from passenger cars (1999/125/EC)
6. Damen K, Faaij APC (2006) A greenhouse gas balance of two existing international biomass import chains: the case of residue co-firing in a pulverised coal-fired power plant in The Netherlands. Mitig Adapt Strategies Glob Chang 11(5–6):1023–1050
7. Edwards R, Larivé J-F, Mahieu V, Rouveirolles P (2005) Well-to-Wheels analysis of future automotive fuels and powertrains in the European context. 2005, EUCAR, Concawe, European Commission Directorate General Joint Research Centre. Version 2a
8. Elam N (2000) Alternative fuels (ethanol) in Sweden. Investigation and evaluation for IEA Bioenergy, Task 27, Atrax Energi AB, Sweden
9. EUCAR, CONCAWE, JRC (2005) Well-to-Wheels analysis of future automotive fuels and powertrains in the European context WELL-TO-WHEELS Report Version 2a
10. Faaij APC (2006) Modern biomass conversion technologies. Mitig Adapt Strategies Glob Chang 11(2):335–367
11. Fargione J, Hill J, Tilman D, Polasky S, Hawthorne P (2008) Land clearing and the biofuel carbon debt. Science 319:1235–1238

12. Guo LB, Gifford RM (2002) Soil carbon stocks and land use change: a meta analysis. Glob Chang Biol 8:345–360
13. Hall DO, House JI (1994) Biomass energy development and carbon dioxide mitigation options. Division of Life Sciences, King's College London. Conference Paper. International Conference on National Action to Mitigate Global Climate Change, Copenhagen, Denmark, 7–9 June 1994
14. Hamelinck CM (2004) Outlook for advanced biofuels. Ph.D.-thesis, Copernicus Institute, Utrecht University
15. Hoogwijk M, Faaij A, van den Broek R, Berndes G, Gielen D, TurkenburgW(2003) Exploration of the ranges of the global potential of biomass for energy. Biomass Bioenergy 25(2):119–133
16. Hoogwijk MM, Faaij APC, Eickhout B, Vries BJM, Turkenburg WC (2005) Potential of biomass energy out to 2100, for four IPCC SRES land-use scenarios. Biomass Bioenergy 29(4):225–257
17. Intergovernmental Panel on Climate Change (IPCC) (2003) Good practice guidance for land use, land-use change and forestry. Institute for Global Environmental Strategies (IGES), Hayama, Japan
18. International Energy Agency and World Business Council for Sustainable Development (2004) IEA/SMP transportmodel. Model and documentation available at: http://wbcsd.org. Cited3Nov 2005
19. International Finance Organization (IFC) (2006) World bank group. Presentation on "Project risks and CDM" January, 25, 2006. Available at: http://www.envfor.nic.in/cdm/presentations/pre6.pdf
20. Intergovernmental Panel on Climate Change (IPCC) (2007) Working group III report "mitigation of climate change", fourth assessment report. Cambridge University Press
21. Krey M (2004) Transaction costs of CDM projects in India—An empirical survey. HWWA-report 238, Hamburg
22. Lewandowski I, Faaij APC (2006) Steps towards the development of a certification system for sustainable bio-energy trade. Biomass Bioenergy 30(2):405–421
23. Little AD (ADL) (1999) Analysis and integral evaluation of potential CO2-neutral fuel chains. Report for NOVEM, Netherlands
24. Macedo IC, Leal MRLV, Silva JEAR (2004) Assessment of greenhouse gas emissions in the production and use of fuel ethanol in Brazil. São Paulo State Environment Secretariat. Also at www.unica.com.br/i_pages/files/pdf_ingles.pdf
25. Michaelowa A, Stronzik M, Eckermann F, Hunt A (2003) Transaction costs of the Kyoto mechanisms. Climate Policy 3:261–278
26. Schlamadinger B, Grubb M, Azar C, Bauen A, Berndes G (2001) Carbon sinks and biomass energy production: a study of linkage, options and implications. Project initiation, coordination and dissemination by Climate Strategies. Climate Strategies
27. Schlamadinger B, Faaij APC, Daugherty E (2004) Should we trade biomass, electricity, renewable certificates, or CO2 credits? IEA Bioenergy Task 38 and 40. Also at www.joanneum.at/iea-bioenergy-task38
28. Schlamadinger B, Edwards R, Byrne K, Cowie A, Faaij A, Green C, Fijan-Parlov S, Gustavsson L, Hatton T, Heding N, Kwant K, Pingoud K, Ringer M, Robertson K,

Solberg B, Soimakallio S, Woess-Gallasch S (2005) Optimizing the greenhouse gas benefits of bioenergy systems. IEA Bioenergy Task 38

29. Searchinger T, Heimlich R, Houghton RA, Dong F, Elobeid A, Fabiosa J, Tokgoz S, Hayes D, Yu TH (2008) Use of U.S. croplands for biofuels increases greenhouse gases through emissions from land use change. Science 319:1238–1240

30. Silveira AM, Victoria RL, Ballester MV,De Camargo PB,Martinelli LA, De Cassia PicolloM(2000) Simulation of the effects of land use change in soil, carbon dynamics in the Piracicaba river basin, Sao Paulo State, Brazil. Pesqui Agropecu Bras 35

31. Smeets EMW, Faaij APC (2008) The impact of sustainability criteria on the costs and potentials of bio-energy production applied for case studies in Brazil and Ukraine. Accepted for publication in: Biomass Bioenergy (in press)

32. Smeets EMW, Faaij APC, Lewandowski IM, Turkenburg WC (2007) A bottom up assessment and review of global bio-energy potentials to 2050. Progr Energy Combust Sci 33:56–106

33. United Nations Development Programme (UNDP) (2003) Energy and environment group, Bureau for development policy. The clean development mechanism: a user's guide, New York, 2003. http://www.undp.org/energy/docs/cdmchapter6.pdf

34. United Nations Framework Convention on Climate Change (UNFCCC) (1997) Conference of the parties, third session, Kyoto Protocol to the United Nations Framework Convention on Climate Change, Document FCCC/CP/1997/L.7/Add.l, Kyoto

35. United Nations Framework Convention on Climate Change (UNFCCC) (2001) Conference of the Parties, seventh session, report of the conference of the parties on its seventh session, held at Marrakesh from 29 October to 10 November 2001. Document FCCC/CP/2001/13/Add.1, Marrakesh, 2001

36. United Nations Framework Convention on Climate Change (UNFCCC) (2002) Conference of the parties 8. Annex to decision 21/CP8. "Draft simplified modalities and procedures for small-scale clean development mechanism project activities." Document FCCC/CP/2002/3/Annex II

37. United Nations Framework Convention on Climate Change (UNFCCC) (2007) Approved baseline and monitoring methodology AM0047 "Production of biodiesel based on waste oils and/or waste fats from biogenic origin for use as fuel"

38. Van den Broek R, M van Walwijk, Niermeijer P, Tijmensen M (2003) Biofuels in the Dutch market: a fact-finding study. NOVEM, Report No. 2GAVE-03.12

CHAPTER 11

Indirect Land Use Changes of Biofuel Production: A Review of Modeling Efforts and Policy Developments in the European Union

SERINA AHLGREN AND LORENZO DI LUCIA

11.1 INTRODUCTION

The use of bioenergy involves use of land for production of, for example, harvest residues, crops or forestry, so increased demand for bioenergy can cause land use changes (LUC), which can have many implications on the economic, social and environmental sustainability of bioenergy. The LUC directly associated with a bioenergy project are referred to as DLUC, for example, when converting one type of land use to a bioenergy plantation. Indirect LUC (ILUC) are the changes in land use that take place as a consequence of a bioenergy project, but are geographically disconnected to it. For example, displaced food or feed producers may re-establish their operations elsewhere by converting natural ecosystems to agricultural land or, due to macro-economic factors, the losses in food/feed/fibre produc-

Indirect Land Use Changes of Biofuel Production – A Review of Modelling Efforts and Policy Developments in the European Union. © Ahlgren S and Di Lucia L. Biotechnology for Biofuels 7,35 (2014), doi:10.1186/1754-6834-7-35. *Licensed under a Creative Commons Attribution 2.0 Generic License, http://creativecommons.org/licenses/by/2.0/.*

tion caused by the bioenergy project may cause an expansion of the total agricultural area, or an intensification of its use.

Although LUC due to increased demand for bioenergy were first discussed in the 1990s (see for example, [1,2]), the debate on indirect LUC caused by bioenergy production intensified with the publication of two studies in 2008 [3,4]. Both studies demonstrated that ILUC could increase carbon emissions following biofuel expansion to such a level that the life-cycle greenhouse gas (GHG) emissions were higher than with fossil fuels. These studies also had impact on policymaking. Within the EU, in 2009 the Renewable Energy Directive (RED) and Fuel Quality Directive (FQD) were introduced with a set of sustainability criteria for biofuels and bi-oliquids used to achieve the Directive targets[a]. One of these criteria is that the use of biofuels must ensure at least 35% reduction in GHG emissions compared with fossil fuels[b]. The threshold will increase to 50% after 2017 (60% for installations that enter operation after 2017). However, the methodology for calculating GHG emissions contained in the Directives does not account for emissions due to ILUC. Only in October 2012 did the EC put forward a proposal to amend the Directives and address GHG emissions generated through ILUC.

One of the major problems with including ILUC in policy is the uncertainty related to the quantification of the GHG emissions [5]. Quantification of GHG emissions due to ILUC is very different from quantification of direct changes, as the theory in ILUC modelling is based on economic market reactions to increasing demand for biofuels, whereas quantifying direct changes relies more on natural science. ILUC are not observable. A wheat-growing farmer in Europe will not see any indirect effects of his or her actions, and it can never be proven that a certain land use in Brazil, for example, is the effect of the European farmer's change from producing wheat for food to wheat for ethanol. The links are complex and impossible to attribute to a certain field.

It is common to use economic equilibrium models to estimate ILUC. These tools are complex optimisation models studying the entire global economy or a specific sector, for example, agriculture. Economic models assume that perfect markets exist and that equilibrium is reached when demand equals supply in the economy [6]. There are several different economic models available, which have been developed and used by researchers for many years. Among the most commonly used are, for example, the Global Trade Analysis Project (GTAP) model developed by Purdue University, the Food Agricultural Policy Research Institute and Center for Agricultural and Rural Development (FAPRI-CARD) model developed by FAPRI together with Iowa State University, and the Modeling International Relationships in Applied General Equilibrium (MIRAGE) model developed by the European Commission French National Institute for Agricultural Research (INRA), the UN and the World Trade Organization. Economic models typically assess changes associated with the implementation of a policy, such as promotion of biofuel. The LUC are often calculated as the difference between scenarios with and without implementation of such a policy. Most economic models are complex and require in-depth understanding of the way they are organised [7,8].

Several alternatives to the economic models have been developed using different approaches. The feature these other methods have in common is that they try to use descriptive methods rather than complex optimisation models. For example, one such model allows reference expert groups to describe likely scenarios of market reaction to increased demand for biofuel [9]. Others use statistics on past LUC to predict future LUC (for example, [10,11]).

The uncertainty in ILUC estimation has been discussed in several papers (for example, [12-14]). There are many explanations for the large variation in modelling results, for example, differences in input data and assumptions. Decisive for the results is the type of land assumed to be affected and the GHG emissions attributed to the LUC, as well as how the emissions are treated over time. LUC, for example, a shift from forest or permanent pasture to annual crops, can lead to large initial carbon losses. However, since the land will continue to produce crops for several years, the carbon losses must be allocated over time, or treated as emission impulses to the atmosphere [8,15,16]. The uncertainty in model results is a problem for policymakers, since it affects the validity of arguments for and against policy support for biofuels. Although the aim of this paper is not to go into details of uncertainty due to underlying assumptions in the models, we return to this in the discussion.

The aim of this paper was to review the recent literature (2009 onwards) reporting efforts to model biofuel ILUC. In our analysis, we evaluated whether results from different types of modelling exercises are comparable and whether there is evidence of result convergence over time. In addition, we reviewed the development of ILUC policies in the EU and compared modelling results with the ILUC values proposed by EU policymakers. We then assessed the possibility of different types of biofuels to comply with the GHG-saving requirements established in the RED and FQD. The paper is intended to give a comprehensive state-of-the-art update of modelling results and policy efforts in the EU, which can be useful to inform future research as well as policy discussions on the topic.

11.1.1 METHODOLOGY FOR LITERATURE REVIEW OF ILUC MODELLING

The literature review was based on a previous summary of studies [17], complemented with a search in Google Scholar in October 2013 using the keywords biofuel, ILUC, model and "g CO_2" for the publication period 2011 to 2013. This yielded 86 hits, of which nine were studies describing models explicitly estimating GHG emissions due to ILUC which were not covered in the previous summary [17].

Most models calculate ILUC on a hectare base, then attribute a GHG emissions factor for LUC and finally allocate the emissions over a number of years and per unit energy of fuel. The allocation of LUC emissions varied in the reviewed literature between 20 and 30 years, most US studies apply 30 years whereas European studies most commonly apply 20 years. In this study, the ILUC data found in the publications were recalculated to a 20-year allocation base in order to be able to compare model results with the ILUC-factors suggested by EU policy-makers, which are allocated over 20 years. Model results are expressed as g CO_2-eq per MJ biofuel. The studies reviewed were divided into two categories: economic (E) and miscellaneous (M) models. The latter including all non-economic modelling, as mentioned in the introduction. The results were analysed qualitatively, that is, no other processing of the data was done except for the time adjustment.

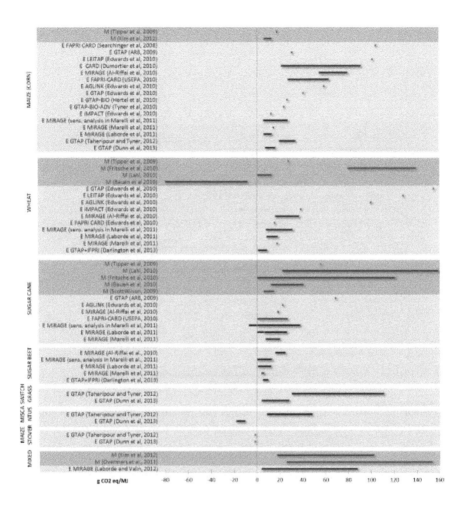

FIGURE 1: Review of modelled greenhouse gas (GHG) emissions due to indirect land use change (ILUC) of ethanol biofuels. All values were recalculated to a 20-year allocation base and are expressed as g CO_2-eq per MJ ethanol. Some studies show results as intervals (illustrated with lines), others as specific values (illustrated with dots). The studies were divided into two categories; economic modelling (E) and miscellaneous (other) modelling (M). FAPRI-CARD, Food Agricultural Policy Research Institute and Center for Agricultural and Rural Development; GTAP, Global Trade Analysis Project; MIRAGE, Modeling International Relationships in Applied General Equilibrium; IFPRI, International Food Policy Research Institute; LEITAP, the abbreviation indicates the extension of the GTAP model developed at the LEI (Landbouw Economisch Instituut) in The Hague; AGLINK, Worldwide Agribusiness Linkage Program.

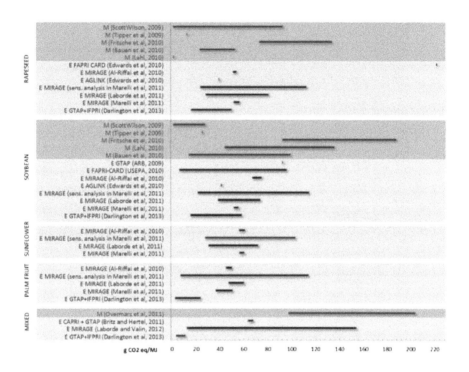

FIGURE 2: Review of modelled greenhouse gas (GHG) emissions due to indirect land use change (ILUC) of biodiesel. All values were recalculated to a 20-year allocation base and are expressed as g CO_2-eq per MJ biodiesel. Some studies show results as intervals (illustrated with lines), others as specific values (illustrated with dots). The studies were divided into two categories; economic modelling (E) and miscellaneous (other) modelling (M). FAPRI-CARD, Food Agricultural Policy Research Institute and Center for Agricultural and Rural Development; GTAP, Global Trade Analysis Project; MIRAGE, Modeling International Relationships in Applied General Equilibrium; AGLINK, Worldwide Agribusiness Linkage Program. Note the different scale of the x-axis compared with Figure 1.

11.2 REVIEW

11.2.1 REVIEW OF ILUC MODELLING RESULTS

The results of the literature review are presented in Figure 1 for ethanol fuels and Figure 2 for biodiesel. The results expressed in g CO2-eq per MJ fuel are arranged by date of publication, type of model (economic or miscellaneous) and maximum values. An additional file shows the full list of references for the studies (Additional file 1).

The review showed that within the selected sample of papers, most modelling was carried out for ethanol, especially with maize as feedstock; that most studies employed economic equilibrium models, and that the majority of the studies were published in 2010 (9 out of 22 studies).

The variation in the results was found to be large. Some of the studies plotted the results as a single number, while others presented a range based on the statistical distribution given by the authors or the sensitivity analysis performed in the studies. This meant that it was not possible to calculate any average values from Figures 1 and 2. Some feedstock displayed more convergence in the results, especially in the results from the economic models. For example, the economic models for sugar cane ethanol yielded ILUC values of between −5 and 25 g CO_2-eq/MJ (excluding one value of 69 g CO_2-eq/MJ in the California Low Carbon Fuel Standard from 2009 [18], which is under revision), whereas the miscellaneous model produced values of between −1 and 159 g CO_2-eq/MJ. However, this does not necessarily mean that the economic models are less uncertain. An alternative explanation could be that economic models do not reflect the full uncertainty associated with quantification of ILUC, as comprehensive sensitivity analysis is often lacking [13], and many of the reviewed alternative models test results using extreme values regarding, for example, carbon stock in soil. Further, as previously pointed out [19], economic models fail to include other important ILUC drivers such as political, cultural, demographic and environmental issues. On the other hand, many of the alternative models risk missing dynamic and interlinked ILUC driving-factors.

Maize stover and sugar beet were identified as feedstocks with low emissions in all studies. We did not analyse the reason for this, but as maize stover is a by-product from maize production, it is not unreasonable for it to have a low ILUC effect. For sugar beet, the high yields obtained with this crop can be one explanation for the low emissions, as higher yields imply lower land use per MJ of produced biofuel and, therefore, less risk of displacement of other agricultural activities (more detailed analysis is needed to explain low ILUC for sugar beet).

The largest variation in results was seen for wheat ethanol and soybean biodiesel. However, over time there was some convergence of results, particularly regarding ethanol from maize, which has undergone much modelling effort. Sugar cane and wheat showed similar patterns. The values reported for biodiesel fuels showed greater variation than those for ethanol. For biodiesel, a few studies reported ILUC values lower than 10 g CO_2-eq/ MJ, whereas for ethanol several studies showed a range of emissions in which the lowest values were below 10 g CO_2-eq/MJ or even close to, or below, zero in some cases.

The review also revealed that only a handful of studies to date have analysed advanced biofuels. Both [20] and [21] modelled switchgrass, miscanthus and maize stover for ethanol production, but reached different conclusions, with [20] reporting lower ILUC emissions for both switchgrass and miscanthus to ethanol. We did not analyse the reason for this, but it can be noted that the two studies used different assumptions regarding, for example, the modelled quantity of biofuels and different models to estimate carbon losses due to the LUC.

11.2.2 EU POLICY TO ADDRESS ILUC OF BIOFUELS

EU policymakers have been struggling for years with the issue of how to deal with the ILUC of biofuels. Already in 2008, during the formulation of the RED and the revision of the FQD, the issue of ILUC took centre stage [22]. According to the final text of the Directives in 2009, the impacts of ILUC had to be investigated further by the EC. In a report released in December 2010, the EC acknowledged that ILUC can reduce the GHG emission-savings associated with biofuels, but it also highlighted the existence of uncertainties and limitations associated with the quantification of indirect emissions in the available models [23]. However, the report did not suggest a concrete approach for tackling the risk of negative ILUC and the related GHG emissions[c].

Work in the following two years was dedicated to reducing the uncertainties and limitations in the scientific knowledge. The European Commission (EC) launched a number of studies on the ILUC of biofuels[d], but only in autumn 2012, nearly two years after the date set by the RED and FQD, did the EC present a proposal to address the risk of negative ILUC [24]. The measures contained in the proposal are to: (i) cap the contribution of conventional (food-crop-based) biofuels to 5% of total motor fuel consumption; (ii) bring forward the introduction of the 60% minimum GHG-saving threshold for new installations to 2014; (iii) encourage the diffusion of advanced biofuels by double or quadruple counting their contribution toward the targets of the RED, and (iv) require Member States and fuel suppliers to report the estimated ILUC emissions, employing feedstock-specific ILUC factors (12 g CO_2-eq/MJ for cereals and other starch-rich crops, 13 g CO_2-eq/MJ for sugars and 55 g CO_2-eq/MJ for oil crops) [24]. These ILUC factors are based on the results of the general equilibrium economic model International Food Policy Research Institute (IFPRI)-MIRAGE-BioF [25] and are introduced only for reporting purposes[e].

The EC proposal is currently being debated within the European Parliament (EP) and Council of the European Union (CEU). The EP adopted a common position on the proposal in September 2013 [26], according to which advanced biofuels should supply at least 2.5% of the energy used for transportation in 2020, and the share of food-crop-based biofuels should be limited to 6% on an energy basis. ILUC factors might be accounted for in the EU sustainability certification system only after 2020. The EC proposal has also been debated within the CEU during 2013. On several occasions the Ministries of Environment and Energy have heavily criticised it, but have failed to produce a common position on the issue. In October 2013, the Irish presidency of the Council advanced a position that calls for a 7% cap on the use of food crops and a minimum share of 2% for advanced biofuels by 2020, but does not mention the introduction of ILUC factors [27]. The negotiation of the proposal will continue throughout 2014, but the chances of adopting a final text before 2015 are now meagre, particularly owing to the parliamentary elections in spring 2014.

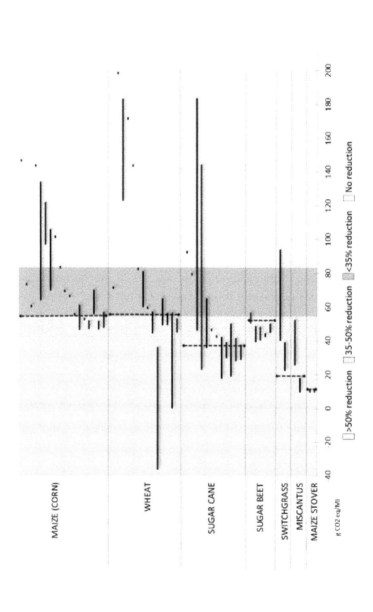

FIGURE 3: Estimates of total GHG emissions for ethanol fuels. Black bars show scientific estimates and include direct emissions as default values established in the Renewable Energy Directive (RED) and Fuel Quality Directive (FQD), and indirect land use change (ILUC) factors from the modelling exercises presented in Figure 1. Dashed black lines show policy estimates and include direct emissions as default values established in the RED and FQD and indirect emissions as estimated in [24].

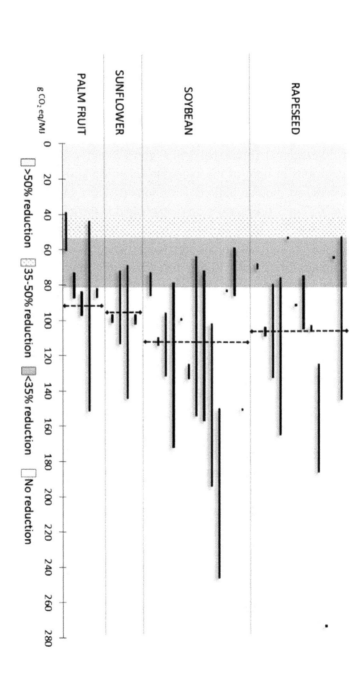

FIGURE 4: Estimates of total GHG emissions for biodiesel fuels. Black bars show scientific estimates and include direct emissions as default values established in the Renewable Energy Directive (RED) and Fuel Quality Directive (FQD), and indirect land use change (ILUC) factors from the modelling exercises presented in Figure 2. Dashed black lines show policy estimates and include direct emissions as default values established in the RED and FQD and indirect emissions as estimated in [24].

11.2.3 ILUC: PUBLIC POLICIES AND SCIENTIFIC KNOWLEDGE

The recent policy developments in the EU can be viewed in light of the scientific knowledge about ILUC provided by the modelling exercises reviewed above (Review of ILUC modelling results). Policy estimates of total GHG emissions (dashed black lines) are compared with the scientific estimates (black bars) in Figures 3 and 4 to evaluate policymakers' decisions in light of the available scientific knowledge. Policy-estimated emissions include indirect emissions (ILUC factors) proposed by the EC in 2012 and direct emissions established in the RED and FQD (default values). The results from the modelling studies reviewed are sorted in order of publication date (oldest first) and thereafter in order of maximum value (highest first), but without distinction into model types. Moreover, Figures 3 and 4 show the GHG-saving thresholds established in the RED and FQD. Although these levels represent the political ambitions with biofuels as climate-change mitigation means, they can be evaluated against the levels of total GHG emissions estimated by policymakers and scientists in order to assess the relative performance of ethanol and biodiesel from different feedstock and, more importantly, how the introduction of the policy values will affect each type of biofuel.

In the case of ethanol fuels, it can be seen that the policy values are generally in line with the results of the latest modelling exercises, in particular for first-generation ethanol fuels from maize and sugar cane (Figure 3). However, in the case of advanced ethanol fuels, the fit between policy values and scientific estimates is less evident (except for the case of maize stover)[f]. Compared with the minimum GHG-saving requirements established in the EU Directives, all types of ethanol fuels will be able to comply with the 35% minimum reduction requirement (assuming limited improvements with direct emissions of wheat ethanol), whereas the 50% requirement will be difficult to fulfil for all but sugar cane ethanol and second-generation ethanol fuels.In the case of biodiesel fuels (Figure 4), it can be seen that the policy estimates contained in the EC proposal do not appear to be in line with the modelling results reviewed earlier (Review of ILUC modelling results), or at least not as much as in the case of ethanol fuels. The EC values for biodiesel are generally higher than the range of values reported in the modelling exercises. Even in cases for which we observed some level of convergence of models results, for example, soybean biodiesel, the EC values are considerably higher. If the proposed values were to be introduced into the EU policy to assess compliance with the minimum saving requirements, none of the (first-generation) biodiesel fuels would be able to fulfil the 35%, let alone the 50%, reduction requirement. Policy estimates of total GHG emissions of biodiesel fuels are (substantially) higher than those of fossil fuels, whereas scientific estimates show lower levels, even though the potential for savings of GHG emissions compared with fossil fuels is limited.

11.3 CONCLUSIONS

Our review of recent ILUC modelling exercises revealed that most studies employ economic equilibrium modelling and focus on ethanol fuels, especially with maize as feedstock, while only a small number of studies to date have modelled advanced biofuels. It also revealed that in spite of some convergence of results over time, particularly for ethanol from maize, wheat and sugar cane, there is still a major variation in the results from the models, especially for biodiesel fuels.

There are many reasons for the variation in the results found in the review. They can be grouped into three major issues: (i) structural components of the models (ii) input data and assumptions and (iii) treatment of carbon stock changes, both concerning the amount of released or sequestered carbon, and how carbon emissions are treated over time.

i) Structural components of the models: economic models (the most commonly used type of ILUC models) were originally developed for quantitative analysis of global economic and political issues rather than assessing ILUC. They adopt different world views and, therefore, have different assumptions about the development of oil prices, and food prices, et cetera [28]. Economic models such as GTAP and MIRAGE (general equilibrium models) study entire economies, whereas others, such as AGLINK (partial equilibrium models), study only specific sectors. Models differ also in their geographical resolution as some study the system at country level, while others aggregate larger, regional areas. Further, the commodity-level resolution varies among models. Whereas some models, for example, IMPACT, study cereals, others, such as CAPRI, can differentiate between different types of cereals. Finally, other issues such as the possibility to model trade of biofuels or the expansion of agricultural land into different types of land uses (only pasture land, only forest, or both) explains some of the variation in model results. For a more in-depth comparison of these issues among models see previous work as examples [12,28-30].

ii) Input data and assumptions: the studies analyse different policies and use different start and end points in time. They assume policy target will be achieved by different ratios of biodiesel and ethanol and only in a few cases include second generation biofuels. An important set of differences concerns assumptions about harvest levels and raw material use per MJ of biofuel and assumptions about the amount and value of by-products, with some studies (for example, IMPACT and LEITAP models reported in [28]) not accounting for by-products at all. In addition, assumptions about how demand for different commodities depends on commodity price (the so-called elasticity factors) are of great importance and vary significantly among the models. Finally, land prices and costs of land conversion are major assumptions influencing the results of ILUC models [8,12,15,16,30-33].

iii) Treatment of carbon stocks changes (including above and below ground biomass): assumptions regarding the type of land that will be converted and the related emissions are important for the variation in results of ILUC models [12]. As these assumptions are usually not included in economic models, additional models need to be added. For example, very high ILUC factors of biodiesel can in some instances be explained by assumptions about ILUC on peat soils in South East Asia [12], see for example, previously reported studies [34].

Our review of the development of EU policies to address the GHG emissions associated with ILUC for production of biofuels showed that EU policymaking suffers from a number of inconsistencies and weaknesses. The approaches officially suggested by policymakers in recent years recognise, to different degrees, the necessity of accounting for ILUC in biofuel support policies. The ILUC factors selected by policymakers for this purpose are very specific in terms of g CO_2-eq per MJ of biofuel, which is clearly at odds with the uncertainty in results emerging from modelling exercises to date. Thus, there is a conflict between the demand from EU policymakers for exact, highly specific values and the capacity of the current models to supply results with that level of precision.

The uncertainty of ILUC estimates may be reduced in the future as better models and better data reduce the epistemological uncertainty associated with lack of knowledge of system behaviour [35]. We observed in our review some convergence of results over time, particularly regarding ethanol from maize (which has undergone much modelling effort), sugar cane and wheat. However, uncertainty will not be eliminated. The modelling results produced to date indicate that due to the complexity of the global economy, no significant reduction in model uncertainty should be expected in the near future [36]. Furthermore, predictions of future changes in complex natural and socio-technical systems are intrinsically uncertain. Future ILUC will be dependent not only on economic reactions, but also on other (unforeseen) factors such as agricultural and trade policies in different parts of the world. This so-called variability uncertainty [35] is not reducible and results in a variety of valid scientific standpoints.

The gap between demand by policymakers for indisputable evidence and final answers and the lack of conclusiveness and definitiveness in the knowledge generated by scientific models is visually displayed in Figures 3 and 4. This gap fuels the abundant criticisms linked to the introduction of ILUC factors in EU policies. Considering that the effectiveness of the introduction of (controversial) ILUC factors on the LUC of biofuel policies has not been modelled, and is probably outside the reach of current models, we must conclude that alternative policy approaches should be further explored. One such alternative approach would be to place a cap on food-crop-based biofuels, as proposed by the EC in 2012 [24]. However, the cap suggested by the EC is a very general measure, which does not reflect the real risk of ILUC of different types of biofuels. It aims at regulating the indirect effects of biofuel promotion without being able to measure them. Hence, it is not an effective measure to reduce the risk of negative LUC. We believe that an overall strategy for land use is more urgently needed than ILUC factors or a cap on food crop-based biofuels.

ENDNOTES

[a]The RED sets a binding target of 10% renewable fuels, including biofuels, in transportation by 2020 for each Member State, while the FQD mandates fuel suppliers to lower the GHG emissions by 6% for each unit of energy from fuel sold by 2020.

[b]The fossil fuel comparator for calculation of GHG reductions is at present 83.8 g CO2-eq/ MJ, but is under revision.

[c]The report identified four options: (i) take no action for the time being, while continuing to monitor; (ii) increase minimum GHG saving thresholds; (iii) introduce additional sustainability requirements on certain categories of biofuels; and (iv) attribute a quantity of GHG emissions to biofuels reflecting the estimated ILUC impact. These options have been evaluated in [25] and [5].

[d]These are available on the EC website (http://ec.europa.eu/energy/renewables/studies/ land_use_change_en.htm).

[e]In a leaked draft version of the EC proposal, ILUC factors were included in the methodology to calculate the GHG balance of biofuels for compliance with the minimum GHG saving requirements.

[f]Note that the EC proposal assigns an ILUC factor of zero to all advanced biofuels.

REFERENCES

1. Marland G, Schlamadinger B: Forests for carbon sequestration or fossil fuel substitution? A sensitivity analysis. Biomass Bioenergy 1997, 13:389-397.
2. Leemans R, van Amstel A, Battjes C, Kreileman E, Toet S: The land cover and carbon cycle consequences of large-scale utilizations of biomass as an energy source. Glob Environ Chang 1996, 6:335-357.
3. Searchinger T, Heimlich R, Houghton RA, Dong F, Elobeid A, Fabiosa J, Tokgoz S, Hayes D, Yu T-H: Use of U.S. croplands for biofuels increases greenhouse gases through emissions from land-use change. Science 2008, 319:1238-1240.
4. Fargione J, Hill J, Tilman D, Polasky S, Hawthorne P: Land clearing and the biofuel carbon debt. Science 2008, 319:1235-1238.
5. Di Lucia L, Ahlgren S, Ericsson K: The dilemma of indirect land-use changes in EU biofuel policy – decision-making in the context of scientific uncertainty. Environ Sci Policy 2012, 16:9-19.
6. Böhringer C, Löschel A: Computable general equilibrium models for sustainability impact assessment: status quo and prospects. Ecol Econ 2006, 60:49-64.

7. Fritsche UR, Sims REH, Monti A: Direct and indirect land-use competition issues for energy crops and their sustainable production - an overview. Biofuel Bioprod Bior 2010, 4:692-704.
8. Nassar AM, Harfuch L, Bachion LC, Moreira MR: Biofuels and land-use changes: searching for the top model. Interface Focus 2011, 1:224-232.
9. Bauen A, Chudziak C, Vad K, Watson P: A causal descriptive approach to modelling the GHG emissions associated with the indirect land use impacts of biofuels. Final report. A study for the UK Department for Transport. E4tech. 2010. [http://www.apere.org/doc/1010_e4tech.pdf]
10. Fritsche UR, Hennenberg K, Hünecke K: Sustainability Standards for internationally traded Biomass. The "iLUC Factor" as a Means to Hedge Risks of GHG Emissions from Indirect Land Use Change - Working Paper. Energy & Climate Division, Öko-Institut, Darmstadt Office; 2010. [http://www.oeko.de/oekodoc/1030/2010-082-en.pdf]
11. Tipper R, Hutchison C, Brander M: A practical approach for policies to address GHG emissions from indirect land use change associated with biofuels. 2009. [Ecometrica and Greenergy Technical Paper - TP-080212-A]
12. Broch A, Hoekman SK, Unnasch S: A review of variability in indirect land use change assessment and modeling in biofuel policy. Environ Sci Pol 2013, 29:147-157.
13. Wicke B, Verweij P, van Meijl H, van Vuuren DP, Faaij APC: Indirect land use change: review of existing models and strategies for mitigation. Biofuels 2011, 3:87-100.
14. Warner E, Zhang Y, Inman D, Heath G: Challenges in the estimation of greenhouse gas emissions from biofuel-induced global land-use change. Biofuel Bioprod Bior 2014, 8:114-125.
15. Khanna M, Crago C: Measuring indirect land use change with biofuels: Implications for policy. 2011. [Agricultural Policy Briefs Dep of Agricultural and Consumer Economics University of Illinois at Urbana-Champaign March 18, 2011 APBR 11–01;] [http://www.farmdoc.illinois.edu/policy/apbr/apbr_11_01/apbr_11_01.pdf]
16. Prins AG, Stehfest E, Overmars KP, Ros J: Are models suitable for determining ILUC factors?. 2010. [PBL, Netherlands Environmental Assessment Agency] [http://www.rivm.nl/bibliotheek/rapporten/500143006.pdf]
17. Di Lucia L, Ahlgren S, Ericsson K: The dilemma of indirect land-use changes in EU biofuel policy - an empirical study of policy-making in the context of scientific uncertainty. Environ Sci Pol 2012, 16:9-19.
18. ARB:Subchapter 10. Climate Change Article 4. Regulations to Achieve Greenhouse Gas Emission Reductions Subarticle 7. Low Carbon Fuel Standard. California Air Resources Board. 2009. [http://www.arb.ca.gov/regact/2009/lcfs09/lcfscombofinal.pdf]
19. Warner E, Inman D, Kunstman B, Bush B, Vimmerstedt L, Peterson S, Macknick J, Zhang Y: Modeling biofuel expansion effects on land use change dynamics. Environ Res Lett 2013., 8
20. Dunn JB, Mueller S, Kwon H-y, Wang MQ: Land-use change and greenhouse gas emissions from corn and cellulosic ethanol. Biotechnology for Biofuels. 2013, 6:51.

21. Taheripour F, Tyner WE: Induced land use emissions due to first and second generation biofuels and uncertainty in land use emissions factors. No 124407. In 2012 Annual Meeting, August 12–14, 2012, Seattle, Washington. Agricultural and Applied Economics Association; 2012. [http://EconPapers.repec.org/RePEc:ags:aaea12:124407]

22. EP: Report on the proposal for a directive of the European Parliament and of the Council on the promotion of the use of energy from renewable sources. Brussels: European Parliament, Committee on Industry, Research and Energy; 2008.

23. EC: Report from the Commission on indirect land-use change related to biofuels and bioliquids. Brussels: European Commission; 2010.

24. EC: Proposal for a Directive of the European Parliament and of the Council amending Directive 98/70/EC and Directive 2009/28/EC, COM 595 final. Brussels: European Commission; 2012.

25. EC: Impact assessment accompanying the proposal for a Directive amending Directive 98/70/EC and Directive 2009/28/EC Commission staff working document. Brussels: European Commission; 2012.

26. EP: European Parliament legislative resolution of 11 September 2013 on the proposal for a directive of the European Parliament and of the Council amending Directive 98/70/EC relating to the quality of petrol and diesel fuels and amending Directive 2009/28/EC on the promotion of the use of energy from renewable sources, First reading. Brussels: European Parliament; 2013.

27. CEU: Progress report on the Proposal for a Directive of the European Parliament and of the Council amending Directive 98/70/EC and Directive 2009/28/EC First reading. Brussels: Council of the European Union; 2013.

28. Edwards R, Mulligan D, Marelli L: Indirect Land Use Change from increased biofuels demand. Comparison of models and results for marginal biofuels production from different feedstocks. 2010. [JRC Scientific and Technical Reports no EUR 24485 EN – 2010. Joint Research Centre, European Commission]

29. Akhurst M, Kalas N, Woods J: Synthesis of European Commission Biofuels Land Use Modelling. 2011. [SCIENCE INSIGHTS for Biofuel Policy, Issue 1, March 2011. LCAworks and Porter Institute (Imperial College London)] [http://www.lcaworks.com/Science_Insights_Issue1.pdf]

30. EC: The impact of land use change on greenhouse gas emissions from biofuels and bioliquids. Literature review. An in-house review conducted for DG Energy as part of the European Commission's analytical work on indirect land use change. 2010. [http://ec.europa.eu/energy/renewables/studies/doc/land_use_change/study_3_land_use_change_literature_review_final_30_7_10.pdf]

31. Cornelissen S, Dehue B, Wonink S: Summary of approaches to account for and monitor indirect impacts of biofuel production. 2009. [Report no PECPNL084225, Ecofys] [https://www.cbd.int/agriculture/2011-121/EU-Ecofys-sep11-en.pdf]. ye

32. O'Connor D: Issues with GTAP iLUC Modelling. 2011. [Presentation at Seminar: Indirect land use change related to biofuels and bioliquids: what option would you chose? Organised by Copa-Cogeca, Brussels 3rd May 2011] [http://www.copa-cogeca.eu/img/user/file/BI_SPEECH/Don_OConnor_BI3592_EN.pdf]

33. Yeh S, Witcover J: Policy Brief: indirect Land-Use Change from Biofuels: Recent Developments in Modeling and Policy Landscapes. Carbon Standards in Agricul-

tural Production and Trade. Sao Paulo: International Food & Agricultural Trade Policy Council 2010. [http://www.its.ucdavis.edu/wp-content/themes/ucdavis/pubs/download_pdf.php?id=1434]

34. Lahl U: An analysis of iLUC and biofuels regional quantification of climate relevant land use change and options for combating it. 2010. [Report no BI(10)8696. BZL Kommunikation und Projektsteuerung GmbH] [http://www.bzl-gmbh.de/de/sites/default/files/iLUC_Studie_Lahl_engl.pdf]

35. Walker WE, Harremoës P, Rotmans J, van der Sluijs JP, van Asselt MBA, Janssen P, Krayer von Krauss MP: Defining uncertainty: a conceptual basis for uncertainty management in model-based decision support. Integr Assess 2003, 4:5-17.

36. Plevin RJ, O'Hare M, Jones AD, Torn MS, Gibbs HK: Greenhouse gas emissions from biofuels. Indirect land use change are uncertain but may be much greater than previously estimated. Environ Sci Technol 2010, 44:8015-8021.

PART IV

CONCLUSIONS

CHAPTER 12

Safe Climate Policy is Affordable: 12 Reasons

JEROEN C. J. M. VAN DEN BERGH

12.1 INTRODUCTION

It is generally felt that a climate policy which stabilizes atmospheric concentrations of greenhouse gases (GHGs) at a 'safe' level will be extremely
expensive, whether measured in terms of monetary costs, reduced GDP
growth or forgone welfare. This is supported by a number of influential
economic cost–benefit analyses of climate policy as reviewed in Kelly and
Kolstad (1999) and Tol (2008a, b). In this paper it will be argued that the
application of cost–benefit analysis (CBA) to climate change and policy
should be judged as being overly ambitious. To avoid the many fundamental and practical problems associated with CBA and the associated
notion of 'optimal' climate policy, it will be argued that a better option is
to adopt a more modest and practical approach, namely examining the cost
of a safe climate policy. This reflects a policy aimed at a stable and safe
level of atmospheric GHG concentrations—thus focusing on mitigation,
not adaptation. The combination of risk aversion, pervasive uncertainty,
and extreme climate change and events motivates such a safe or precautionary approach as a rational alternative to an optimal climate policy. In

Safe Climate Policy is Affordable: 12 Reasons. © *van den Bergh JCJM.* Climatic Change *101 (2010),*
DOI 10.1007/s10584-009-9719-7. Reprinted with permission from the author.

fact, (avoiding) extreme climate change may be regarded as the ultimate reason for us to worry about and respond to climate change. Even two strong advocates of using CBA to analyze climate change, Tol and Yohe (2007, pp. 153–154), state: "A cost–benefit analysis cannot be the whole argument for abatement. Uncertainty, equity, and responsibility are other, perhaps better reasons to act."

It will be argued here that the cost of climate policy has so far been approached from too narrow a perspective. This will involve a discussion of fundamental problems associated with applying CBA to climate change and policy. Spash (2007) concludes that cost-effectiveness studies are not much better than CBA's. Indeed, studies attempting to assess the monetary cost of climate policy make many debatable assumptions as well. Nevertheless, the shortcomings are less serious than in the case of climate CBA studies because the monetization of climate damage is avoided. Since some of the shortcomings of CBA's and cost assessments of climate policy cannot be resolved, one cannot hope for a single model analysis of climate policy to provide the definite insight about its cost let alone its optimality.

This paper will therefore offer an alternative approach consisting of assessments of the cost of climate policy from a range of complementary perspectives. Together, these aim to avoid or surpass the limits of existing CBA and policy cost studies. The alternative approach can be seen as trying to determine the economic and social costs of a safe or reasonably safe—given all sorts of uncertainties involved—climate policy by considering a range of perspectives to somehow bound the "cost space". The focus on a safe or precautionary climate mitigation policy can be regarded as the outcome of a qualitative risk analysis, as will be discussed in Section 3. Twelve perspectives on the cost of climate policy are offered. Together they deliver quite an optimistic conclusion, namely that climate policy is not excessively expensive and is certainly cheaper than suggested by most current studies. In other words, our global society can afford to invest in a safe climate policy. This should serve as relevant information for all politicians who fear severe economic consequences from stringent regulation of GHG emissions.

The remainder of this article is organized as follows. Section 2 briefly argues the failure of cost–benefit analysis of climate policy. Section 3 presents the main arguments in favor of a safe, precautionary approach to climate policy. Section 4 discusses the meaning of the cost of a safe climate policy and reviews the methods and assumptions that have been used to produce the main cost estimates. Together, Sections 2 and 4 show that the current economic approaches to assessing the (net) costs of climate policy have severe limitations. As a result, they are prone to generating inaccurate estimates of these costs, so there is a need for an alternative approach, as offered here. Section 5 contains the main thrust of the paper. It presents the new approach consisting of 12 perspectives on, and interpretations of, climate policy costs that move beyond current model assumptions and limitations. Section 6 provides conclusions.

12.2 THE FAILURE OF COST–BENEFIT ANALYSES OF CLIMATE POLICY

The history of climate CBA shows enormous variation in estimates. For example, whereas early studies (e.g., Nordhaus 1991) excluded adaptation to and benefits of climate change, later studies did take them into account and arrived at lower climate damage costs. Despite variation, most climate CBA studies share many basic assumptions. These have received considerable criticism, much of which is difficult to resolve (e.g., Ayres and Walters 1991; Daily et al. 1991; Broome 1992; Barker 1996; Azar 1998; Neumayer 1999; Spash 2002; DeCanio 2003; van den Bergh 2004; Padilla 2004; Ackerman and Finlayson 2007;Maréchal 2007; Gowdy 2008; Tol 2008b; Ackerman et al. 2009; and various responses to the Stern Review). Criticism has been directed, among others, at the assumed behavior of economic agents, the social welfare objective used, the treatment of small-probability-high-impact scenarios, discounting and social discount rate values, monetary valuation of a human life, and the neglect or incomplete treatment of certain cost categories.

A main criticism is that the analysis of climate policy should not be conceptualized as a problem suitable for quantitative cost–benefit analysis but as one of risk analysis, since the cost of climate damage cannot be assessed with any acceptable degree of certainty (e.g., Azar and Schneider 2003; van den Bergh 2004; Stern et al. 2006). Weitzman (2007, p. 703) says about this: "The basic issue here is that spending money to slow global warming should perhaps not be conceptualized primarily as being about consumption smoothing as much as being about how much insurance to buy to offset the small chance of a ruinous catastrophe that is difficult to compensate by ordinary savings." The latter means that social welfare losses due to extreme climate change cannot be reversed or undone through adaptation. This view is the motivation for the approach adopted in this paper, namely an assessment of the cost of a (reasonably) safe climate policy. In this context, the treatment of extreme climate change and climate events characterized by a combination of small probabilities and large impacts has been argued to not go together well with an expected value approach to cost–benefit analysis. This specific, fundamental criticism is addressed in more detail in Section 3, which will result in an extended argument in favor of a precautionary approach to climate mitigation policy.

Several other fundamental issues can be raised about CBA-style evaluations of climate change. Woodward and Bishop (2000) argue that current economic analysis of climate policy falls short as it focuses on economic efficiency, while the underlying concerns about climate change are driven by an intergenerational allocation of economic endowments. Using a simple model, they show that efficiency does not guarantee environmental sustainability. Tol (2008b) mentions population as being endogenous to climate change since the latter will directly influence mortality and migration and indirectly affect long-term birth, mortality, and migration rates through climate impacts on poverty and economic development. As a result, welfare optimization may involve ethically debatable implications for population size. More recently, Llavador et al. (2008) reject discounted utilitarianism as a normative criterion for intergenerational public decision making, and instead examine climate strategies under an intergenerational maximin criterion and maximization of a quality of life (human development) indicator. Indeed, there is an extensive literature on happiness and economics which finds that income or GDP is generally not an accurate indicator of welfare. This suggest that the focus on GDP growth in climate– economy models and the resulting interpretation of GDP losses (or foregone GDP growth) as a measure of the cost of climate policy is misplaced. This will be discussed in more detail under Section 5.4 in Section 5.

Perhaps the most important shortcoming of current economic studies of climate policy relying on CBA is that they incompletely account for extreme and irreversible climate scenarios, such as (Easterling et al. 2000; Reilly et al. 2001; Bryden et al. 2005; Royal Society 2005): extreme low or high temperatures; a slow-down or halting of the global thermohaline circulation, of which the Gulf Stream is a part; an extreme increase of the world's mean sea-level over centuries due to the collapse of the ice sheets on Greenland and West Antarctica; 'runaway dynamics' caused by positive feedback mechanisms in the biosphere, such as substantial emissions of methane (with a much higher warming potential than CO_2) from permafrost regions; changes in climate subsystems such as the 'El Niño Southern Oscillation'; acidification of the oceans due to high atmospheric CO_2 concentrations, meaning a deterioration in the living conditions for marine organisms with yet unforeseen effects; and extreme weather events, notably extreme rainfall, an increased probability of heat waves and droughts, and an increased intensity of hurricanes due to warmer seas. If, moreover, such changes take place rapidly, then insufficient time for adaptation will contribute to higher damage costs. According to Weitzman (2007), "There is little doubt that the worst-case scenarios of global-warming catastrophes are genuinely frightening." The omission of these extremities from CBAs is incomprehensible given that the ultimate reason for studying climate change is—or in any case should be—a concern for extreme events which will fundamentally alter the environmental conditions for humans and the rest of the biosphere. In fact, studies that have incompletely taken into account extreme events should not be taken too seriously—they really involve nothing more than toy models—and the respective authors should be modest about the policy implications of their analyses (see also Azar and Lindgren 2003). In particular, studies omitting extreme events will underestimate the cost of climate change, or the benefits of climate policy, and therefore be biased against safe climate policy. The omission of worst-case scenarios from cost–benefit analysis is not so much a fundamental argument against cost–benefit analysis but rather a serious omission by the modelers. The reason is, of course, that most worst-case climate change scenarios cannot be accurately quantified.

The differential treatment of extreme climate events offers one explanation for the wide range of damage cost estimates of GHG emissions that one can find in the literature (Tol 2005; Fisher and Morgenstern 2006). Tol (2008a) performs a metaanalysis of them, suggesting that the most reliable estimate cannot be the outliers, thus explicitly questioning the high damage estimates used in the Stern Review. However, a meta-analysis assumes that all studies are equally valuable unless one weights studies, for instance, by giving a relatively high weight to more recent studies using updated information. But since Tol does not apply such a weighting scheme, the outcome of his analysis is dominated by the large share of (older) studies which neglect or incompletely address extreme climate change scenarios and events. The meta-analysis thus hides the fundamental shortcomings of the primary studies, even though it gives the impression of being an objective aggregation.

Other limitations and weaknesses of CBAs of climate policy have been welldocumented. Tol (2008b) lists the many imperfections in a refreshingly critical and honest account of climate damage cost studies. In particular, he notes the neglect in existing studies of the impact of climate change on human conflict, large-scale biodiversity loss, economic development, and human population/demography. Most models take immediate adaptation for granted by assuming rational behavior by economic agents. A general shortcoming is the neglect of any impacts beyond 2100 in many studies. An entirely different concern is that damage cost estimates for developing countries are of lower quality than those for developed countries (and many are extrapolations from earlier studies, often for the USA). This is especially problematic since developing countries will not only suffer severely from climate change but also be less able to undertake protection or adaptation. In addition, to estimate the costs of illness, accidents and human mortality, which comprise a considerable share of the costs of climate change, estimated 'values of a statistical life' have been used. However, these are problematic at a global scale in view of the immense economic and cultural heterogeneity as well as a heavily skewed international income distribution. Moreover, the environmental changes to be valued under extreme climate change scenarios are large, which creates the problem that their monetary valuation is inconsistent with a necessary condition, namely that the change to be valued is small compared to income. The latter follows from monetary valuation being based on the theoretical idea of income compensation or equivalence (e.g., Johansson 1987).

Next, over long-term horizons, such as in climate change analysis, CBA is extremely sensitive to discounting and particularly the choice of (social) discount rate. This in combination with the fierce, long-standing debate over the "correct" social discount rate serves as an important reason for many observers to question the robustness of CBA studies of climate policy. In fact, a large part of the variation in results of studies that have undertaken a quantitative CBA of climate policy is due to this discount rate sensitivity. The debate on intergenerational discount rates was revived by the Stern Review (Stern et al. 2006). The social discount rate (r) interpreted as an interest rate is generally defined as the sum of two elements, namely the pure rate of time preference (δ) and the average growth rate of per capita consumption (g) multiplied by the elasticity of marginal utility of consumption (η): this results in the "Ramsey formula" $r = \delta + \eta g$. The pure time preference δ (what Quiggin (2008) called the inherent discount rate) was set by the Stern Review team at nearly zero to reflect intergenerational equity. The exact value chosen was 0.1%, corresponding to a 90% probability of the human race surviving 100 years (implying a chance of human extinction of about 0.1% per year). The probability of human extinction equal to 10% in 100 years is likely to be an upper bound to the real value, suggesting that the pure rate of time preference of 0.1% is also an upper bound. Respected even prominent economists can be found to support the nearly zero value of δ (Dasgupta 2007; Cline 2007; Quiggin 2008) and to criticize it (Tol 2006; Nordhaus 2007; Weitzman 2007). The value $\eta = 1$ adopted in the Stern Review has been criticized as well (Dasgupta 2007; Nordhaus 2007). Stern (2007, p. 140) has said he is prepared to raise it to 1.5 and notes that this value receives support from Nordhaus; since the damage costs then are still much larger than the policy cost, this change will not alter the conclusions. Quiggin (2008) nevertheless argues that $\eta = 1$ is not a bad compromise given the variation of estimates in the literature, and that the criticism by Weitzman, Dasgupta, and Nordhaus implicitly sticks to the old assumption of expected utility behavior, even though this has been refuted as an explanatory decision-under-uncertainty model by advances in behavioral economics (see also Section 3). All in all, the criticism on the social discount rate by the Stern Review is not convincing. Note in this respect the range of social discount rates used in the various climate CBA studies. The Stern Review uses 1.4% per year

(δ = 0.1% per year, η = 1, and g = 1.3% per year); with η = 2 this would become 2%. Nordhaus in his various DICE and RICE models has used rather complicated procedures involving endogenous and regional discount rates. Roughly, this results in implicitly assuming a social discount rate in the range of 3–5%. Nordhaus and Boyer (2000) apply declining discount rates which doubles the social cost of carbon dioxide emissions compared to the earlier DICE calculations. Generally, declining discount rates mean that climate policies will more easily pass a cost–benefit test (Guo et al. 2006). Most importantly, as noted by Arrow (2007), even with a much higher social discount rate than the one resulting from the Stern Review's assumptions, and well above the value range accepted by most economists (3–6%), the cost–benefit argument for stringent climate policy remains valid.

Of course, there are several fundamental objections to be made against discounting as formalized in the Ramsey formula above. Three important ones are as follows (for others, see Ackerman et al. 2009). A first, longstanding objection is ethical in nature and was already mentioned by Ramsey himself. It recognizes that long term discounting, even with small discount rates, effectively means giving very large weights to generations early in time and very small weights to generations distant in time. Many economists and philosophers have expressed that equal weighing of all generations, meaning the use of a zero discount rate, is the only ethically defensible approach. A second objection is that the formula does not account for the "diversity of uncertainty" in the sense that in reality one typically finds multiple interest rates associated with assets that show varying degrees of risk, rather than a single market discount rate (Ackerman et al. 2009). I want to propose a third fundamental consideration as well. Social discounting means imposing a feature of human individuals, who show time preferences or impatience, upon a society. This comes down to treating a society as analogous to an individual. But unlike an individual, a society does not have a finite life as it always includes multiple, overlapping generations, so that one can regard it as continuous and immortal. This holds especially true when the society aims for (environmental) sustainability, which is in fact the motivation for the evaluation of potential climate policies. A related difference is that whereas individuals discount their own future utility over their lifetime, social discounting by the current generation discounts utility of other, future generations over time periods beyond the current generation's lifetime. These differences suggest that applying a positive time preference to societal, intergenerational decisions can be seen as employing an erroneous analogy (van den Bergh 2004). Nordhaus, Weitzman, Tol, and various others do not want to give credit to such fundamental objections against social time preference discounting and instead harshly judge the Stern Review as representing a "decidedly-minority paternalistic view", "lowest bound of just about any economist's best-guess range" and "nonconventional assumptions that go so strongly against mainstream economics". But these are rhetorical statements reflecting the fact that economists, just like ordinary people, are prone to conformist behavior. But conformism does not in any way guarantee truth. Moreover, speaking of mainstream economics in relation to climate policy analysis does not do justice to the fundamental criticism of the suitability of CBA as a method to evaluate climate policies, as summarized above. One can indeed interpret the fierce attacks by Nordhaus and Weitzman on the Stern Review as a "historical accident", to use a term from the literature on path-dependence: if Cline and Stern had been the dominant players in the field, and Nordhaus, Mendelsohn, Tol, and Yohe had arrived on the scene late, they would have likely been the ones receiving fierce criticism for making unorthodox assumptions.

Nordhaus, Weitzman, and others particularly refer to the gap between the Stern Review's discount rate choices and market interest rates. However, as Stern has noted in his responses (e.g., Stern 2007), market interest rates cannot serve as a guide for a "prescriptive or even a descriptive account of value judgments", as they are the result of short-term decisions by many individual consumers and producers on investment, saving, and consumption motivated by personal gains. Moreover, there are many rates of return which vary significantly (Brekke and Johansson-Stenman (2008) mention an average range of 0.4% to 8.8% for the USA during the period 1926–2000), market interest rates vary over time, and empirically estimated implicit discount rates obtained through stated choice experiments vary considerably as well (Frederick et al. 2002). This all suggests that any ethical judgment would vary over time or between the specific markets or investment assets taken as a basis, thus implying arbitrary discrimination among individuals living in different periods or born at different moments in time. In addition, the market interest rate is affected by failures of financial markets, which are due to myopic behavior, asymmetric information, market power, environmental externalities, etc. Witness in this respect the current worldwide financial crisis, which has consequences for market interest rates. Next, various proponents of a low discount rate have argued that if the market serves as a guideline, one should not focus on risky investments in stocks but on safe or "risk-free" investments, such as money market funds and low-risk government bonds. These typically give a very low return with an order of magnitude of 1% and 2%, respectively. These numbers are surprisingly consistent with the Stern Review's social discount rate of 1.4 (η = 1) and 2% (if η = 2 would have been used), as mentioned above.

Finally, low discount rates are consistent with certain stated preferences and theoretical findings. Based on the results of a survey among 2,160 economists, Weitzman (2001) finds that even if every individual believes in a constant discount rate, the wide spread of opinion on what is the appropriate social discount rate causes it to decline significantly over time. Extrapolation of this finding supports a zero longterm or intergenerational discount rate. Other support for a low or zero discount rate comes from the observed tendency of humans to hyperbolically discount, i.e. to use a decreasing discount rate as the time horizon increases. Different theoretical explanations involving evolutionary history, commitment, and self-control have been offered (Frederick et al. 2002; Dasgupta and Maskin 2005; Fudenberg and Levine 2006). Given bounded rationality in intertemporal decision-making by individuals, Brekke and Johansson-Stenman (2008) think that it makes sense to use a substantially lower social discount rate than the average return on investments. To support this, they wield arguments relating to prospect theory (Kahneman and Tversky 1979), notably status quo (climate damage is a loss, not a foregone gain) and nonlinear responses to (subjective) probabilities associated with climate risk (Botzen and van den Bergh 2009). Finally, it should be noted that the debate on discounting and the choice of discount rate are not just important for CBA but also for assessing the costs of climate policy. The reason is that these costs are not occurring in a single point in time but extend over a long period of time.

All in all, there are many reasons for not having much confidence in CBAs of climate policy. Regardless of where one precisely stands in the debate on using CBA for making choices about climate policy, one has to admit that there are many elements that can and will be disputed. The extensive and fierce debate following the Stern Review illustrates this. While Stern has been able to put climate change on the retina of many economists and government officials precisely because it was seen as an economic study, the CBA part received severe criticism. Stern (2007) has clearly stressed that he purposefully wanted to deviate from certain incorrect assumptions of earlier climate CBA studies, and that he regards climate change as a problem that really requires a risk analysis rather than a CBA, which he merely regards as one input into the debate. One can at least be positive about the Stern Review because it initiated a much-needed fundamental debate on the very young but already (policy-) influential research field of climate economics. Possibly, many economists previously not working on environmental issues have become aware of the combination of economic research challenges and political relevance that climate economics offers. An injection of new ideas and expertise is very much needed, as so far the debate has been too dominated by a small group of like-minded individuals adopting very similar assumptions. As the field of innovation studies teaches us, a diversity of approaches enhances the pace of innovation, which one would hope is the fate of climate economics.

CBA is an attractive and reasonable evaluation method for well-bounded problems (local, sectoral) with limited time horizons, non-extreme and manageable uncertainties, reversible scenarios, and limited income inequality. But its application to global, long-term climate change and policy questions runs into severe problems. Here CBA is not merely stretched to its extreme but breaks down. This does not mean that one has to reject qualitative-type of CBA thinking. Indeed, it is difficult to escape thinking in terms of trade-offs between qualitative costs or the disadvantages and benefits or advantages of any choice. Such a qualitative, conceptual approach is in fact needed to support a precautionary approach to climate policy. But unlike the quantitative CBA approach, its qualitative counterpart expresses clearly that specific, detailed statements about the social optimality of choices in the context of climate policy are overly ambitious.

12.3 ARGUMENTS FOR A SAFE, PRECAUTIONARY APPROACH TO CLIMATE POLICY

If it can be argued that a safe climate policy means considerably lower net costs than the absence of such a policy, it is rational to be in favor of such a policy. This represents a kind of cost-effectiveness combined with precaution, given the uncertainties involved, aimed at avoiding extreme damage costs due to climate change. As a guide we can take Nordhaus and Boyer's (2000) estimate of 10% and the Stern Review's estimate of almost 20% potential GDP damage cost of extreme climate change (Stern et al. 2006). As noted in Section 2, considerably lower damage costs require the omission of relevant extreme climate events and scenarios. If we compare these figures with climate policy cost estimates in the IPCC range of 1–4% of global GDP (Section 4), then safe climate policy is clearly seen to be socially efficient. The slogan used by some environmental NGOs is surprisingly appropriate: 'the most expensive climate policy is doing nothing'.

The combination of small probabilities and large impacts associated with extreme climate change and climate events does not go together well with an expected value approach to cost–benefit analysis, and moreover does not reflect the way humans generally tend to evaluate such problems (Kahneman and Tversky 1979; Diecidue and Wakker 2001; Weitzman 2007; Botzen and van den Bergh 2009; Quiggin 2008). This can partly be understood through different treatments of riskaversion in expected and non-expected utility approaches (Cohen 1995). Lowprobability, high-impact scenarios have a small expected value compared to more certain changes associated with less extreme costs, and as a consequence receive a relatively low weight in CBA analysis. This effectively means a risk-neutral or riskloving approach. Nevertheless, one may perceive such costs as very undesirable and hence place a considerable value on preventing low-probability, high-impact events from occurring, especially when such events are irreversible and involve the loss of non-substitutable goods or services, as is the case with climate change.

In line with this view, Loulou and Kanudia (1999) and van den Bergh (2004) have proposed studying climate change using a precautionary principle formalized via a minmax regret goal. This represents more risk aversion than an expected value approach and less risk aversion than, for example, maximin net benefits. Tol (2008b, p. 10), a fervent believer in climate CBA, supports the precautionary approach to climate policy evaluation implicitly by stating that in view of the strongly right-skewed distribution of climate change damage costs (median \$14/tC, mean \$93/Tc, 95 percentile \$350/tC; Tol 2005): "The policy implication is that emission reduction should err on the ambitious side". Dietz et al. (2007, p. 250) make a convincing plea for precaution in climate policy as well: "Those who deny the importance of strong and early action should explicitly propose at least one of three arguments: (1) there are no serious risks; (2) we can adapt successfully to whatever comes our way, however big the changes; (3) the future is of little importance. The first is absurd, the second reckless, and the third unethical."

Environmental economists have long thought about uncertainty, irreversibility and precaution, which has given rise to option value theory (Arrow and Fisher 1974). But surprisingly they have refrained from systematically applying it to the most relevant case of irreversible environmental change, namely climate change (an exception is Schimmelpfennig 1995). In brief, this would mean that the foregone benefits of a certain 'preservation scenario' (i.e. safe climate policy) are included as a cost category of the 'development scenario' (i.e. no policy, leading to climate change). The resulting option value can be interpreted as the value of flexibility to either accept climate change at a later date or not, where the flexibility is due to investing in GHG emissions reduction to avoid the irreversible build-up of greenhouse gases in the atmosphere. Ha-Duong (1998) applies the notion of quasi-option to climate policy, which states that precaution allows for learning about climate change in terms of risks, costs, and adaptation opportunities. Admittedly, a main weakness of applying (quasi-)option value theory to climate change policy is that it takes expected utility theory as a basis, which, as argued above, is problematic in view of the lowprobability, high-impact scenarios associated with climate change.

Gollier et al. (2000) have shown the precautionary principle to result from a rational decision formalized as dynamic optimization under uncertainty and irreversibility involving Bayesian updating/learning. The conditions for precautionary action turn out to depend on risk aversion and "prudence". The latter is captured by the third derivative of the utility function and reflects the degree to which an individual increases his savings in response to an increase in uncertainty about future revenues (Kimball 1990). Other approaches than expected utility maximization and minimax regret to support a precautionary policy are maximin utility and nonlinear methods like prospect theory or rank-dependent utility theory, which one can characterize either as rational or boundedly rational (but not irrational) approaches. Although experts seem not to entirely agree on the best theoretical approach to address decisions in the face of low-probability, high-impact scenarios, a defensible approach seems to be to give relatively more attention or weight to extreme case scenarios, which comes down to a kind of minimax regret approach.

In the face of extreme uncertainty a quantitative analysis will not necessarily be able to offer more informative insight than a mere qualitative analysis. The reason is that the extreme uncertainty does not disappear by adding more quantitative sophistication to the method of analysis or by reducing uncertainty to (subjective) risk. All existing models that include uncertainty somehow apply arbitrary probability distributions to extreme climate events and changes (surveyed by van den Bergh 2004). These models regard investments in emissions reduction as a decision on risky investments, but they insufficiently reflect the irreversibility of climate change, the extreme uncertainty (content and likelihood) associated with certain scenarios and events, and the non-insurability against extreme climate change and events due to risks being highly correlated for all regions in the world.

A somewhat different way to understand the rationale behind a precautionary approach to climate policy is based on comparing the likelihood and features of climate and economic instability. This represents a kind of risk management view, which conceptualizes climate policy as the outcome of a trade-off between the risks and costs associated with natural and economic instabilities. However, these two risks are neither on equal par nor symmetric. One may even go as far as to say they are of a different order and thus simply incomparable. This can be reasoned as follows. With a given global environment under a stringent climate policy, humans cannot predict economic changes with certainty, but they can guide and control them within boundaries. Economic stability can then be maintained. For example, if a stringent climate policy turns out to create too high economic costs and too much instability, the policy may be altered or adapted. However, under extreme climate change—due to a lax or lacking climate policy—one has to reckon with macro-scale risks, with catastrophic and irreversible changes in the coupled climate-biosphere system which cannot be controlled by any public policy, even though impacts may in some cases be ameliorated by climate adaptation policies. Governments will then be unable to avoid extreme impacts on the world economy, and economic policy will have a very hard time stabilizing economic responses to extreme climate change. In fact, a severe climate crisis may very well stimulate an unprecedented economic crisis. All in all, economic adaptation and policy under stable natural, climate conditions, enhanced by a stringent climate policy, are easier and safer than responding to unstable natural conditions resulting from a lax climate policy. This is consistent with the view of Azar and Schneider (2003, p. 331): "Thus, we do not see costs and benefits in a symmetrical cost–benefit logic, but rather as an equity problem and a risk management dilemma." The Stern Review also shares this standpoint, and many other observers have made similar statements.

The extensive literature on resilience and ecosystem functioning also suggests that we should be extremely careful in tinkering with the biosphere through humaninduced climate change, as this may cause discrete, structural changes in all kinds of ecosystems (freshwater, marine, rangeland, wetland, forest, arctic) when certain critical thresholds of GHG concentration in the atmosphere are surpassed (Holling 1986). The risk of extreme events or disasters, as documented in Section 2, is relevant here, as many of them will considerably affect basic conditions for many ecosystems. In addition, the uncertain synergy between biodiversity loss and climate change is relevant. Biodiversity supports the stability of ecosystem functions and related services to humans, while biodiversity loss is being enhanced by climate change. Against this background, some have even denied the relevance of normal scientific analyses of complex issues like climate change and climate policy on the basis of the climate system being complex and able to show catastrophic behavior (Margolis and Kammen 1999; Rind 1999). Add to this the other dimensions of global change that may interact with climate change in nonlinear and unknown ways, such as land use, deforestation, water use, destruction of wetlands, acid rain, acidification of the oceans, and human control over a sizeable portion of primary production. Complexity implies that causal connections between a multitude of potential factors and effects cannot be identified, let alone be quantified. Against this background, a 'post-normal science' has been pleaded for, characterized by "uncertain facts, values in dispute, high stakes and urgent decisions" (Funtowicz and Ravetz 1993). The climate problem meets all four characteristics.

The foregoing set of considerations suggests that the implementation of a precautionary principle in climate policy emerges as a rational strategy. Neither decisionmaking based on quantitative CBA nor waiting until more information is available are convincing strategies. An often-heard argument against the precautionary principle is that climate policy means that alternative public goals have to be sacrificed (Lomborg's simplified view—see note 2). But whereas, for instance, less health care and education can indeed reduce growth and welfare, they are unlikely to cause extreme and discrete changes at a global scale. For this reason, climate policy needs to be treated as fundamentally different from many other areas of public policy.

12.4 COST ASSESSMENTS OF CLIMATE POLICY

Next we consider less ambitious studies that use models to assess the cost of climate policy. To begin with, it should be noted that the notion of policy cost is not very clear as both the policy scenario and the benchmark (status quo) are a matter of choice and surrounded by a considerable degree of uncertainty. The best starting point is undoubtedly IPCC (2007), as it synthesizes the then available primary studies on the cost ofmitigation options. It aims for stabilization in the range of 535–590 CO_2 equivalent ppmv and reports cost estimates ranging from slightly negative to 4% of global income. IPCC makes a distinction between market and economic mitigation potentials, where the first type is based on private costs and discount rates given current market conditions and prevailing policies, while the second type involves social costs and discount rates under appropriate policies to remove market failures. IPCC notes that an evaluation of primary cost-effectiveness studies indicates that a global carbon price in the range ofUS\$20–80/tCO_2e (CO_2-equivalent) by 2030 would be able to realize a stabilization of approximately 550 CO_2e by 2100. It is, however, still debatable how safe the goal of stabilizing greenhouse gas concentrations in the atmosphere at 550 parts per million (ppmv of CO_2e) is, which might cause a 2 C higher temperature 100 years from now. While the pre-Industrial Revolution concentration was 280 ppmv CO_2e, the current level is about 385 ppmv CO_2e. We are quickly approaching the relatively safe concentration of 450 ppmv, meaning that (cheap) opportunities to stabilize at a safe level are become scarcer. It is even being debated whether anything beyond the current 380 ppmv is safe; 450 ppmv may lock us into a 3°C temperature increase (Lynas 2007). An important uncertainty factor is the residence time of CO_2 in the atmosphere: it is highly uncertain and estimate have increased considerably during recent years. According to Montenegro et al. (2007, p. 1), "About 75% of CO_2 emissions have an average perturbation lifetime of 1800 years and 25% have lifetimes much longer than 5000 years." Matthews and Caldeira (2008) even suggest that in order to stabilize atmospheric concentrations, anthropogenic emissions will have to be reduced to zero for a period ranging from decades to possibly centuries.

Various studies offer systematic comparisons and meta-analyses of cost estimates of climate policy (Repetto and Austin 1997; Barker et al. 2002, 2006; Fisher and Morgenstern 2006; Hawellek et al. 2007; Söderholm 2007). This simultaneously involves identifying critical assumptions and weaknesses of existing economic analyses of the costs of climate policy. Söderholm (2007) distinguishes between direct, partial equilibrium, general equilibrium, non-market, and policy design costs. Different studies have emphasized particular costs while ignoring or simplifying others. Systems engineering or bottom-up models stress the cost of behavioral change and changes between discrete technologies. Top-down models such as general equilibrium and neoclassical growth models describe continuous production (or cost) functions and focus on interactions between aggregate markets. The two model approaches are in this sense rather complementary, explaining the existence of hybrid models. The indicators of the cost of policy reported in different studies vary: total direct compliance costs, the carbon price required to comply with a given emissions reduction, the loss in GDP (country or world), and equivalent variation (the income change that would cause the same utility change as the climate policy).

The estimated costs of climate mitigation policy show a great deal of variation. This is due to a number of assumptions which differ between studies: the assumed required emissions reduction (trivial, but it nevertheless makes straightforward comparisons impossible); the structural characteristics of the models, notably substitution possibilities (fuels, products), level of technological detail, assumptions regarding technological progress (and whether it is exogenous or endogenous); the design of climate policy (e.g., market-based instruments like taxes or tradable permits, or command-and-controlmeasures); and the inclusion of non-market costs and benefits.

Technical progress emerges as a crucial factor, and therefore the specification of the relation between policy and technical change is important. This requires models with endogenous (or policy-induced) technological change. Models without this feature are likely to overestimate policy costs. Bottom-up models account for endogenous technological change through the use of learning curves. Neoclassical economic models see technological change as due to investment in R&D. The learning curve approach generally produces much lower policy cost estimates. Underrepresentation of knowledge spillovers across different sectors of the economy may further lead to overestimating climate policy costs. On the other hand, the existence of market failures associated with R&D as well as path dependence and technical lock-in may mean that policy costs are underestimated. Finally, differences in cost estimates can be due to different assumptions regarding the use of carbon tax revenues (if policy is a tax, charge or levy): revenue recycling in the most neutral way (lump-sum rebates) or reducing other types of taxes (notably on labor) to correct market distortions and thus improve welfare.

Despite the shortcomings identified, let us consider a few recent studies in more detail to illustrate a few more aspects of estimating the cost of climate policy. A study undertaken for the USA by a renowned team of energy-CGE modelers (Jorgenson et al. 2008) finds that the US economy can easily accommodate a stringent climate policy. By 2020, reductions in real GDP are in the range of 0.5% to 0.7%, and by 2040 they are 1.2%. Spread over 34 years, this loss entails a negligible slowdown in economic growth. Evidently, energy prices are most affected, and coal the most. While the production side of the economy feels the negative effect, con-

sumption is much less affected. By 2020, foregone consumption is in the range of 0.1% to 0.2% of baseline levels and, by 2040, the loss rises to 0.5%. In dollar terms, policy costs are $33 per household in 2010, $158 per household in 2020 and $672 per household in 2040. The analysis assumes lump-sum transfers of permits and tax revenues, which means that the effects may be even smaller (double dividends). The same holds for the omission of induced technical change. The authors think that its contribution is considerably smaller than that of substitution and economic restructuring.

Bollen et al. (2004) have studied the cost of post-Kyoto climate policy for The Netherlands. They assume that developed countries reach emission reduction levels of 30% below 1990 levels, which is the European Union's goal, consistent with an increase in average world temperatures of no more than two degrees Celsius above the pre-industrial level. Note that the 2°C temperature rise is surrounded by uncertainty (Meinshausen 2006) and that it implicitly assumes that the emission patterns of countries not bounded by the Kyoto protocol do not rise sharply. Related to the latter, an important cost factor is whether developing countries participate. If they do, the estimated costs will be 0.8% of the real national income (assuming a high economic growth rate). If only industrialized countries participate, many cheap options will disappear, leading to a cost ratio of 4.8%. The second important cost factor is the growth rate. If moderate instead of high growth is assumed, the cost falls to 0.2% of the National Income, as the economy and with it GHG emissions rise more slowly. A third factor is the policy instrument: a system of tradable emission rights will realize the minimum cost of mitigation (although only under the assumptions of a general equilibrium system, with fully rational agents, perfect information, no strategic behavior, and clearing markets). The results further assume inclusion of the United States and Australia, even though these countries at the time of the study had not ratified the Kyoto Protocol (although Australia has in the meantime). The analysis ignores the costs of adapting the economic structure and the transaction and enforcement costs of climate policy.

The Stern Review (Stern et al. 2006) assesses the mitigation costs based on the annual cost of cutting total greenhouse gas emissions to 75% of current levels by the year 2050, leading to an atmospheric stabilization level of 550 ppmv CO_2e. The PAGE2002 model was used giving an estimate of the associated mitigation costs, yielding an average of 1% of global GDP, with a range of ±3%. The wide range reflects the uncertainties related to technological innovation, scale of mitigation needed, flexibility of policy, and fossil fuel extraction costs. Cost estimates for mitigation over time can vary considerably based on the level of technological change in low-carbon technologies and the improvements in energy efficiency expected in the model, to the extent that it is not even certain which technologies will have the most market potential and lowest social cost. In addition, the timing of mitigation has an impact on the cost estimates. Delaying the emission reductions will most likely cut the costs of mitigation. However, delaying action may imply very high damage and adaptation costs (Stern et al. 2006). Finally, some relatively cheap strategies to reduce emissions are not considered in most economic studies, such as reducing emissions from deforestation and forest degradation (REDD; Lubowski 2008).

McKinsey and Company (2009) suggests that if immediate action is taken it will be possible to maintain global warming below 2∘C at an overall cost of less than 1% of global GDP. This is based on a study that constructed a global cost curve (version 2.0), including more than 200 opportunities for reducing GHG emissions across ten sectors and 21 world regions. According to the report's most optimistic scenario, investing in a transition to a low-carbon economy is potentially as 'cheap' as 0.5% of global GDP. This, however, involves an assumption that approximately 40% of the global potential reduction of GHG emissions can be achieved by energy conservation and more energy-efficient technologies in transport, daily household activities and buildings. This may be overly optimistic as it seems to be based on a partial analysis, having overlooked the rebound effects of energy conservation. There is increasing evidence that optimism about energy conservation is unwarranted, since rebound effects are much larger than often thought. Rebound may even be higher than 100%, but this is impossible to prove due to the limited system boundaries and time horizons used in energy analyses as well as the difficulty of proving causal effects of energy conservation (Sorrell 2007, 2009). From the perspective of rebound risks, renewable energy has a major advantage over energy conservation. That is why it is given prominent attention in Section 5.1 in Section 5.

The previous discussion suggests that cost estimates of climate policy in the literature are clouded by uncertainty and debatable assumptions. Abroad view going beyond single model limits can improve our understanding of the costs of climate policy.

12.5 TWELVE REASONS WHY A SAFE CLIMATE POLICY IS AFFORDABLE

The section below presents twelve new, complementary perspectives on the cost of climate policy.

12.5.1 PERSPECTIVE 1: EXTRAPOLATING LEARNING CURVES FOR RENEWABLE ENERGY

The easiest way to reason about the cost of climate policy is by considering a most likely definite solution to the core problem, that is, the emission of greenhouse gases, notably carbon dioxide. Renewable energy really offers the only definite solution, as it can in principle support the supply of electricity and other types of energy carriers in a carbon-free way. Of course, this requires the equipment and indirect support of renewable energy themselves to be produced with renewable, carbon-free energy. In order to allow for the wide-spread adoption of renewable energy, it needs to produce electricity at market-competitive prices.

Within renewable energy, one can identify wind turbines, water power, biomass energy (including biofuels), concentrated (solar) heat power, and solar photovoltaics (PV) as the main candidates for future dominance. However, which technology will ultimately emerge as the most attractive is uncertain. Wind is close to being competitive but limited in application because of visual hindrance and noise problems. Still, it is expected to undergo a large growth before saturation. The early optimism about biofuels has been severely tempered in the last few years. The net energy and de-carbonization effect of (first-generation) biomass and biofuels have recently been questioned. It has even been claimed that some biofuels produce more CO_2 emissions than they save because, among other reasons, they involve clearing natural vegetation (Fargione et al. 2008; Searchinger et al. 2008). In addition, concern has been raised over the potential impact on food production and prices, as well as over the unsustainability of bio-fuel agriculture. Many observers plea for a second generation focusing on waste biomass and non-edible parts of plants, but this still requires a long process of R&D and learning, while it is not expected to provide a major contribution to the energy supply, simply because suitable waste streams are limited. Concentrated solar (heat) power (CSP) seems to be a neglected technology, which is surprising since it is a fairly simple technology which has the advantage that there are few surprises in further development and application. Moreover, one square km of desert with CSP can generate about 100 times the amount of electricity produced from biofuel crops grown on 100 km^2. Perhaps the main barriers are related to requiring inter-national cooperation and infrastructure, such as high voltage direct current cables under the sea between Europe and North Africa, along with con-tinuing Western energy dependence on unstable regions. All in all, there is much to say in favor of Solar PV: it is characterized by decentralized, lo-cal applications, can be easily integrated into buildings, produces directly electricity, and there are diverse sub-technologies which suggest a great deal of learning potential and many options in the face of uncertainty. Compare this, for instance, with nuclear fusion, of which there is only one large-scale experiment worldwide, namely the ITER reactor constructed at Cadarache in southern France.

Energy conservation, including increased energy efficiency improve-ments, is generally seen as a cheap strategy to achieve GHG mitigation. However, this is not always the case if it requires costly investments in new capital and R&D. In deciding between investments in energy conservation and renewable energy, one should take into account which learning curve is steeper, i.e. where investment can lead to substantial cost reductions. One should further recognize that energy conservation can come hand in hand with large rebound effects (Greening et al. 2000; Sorrell 2007, 2009). If new, energy-efficient technologies considerably improve the productivity of energy intensive industries, economy-wide rebound effects can exceed 50% (Sorrell 2007) and in the long run energy consumption might even increase—what Polimeni et al. (2008) call the "Jevons paradox". Based on a panel of German household travel diary data, Frondel et al. (2008) find that between 1997 and 2005 cars' higher energy efficiency caused 57% to 67% more travel. Rebound involves substitution in consumption (saving energy involves shifting expenditures, like to air travel), price effects (less market demand), coupled markets (imperfect substitutes, such as transport and telecommunication), substitution of inputs (more capital or materials) causing indirect energy use effects, and even sector restructuring with consequences for communication, transport, and international trade with associated (embodied) energy uses. The policy lesson is that rather than stimulating voluntary energy conservation, it would be more effective to focus policy on avoiding GHG emissions with adequate price corrections for energy-related externalities. This would then prevent leakages and thus rebound in terms of both energy use and related externalities. Currently, for instance, the lack of good regulatory climate for air traffic offers a clear route for rebound effects. With adequate externality and thus energy pricing, the economic system can endogenously find the best route to energy sustainability in terms of combinations of energy conservation and renewable energy over time, something that cannot be accomplished by stimulating voluntary energy conservation.

In view of the foregoing discussion, the case of solar PV will be considered here in more detail as a basis for assessing the cost of climate policy (for a similar exercise for wind see Nemet 2008). van der Zwaan and Rabl (2003, 2004) have analyzed scenarios of the price and cost of solar PV on the basis of experience or learning curves. Such curves convey that overall production costs tend to decline with an increase in cumulative production. It is true that overall costs not only capture learning and innovation (R&D) effects but also change in market prices of inputs (notably material inputs). The latter may sometime increase which can (temporarily) reverse the normal, negative relation between cumulative production and costs. Nevertheless, generally speaking learning curves are seen as quite a robust tool to examine the long-run cost behavior of technologies. For solar photovoltaic (PV) energy, a most likely or middle scenario delivers an estimate on an order of magnitude equal to US$60 billion associated with a cumulative production of about 150 GW_p (note: in 2004 cumulative production was about 1 GW_p). This amount of money represents an extra expenditure over the investment in fossil fuel electricity, which is needed to make solar PV competitive with electricity produced from fossil fuels (van der Zwaan and Rabl 2004, Table 2, progress ratio 0.8). If learning is favorable, then US$30 billion (at 50 GWp) is a better estimate, while if learning is slow the cost may rise to US$300 billion (at 1000 GW$_p$).

van der Zwaan and Rabl (2004) provide several arguments for why these figures are reliable. PV cost decreases have been following the learning curve model rather well; PV is already competitive in certain niche markets, and the PV market has expanded by 15% annually over the over the past two decades, which serves as a good basis for cost reductions. Furthermore, sharp rises of the price of oil, as witnessed in 2008, would-make it easier for solar PV to become competitive, although a potential negative consequence is a worldwide rise in electricity production with coal. Evidently, it is important that good choices are made in terms of diversity of solar PV sub-technologies and applications (van denHeuvel and van den Bergh 2009; van den Bergh 2008a, b). Nemet (2008) and van der Zwaan andRabl (2004) warn that projected costs may bemay be very sensitive to certain parameters. In addition, using a learning curve may overestimate the cost decrease if rapid investment in solar PV capacity takes place, simply because then the learning time is shorter and the capacity to spend the R&D funds will be limited—notably, too few researchers with adequate education and expertise will be available. Finally, studies show that cost reductions are not a free lunch but require investments in private and public R&D and public policies like subsidies (Messner 1997).

It is often argued that we cannot hope—not even in the very long run—for the energy supply to rely totally on renewable energy sources, mainly because of their volatility due to natural fluctuations in sunshine and wind. There are many ways to resolve this, including energy storage, creating over-capacity, and combining energy sources with complementary variation patterns. Once solar PV electricity has become competitive, we can resolve many of the remaining barriers. It was argued here that the expected cost of bringing the price of solar PV electricity down is not excessive. The presented cost range will gain more meaning in the subsequent perspectives.

12.5.2 PERSPECTIVE 2: GLOBAL CLIMATE POLICY COST NORMALIZED BY OECD GDP

Here the cost of worldwide climate policy will be normalized by the GDP of OECD countries. This can be justified on the basis of their historical contribution to climate change (Botzen et al. 2008) as well as their currently high incomes relative to the rest of the world, i.e. historical and intra-generational fairness. We can then take the range of 1–4% suggested by a survey of studies by IPCC (2007) as one basis for a climate policy cost estimate (see Section 3). The second estimate can be drawn from the previous section, where the cost of public support to make solar PV competitive was estimated to be in the range of US$30 billion to US$300 billion with a best, middle estimate of US$60 billion. These costs result in only 0.17% (with an uncertainty range of 0.08–1.65%) of the joint GDP of the 30 OECD countries in 2007 (which was US$ 36,316 billion; OECD 2008). An equal distribution would simply come down to 60/30 = US$2 billion per country, which is not a shocking figure. If the investment were spread over the course of ten years, then it would amount to only US$200 million per country per year (over 10 years) or on average 0.017% of GDP (with an uncertainty range of 0.008–0.17%). In the worst case scenario, this would imply a cost to a family with a net income of € 25,000 about € 40; in the most likely case this would be € 4, and in the most favorable case € 2, over a 10-year period.

An alternative is to allocate costs proportional to country GDP or country per capita GDP, which would simply mean higher absolute costs for some and lower absolute costs for other OECD countries. Of course, if non-OECD countries share in the costs, the burden will be spread over a larger base and result in lower figures per country and individual. In any event, the aim here is just to show the magnitude of the cost rather than suggest any fair distribution.

In 2007, OECD income was about 55% of world GDP (about US$66 trillion). If OECD would carry all the cost of climate policy, and taking the climate policy cost range identified by IPCC (1–4%), this would lead to an average cost for OECD countries equal to 1.8–7% of GDP. This is significantly higher than the estimates based on public support of solar PV. Why is that so? First, the 4% is quite a high estimate, and it is likely that the 1% estimate is a more reasonable order of magnitude, yielding 1.8% for the OECD countries. This is, however, still about 100 times larger than the yearly middle estimate and ten times the yearly upper end estimate (assuming a 10-year investment period to make solar PV competitive) of the cost of public support of solar PV. One important reason is that climate policy initially will indeed be more expensive as solar PV is still maturing, meaning that it can not make a significant contribution to reducing GHG emissions. However, according to the scenario sketched under Section 5.1, after a 10-year period solar PV should fairly quickly take over the market and provide the major means of reducing GHG.

Therefore, during the first ten years one should expect a relatively high cost of 1.8% and subsequently a rapid drop in the cost of climate policy to 0.017% (with an uncertainty range of 0.008–0.17%). This pattern should not come as a surprise, as it simply reflects an initial investment in R&DDD and then enjoying the returns on this investment. This is consistent with the suggestion by Sandén and Azar (2005) that we need to enter a decade of experimentation with low carbon technologies.

12.5.3 PERSPECTIVE 3: DELAYED GDP GROWTH

If it is true that climate policy will cost about 1% of GDP per year, then given that economic growth in many countries has historically been around 2% on average, and in some countries higher, this would mean that net growth, after discounting the cost of climate policy, would still be positive, and that one would reach a certain level of income with a delay.

A related perspective on the cost of climate policy was proposed by Azar and Schneider (2002). They take as a starting point studies suggesting that the absolute cost of reaching what is regarded by the IPCC as "safe" concentrations of CO_2 is in the range of 1 to 20 trillion US$. Although this may seem impressive, it turns out to imply only a few, namely one to three, years' delay in achieving a specific level of income in the distant future. The delay evidently depends on income growth. Global income during the twenty-first century is expected to increase about tenfold (on average 2.35% per annum). Azar and Schneider (2002, p. 77) calculate that "if the cost by the year 2001 is as high as 6% of global GDP and income growth is 2% per year, then the delay time is 3 years. . . ". This 3-year delay is moreover easily dominated by random noise given the uncertainties involved in GDP movements over a period of one century. That is, uncertainty over such a long time horizon might translate in a variation of the final GDP level (i.e. after one century) which exceeds the 6% figure. This all means there is little reason to worry about the long-term negative effects of climate policy on the economy. In other words, seen in a long-term perspective, the costs of a stringent climate policy are marginal in economic terms. Aznar and Schneider further note that ". . . the global economy is expected to be an order of magnitude larger by the end of this century. . . we would still be expected to be some five times richer on a per capita basis than at present, almost regardless of the stabilization target."

12.5.4 PERSPECTIVE 4: HAPPINESS INSTEAD OF GDP

As was made clear in Sections 2 and 4, the economic evaluation of climate policy is often cast in terms of lost GDP. This seems attractive, as the economic and welfare impact is captured in a simple, aggregate number. However, this neglects the fact that the implicit assumptions and judgments about the relationship between wellbeing, happiness, and GDP have been staunchly criticized (van den Bergh 2009). In fact, there is quite extensive literature on this topic involving two approaches. From the angle of traditional microeconomic and welfare theory, notions like negative externalities, inequity, non-substitutable or lexicographic needs, informal activities, unaccounted resource use, and environmental degradation are stressed (Mishan 1967; Nordhaus and Tobin 1972; Hueting 1974; Sen 1976; Daly 1977; Dasgupta 2001). In addition, a steadily rising number of empirical studies in economics, sociology, and psychology assessing subjective well-being and happiness have questioned the use of indicators like income and GDP as proxies for social welfare and progress. There is much support for the view that beyond a certain threshold, which has been passed by most rich countries, average income increases do not translate in significant rises in well-being. In particular, this research indicates that somewhere between 1950 and 1970, the increase in welfare stagnated or even reversed into a negative trend in most industrials (OECD) countries, in spite of steady GDP growth (Blanchflower and Oswald 2004). This so-called "Easterlin Paradox" (Easterlin 1974) is supported by the 'Euroba-rometer surveys', the half-yearly opinion polls of the inhabitants of the EU member states, as well as by aggregate indicators of sustainable income based on GDP corrections, notably the ISEW and (derived) GPI indica-tors (Daly and Cobb 1989; Lawn and Clarke 2008). The income level at which de-linking occurs between GDP and (subjective) social welfare has been estimated to be approximate $15,000 (Helliwell 2003). Of course, one should not expect a rigid threshold to apply generally for all countries, cultures, and times. Nevertheless, the various empirical findings provide evidence for a stabilization of happiness and social welfare in spite of continued GDP growth. Layard (2005) also provided support by showing that countries with high incomes show little variation in average report-ed happiness. At best, the country comparison clarifies that happiness is characterized by diminishing returns on increases in GDP per capita. This means, not surprisingly, that for poor, developing countries the correlation of income and well-being is higher than for rich countries.

Three stylized facts assessed by happiness research can explain the ob-served de-linking of income and happiness (Frank 1985, 2004; Ng 2003; van Praag and Ferrer-i-Carbonell 2004; Kahneman et al. 2004). First, in-come and income growth contribute considerably to happiness if people are poor or countries are in a low development phase, as extra income will be mainly spent on basic needs. Second, although people may enjoy short-term or transitory increased happiness effects, ultimately they will adapt or get used to a higher income and changed circumstances in vari-ous other dimensions (Frederick and Loewenstein 1999). One explanation for this is that our senses can only handle a limited amount of stimuli, and ultimately satisfaction or boredom ensues. Since most people are not aware of the phenomenon of adaptation, they continue striving for 'more'. This is reflected by a range of terms used by different researchers: 'ad-diction', 'hedonic adaptation', 'hedonic treadmill', and 'preference drift'. Third, people compare their situation with that of others in a peer group, so their welfare has a relative component. This is associated with status-seeking and rivalry in consumption. In addition, studies have consistently found that income-independent factors greatly influence individual welfare or happiness, the most important ones being health, having a stable family (partner, children), personal freedom (political system), and being employed. Certain studies reported below also point out the relevance of environmental and climate factors. Note that some of these findings from rigorous econometric studies of subjective well-being data were already hypothesized in older writings (Hirsch 1976; Scitovsky 1976).

Happiness research further suggests that there are limits to improving happiness through income since happiness is to a large extent based on unobservable or not easily observable factors which may be summarized as a pessimistic or optimistic attitude towards life in general. Indeed, the causality may be often opposite in the sense that optimistic individuals are found to be relatively happy and successful in life on average and are thus capable of earning a relatively high average income (Ferrer-i-Carbonell and Frijters 2004). Another relevant consideration is that high incomes generally come with many working hours. But happiness evidently depends also on leisure, which is implicitly valued negatively if one employs GDP as a progress indicator, since it has an opportunity cost in the sense of forgone production opportunities. The OECD (2006) adjusts GDP by valuing leisure at GDP per hour worked (somewhat debatable), and finds that the result (in per capita terms) leads to a quite different ranking of countries than according to GDP per capita. The Netherlands leads the OECD countries in this ranking, which can be explained by the fact that the inactive part of the working force is relatively large and parttime working is very common there (de Groot et al. 2004). Note that because being (un)employed is also an important factor in happiness, the OECD adjustment represents merely a partial correction.

The implication of the foregoing stylized facts is that absolute individual income at best imperfectly, and beyond a certain threshold hardly, correlates with individual welfare (Easterlin 2001; Frey and Stutzer 2002; van Praag and Ferrer-i-Carbonell 2004; Ferrer-i-Carbonell 2005; Clark et al. 2008). Relative income turns out to be critical. But at the societal level, relative income changes are largely a zero-sum game: what one wins another loses. In conclusion, GDP (per capita) increases are neither a necessary nor sufficient condition for improving individual well-being and social welfare (for more detailed argumentation, see van den Bergh 2009).

Therefore, using effects on happiness instead of GDP as a criterion for judging climate policy is likely to provide quite different conclusions. Three considerations are relevant here. First, although climate policy may lead to a slower pace of economic growth, the foregoing discussion suggests that this translates into a smaller or even insignificant loss in happiness terms, depending on which country or group of people is considered. Secondly, climate policy aimed at preventing extreme events implies avoidance of serious reductions in happiness, given that happiness directly depends on climate, i.e. it involves direct non-market effects on individuals and households. This means that the economic and welfare effects of climate change measured in GDP terms may underestimate the real impact on happiness. Especially extreme climate events are not easily captured by GDP or other monetary cost terms, as argued in Section 2. Extreme climate change will have a profound impact on local and regional sea levels, temperatures, and weather patterns. This can in turn cause extreme effects on resource availability (notably clean water), human health, human security, vulnerability of poor people in regions with low productivity (Sahel countries), migration, and violent conflicts. It is virtually impossible to costaccount for these, even though it is clear that human happiness and basic needs are then seriously at stake. Third, although climate change may not affect the happiness of people in Western countries much, for people in poor countries it may mean that their basic needs will come under threat, which is likely to create severe and structural losses in happiness. In addition, richer people and richer countries can more easily adapt to climate change so that they can restore or approximate their old happiness levels. This is because rich countries are characterized by high levels of wealth (financial reserves), high average education, good access to modern technologies, and a generally high capacity for collective action.

Although no serious climate policy study has employed a happiness type of criterion or goal, a few studies have examined the impact of climate conditions on happiness. Rehdanz and Maddison (2005) start from the view that climate affects the daily life of humans in various ways: through heating and cooling requirements, health, clothing, nutritional needs, and recreational activities. Therefore, they expect individuals to have a clear preference for particular climate conditions. Based on a panel of 67 countries and using self-reported levels of happiness in relation to climate variables like temperature and precipitation, they use multiple regression analysis to show that climate variables have a highly significant effect on happiness. The authors find that high-latitude countries generally benefit from climate change raising temperatures, while countries already characterized by very high summer temperatures would most likely suffer losses. Other studies with similar findings are Frijters and van Praag (1998), who focus on well-being in Russia in 1993 and 1994. They examine how climate conditions in various parts of Russia affect the cost of living and wellbeing. Maddison (2003) applies the hedonic pricing method assuming that individuals can freely migrate in response to geographical conditions, including climate-related ones. Using data for 88 countries, a 2.5°C increase in mean temperature is found to benefit individual well-being in high latitude countries whereas it will lead to losses in low latitude countries. Rehdanz and Maddison (2004) perform a similar type of study to assess the amenity value of climate in Germany. They find that German households are compensated for climate amenities mainly through hedonic housing prices. House prices turn out to be higher in areas with higher January temperatures, lower July temperatures and lower January precipitation.

Welsch (2006) examines the relationship between pollution and happiness using subjective well-being panel data for ten European countries combined with air pollution data. Pollution is found to play a statistically significant role as a predictor of inter-country and inter-temporal differences in subjective well-being: $750 per capita per year for nitrogen dioxide and $1,400 for lead emissions. A related study is Luechinger (2007). Other studies supporting the relevance of climate and environmental conditions on happiness are Ferrer-i-Carbonell and Gowdy (2008) and Brereton et al. (2008). The shortcoming of many of the previous studies of the link between climate and happiness is that they consider small temperature changes or differences and give no attention to large changes or even extreme climate change or events.As a result, these studies may deliver an overly optimistic and insufficiently representative general picture of how people's happiness responds to climate change.

In translating the results of such studies, one might take into account the fact that the projected temperature change is largest for higher latitudes, so that happiness effects may be larger here; at lower latitudes with already high temperatures, the change is projected to be lower.

Cohen and Vandenbergh (2008) consider the lessons that can be learned from happiness research for climate policy, focusing on consumers. Taxes on pollutive consumption with a positional good character has two benefits: it reduces the status externality due to reduced consumption of such goods (Ireland 2001), and it reduces the total pollution associated with the consumption. Layard (2005) suggests taxing income to stimulate leisure and temper "status games" with respect to income and consumption. This may reduce status effects and pollution related to goods consumption equally, although this will depend on the shift in consumption (e.g., more holidays to distant countries will give rise to increased air traffic with associated GHG emissions). Brekke and Howarth (2002) have studied the interaction between status and environmental externalities and even apply this to the context of climate change. They find that ignoring status signalling in the analysis of public policy, in particular consumption and income taxes, will lead to significant biases in optimal tax rules. In addition, the inclusion of status in a climate-economy model shows that traditional policy analysis overvalues consumption and undervalues a stable climate.

According to Cohen and Vandenbergh (2008, p. 9), "Economists have never argued that money and economic wealth are all that matters. Instead, their starting point has always been 'utility maximization' which includes individual leisure activities, health, family situation, and other components." Two comments are in order here. First, standard microeconomics and its application to environmental policy theory, labor economics, and many other areas of applied economics generally employs utility functions which do not take the phenomena of adaptation and relative welfare into account. Second, empirical macroeconomics and political pleas for economic growth are completely uncritical of GDP information and assume that GDP growth is equivalent to (social) welfare growth or human progress. This is inconsistent with both standard microeconomics and enlightened microeconomics incorporating insights from empirical happiness research. Applications of economics to climate change and climate policy need to start taking the lessons of happiness research into account. The general implications outlined above are that we should worry more about climate change, and that safe climate policy becomes a more attractive option than under CBA-GDP types of evaluations.

A provision to the above arguments is that people may adapt to a changed climate in the sense of being initially (negatively) affected in their happiness, while later slowly recovering their old happiness level. However, such adaptation is difficult to imagine for extreme climate change and events. Finally, note that adopting a happiness approach may also affect the discount rate debate. The reason is that one would then be less inclined to discount as this would mean that the happiness of a person in the future would be valued less than that of a person living now. When more general, abstract notions like costs and benefits are employed instead, as in CBA studies, specific people and their happiness disappear from the picture, making the case for discounting easier to defend.

The happiness perspective has been given relatively much attention here. The reason is that it may well be the major alternative perspective to CBA on the cost of climate policy, one that is urgently needed in order to arrive at an accurate picture of what we really gain and sacrifice if we undertake a stringent, safe climate policy worldwide.

12.5.5 PERSPECTIVE 5: COMPARISON WITH LARGE PUBLIC INVESTMENTS: IRAQ WAR, FINANCIAL CRISIS, MILITARY R&D, AND SECTORAL SUBSIDIES

The cost of climate policy or more particularly of making solar PV a competitive technology might be seen as a large public project. This suggests a comparison with other public projects. Four large 'projects' will be considered, namely the Iraq war, combating the financial crisis, R&D investment in the military sector, and expenditures on subsidies to economic sectors.

Stiglitz and Bilmes (2008) have estimated the cost of the Iraq war to the United States to be at least US$3 trillion (3,000 billion). This excludes the cost to the rest of the world (notably the UK and Iraq, with an estimated 40–100,000 casualties). The Iraq war comes out, then, as the second most expensive war in history, after the Second World War, which cost about $5 trillion (in 2007 dollars adjusted for inflation). The cost estimate for the Iraq war is much higher than the official number given by the Bush administration as this excludes relevant cost categories. The broad categories in the Stiglitz/Bilmes figure includes both budgetary costs (notably military operations, health care, and disability compensation) and economic costs (notably loss of lives, welfare effects relating to oil prices, and interest payments). Hartley (2006) has suggested a figure of a similar magnitude, at least about US$1 trillion up to 2007, though including the cost to civilians and of reconstruction in Iraq. He argues that the economic costs of war receive far less attention than political, moral, legal (UN), and military (safety) considerations. In line with this, he makes the interesting remark that the US could have bribed Saddam Hussein by offering him and his family US$20 billion to leave Iraq, giving the Iraqi people US$50 billion, and on top of that save US$30 billion given that the cost of the war was ex ante (grossly under-) estimated at US$100 billion. The main message here is not that outlays on certain wars have been too large (that would be the theme of another paper), but that democratic societies have clearly shown a willingness to spend large amounts of money to avoid low-probability, high-impact catastrophes in the social realm.

Another interesting comparison is with the financial crisis in 2008/2009. The USA decided overnight to reserve US$700 billion to stabilize the US banking system. Governments in Europe are likely to have reserved a similar amount. For example, The Netherlands created a e20 billion fund to stabilize the financial sector, while it acquired Fortis Bank Nederland (Holding) N.V. for a total sum of EUR 16.8 billion. The Belgium government spent e4.7 billion on Fortis Bank Belgium, and Luxemburg e2.5 billion on Fortis Bank Luxemburg. The UK spent about e44 billion to take a majority share in four large British banks to rescue them, namely HBOS, Royal Bank of Scotland, Barclays, and Lloyds TSB. In total, OECD countries may have invested more than US$2 trillion (2,000 billion) to stabilize the financial system. The urgency of this was evident in view of the threats the financial crisis posed on the world economy. One may argue that some of the guarantees offered by countries in response to the financial crisis are in fact only creating reserves or represent investments in (shares of) banks rather than being effective spending, but nevertheless the countries or at least their governments were willing to set aside so much money in response to a threat without the support of any cost–benefit analysis or any other type of pseudo-welfare optimization. Similarly, if the threats posed by climate change were recognized and translated into a similar investment in GHG emissions reduction and renewable energy, the most serious risks associated with climate change might be avoided. An important barrier may be that whereas we have negative experiences with financial crises in the past and are determined to avoid new ones, we humans lack similar experiences with extreme climate change in the past.

So governments worldwide are investing roughly US$5 trillion in the Iraq war and countering the financial crisis jointly. We can compare this with the range of climate policy cost estimates, i.e. 1–4% of world GDP (US$66 trillion in 2007), or 0.7–2.7 trillion US$, which is only 14–54% of the aforementioned public investments. If one focuses on the cost range of making solar PV competitive, i.e. US$30 billion to US$300 billion with a middle scenario estimate of US$60 billion (Section 5.1 in this section), then as a proportion of the current investments in Iraq and the financial crisis this comes down to a central estimate of about 1% and a range of 0.6–6%. In other words, if these percentages of current public investments would be diverted to renewable energy, we would very likely solve the problems of energy scarcity and climate change. If the cost of making solar PV competitive is compared only to the cost of the Iraq war, then the assessed central estimate of US$60 billion and the higher end estimate of US$300 billion result in only 2% and a uncertainty range of 1–10% of the expenditures on the Iraq war.

A third relevant comparison is with current expenditures by countries worldwide on military research, which is estimated to be roughly US$140 billion. Of this, the largest single investor, the USA, spends about US$85 billion per year (Brzoska 2008). Earlier, between 1953 and 1970, America spent about 1% of its GDP on military R&D. Later, this percentage dropped and reached a minimum of 0.45% in 1979 (Roland 2001). Of course, the tremendous increase of GDP means that the absolute value of the expenditures on R&D has increased steadily over time. These investments in military R&D are especially interesting as they suggest that governments are willing to undertake enormous investments in R&D even when there is no clear problem to be solved. Similarly, safeguarding us from the effects of extreme climate change might be responded to with an investment in R&D in renewable energy at a similar scale. In two years, the world spends almost US$300 billion on military R&D, which is equivalent to the upper-limit estimate of the investment needed to make solar PV competitive. Per year, the world invests more than twice the middle estimate (US$60 billion). If the solar PV investment were to be spread over 10 years, then it would equal 5% of world expenditures on military research in the same period.

A final comparison is with current expenditures worldwide on direct (on-budget) and indirect (off-budget) subsidies to economic, resource-based sectors like agriculture, energy, and transport. In the period 1994–1998, more than US$1 trillion was spent worldwide on subsidies to these sectors (van Beers and van den Bergh 2009). Important subsidies include market price support, output payments, and input subsidies. For OECD countries, off-budget subsidies to the agricultural sector amounted to US$318 billion in 2002, which is 1.2% of the total OECD GDP. A lower bound estimate of energy subsidies (excluding external costs) for the EU-15 countries is US$37 billion. In the transport sector, most off-budget subsidies relate to road transport infrastructure, and for the world as a whole these are estimated to range from US$225 to 300 billion. The GATT and WTO trade rounds in the last 20 years led to subsidy reform but at a very slow pace, so the absolute and relative level of subsidies is still quite high. Although economists have written critically about subsidies, noting the harm they cause to social welfare and the environment, politicians have not been eager to systematically evaluate the net benefits of subsidies. The size of subsidy flows has additional relevance here, since an effective climate policy requires removal of the most environmentally harmful subsidies.

Evidently, the comparisons made under the current perspective do not deliver a sufficient argument for safe climate policy, but they should be seen in addition to some of the other perspectives offered in this section. Nor is the intention here to argue that precedent—notably, wasting money on public projects and wars—serves as an argument. Simply, for those stating that climate policy is (too) expensive and will bring about economic disaster, it is good to see things in a broader context.

12.5.6 PERSPECTIVE 6: THE CURRENT COST OF ENERGY IS FAIRLY LOW

Here it is argued that current fossil fuel-based energy (gasoline and electricity) is cheap, too cheap in view of associated negative externalities.

The latter is especially true if the cost of CO_2 reflects extreme climate events and scenarios. Current studies estimate this cost to have an order of magnitude equal to €200–450 per ton emitted CO_2 (Kuik et al. 2008; see also Section 2). This is even likely to be an underestimation as the studies on which it is based only partially address potential catastrophic scenarios.

While the cost of energy sources has fluctuated over time, in part due to instability of the OPEC cartel and conflicts in oil-rich regions, energy efficiency improvements in electricity generation and light production have caused a structural trend of falling energy costs in production, household consumption, and transport. Fouquet and Pearson (2006) show for the UK that in 2000 the cost of lighting was 1/3,000 of its 1,800 value, while during the same period income (purchasing power) had increased 15-fold. Of course, the falling cost of energy services (light, manufacturing, transport, and more recently various household appliances) has come with rebound effects (see Section 5.1) and an increasing demand for such services due to sustained increases in income as well as product and process innovations over time. Fouqeut and Pearson document this by noting a 25,000-fold increase in lighting consumption between 1,800 and 2,000. If one regards the share of energy cost in total income as a useful measure of the cost of energy, then the following picture emerges from these findings: the share for light services alone dropped between 1,800 and 2,000 by a factor $25,000/(3,000 \times 15) = 5/9$. For other uses of energy, the story is more complicated as the energy output is not a homogeneous service. Nevertheless, energy intensity defined as energy input per monetary output has dropped by more than 30% since 1970.

The falling cost of energy in various areas can be observed by considering the share of energy cost in total national income. The ratio of (all) energy expenditures to GDP since the 1970s shows a pattern that starts at around 8%, increases to about 14% in the early 1980s and then drops again to levels below those of 1970 and recently increases again (EIA 2008). This illustrates that—in any case, until recently—the cost of energy can be judged as fairly low. Even though energy is the fundamental input to all human economic activity, roughly 90% of income is spent on things other than energy. Moreover, continuous GDP growth and an almost constant share of energy costs in it suggest that the disposable income after energy expenditures has increased over time. A disadvantage of the aggregate approach to measuring energy expenditures as a share of GDP is that it hides income inequality. Generally, low income families spend a larger part of their income on energy, and they will also see a relatively rapid increase in the cost share when energy prices rise. The shares can differ between low, middle, and high incomes from 15%, 5% and 2%, respectively. This suggests that for some people, energy use may represent a considerable expenditure, while for many it does not. Roberts (2008) regards households as "undoubtedly fuel poor" when they are spending more than 10% of their income on energy just to meet basic requirements. This 10% threshold may reflect, however, that we take a very low share of energy cost in income for granted simply because this is a historical fact. Income inequality does suggest, though, that a serious climate policy raising energy prices might need to be complemented by an income redistribution policy (e.g., as part of shifting taxes from labor to energy).

Another indication that the cost of energy is not very high or even low is that the long-term average oil price (US crude oil prices adjusted for inflation in 2006 US$), if calculated from 1869 to 2007, equals $21.66 per barrel for world oil prices, while during the same period 50% of the time world prices were below the median oil price of $16.71 per barrel. For the post-1970 period, equivalent indicators are $32.23 and $26.50 (http://www.wtrg.com/prices.htm).

In addition, the sharp increase in the oil price in 2007–2008 did not give rise to serious, sustained social unrest. This supports the belief that the cost of energy is not perceived as very high, and since then it has even come down a lot. This all means that there is room for safe climate policy, which will undoubtedly increase the price of energy. Admittedly, future rises in the oil price weaken this argument. Note, however, that high oil prices are no substitute for climate policy as they are likely to stimulate a worldwide shift to coal, the combustion of which contributes considerably more to enhanced global warming than the combustion of oil (per unit of useful energy generated). Furthermore, if the price of fossil fuel energy goes up due to climate policy, this will also increase the cost of renewable energy since the production of the latter depends on inputs of fossil fuel energy. In other words, the environmental gains of (endogenous) increases in fossil fuel prices should not be overestimated.

All in all, higher energy prices are feasible. It is likely that the economic system and the consumer (in developed countries) can handle a fair amount of increase in energy costs due to climate policy without serious social repercussions.

12.5.7 PERSPECTIVE 7: STIMULATING A FUNDAMENTAL SOCIAL–TECHNICAL TRANSITION

Combating climate change is not about installing a one-time solution with a fixed cost. It is better conceptualized as balancing investment and R&D for many years to come: too much R&D will mean waiting too long for effective investments in reducing emissions; too little R&D will mean investing too quickly in less than mature—environmentally ineffective or overly expensive—technologies. The right balance has been cast in the literature as the exploration-versus-exploitation (March 1991) and optimal diversity problem (van den Bergh 2008b). The long-term dimension of investments and R&D is important, as R&D will take time before new technologies can diffuse worldwide on a significant scale. In the meantime investments will go to carbon-intensive fossil fuel technology and infrastructure+. The long-term perspective is also needed for making good choices about nuclear and carbon capture and storage: are they suitable as transition technologies or should we intend to move them further on their learning curves? A long-term angle will further affect choices between costly investments in energy efficiency improvements and renewable energy. Details of all these options were already discussed under Section 5.1.

Climate change policy is not a simple, one-dimensional policy or an instrument with a clear cost, rather a complex process of multilevel andmultidimensional change involving the unlocking of a dominant, undesirable system of fossil fuel technologies and infrastructures, and changing institutions, incentives, knowledge bases, and international cooperation. Very likely, a mixture of general policy principles is needed, notably: regulation of externalities (environment foremost), resource policies (prices reflecting real scarcity), innovation policies (including public investments), and specific unlocking policies (e.g., subsidy programs). A growing group of researchers is calling this approach a "social-technical transition to sustainability". It recognizes that in due time a stringent climate policy will lead to structural changes in the economy, including technological innovations and alterations in sector structure, demand side patterns, products types and designs, and institutional arrangements. Such qualitative changes are not well captured in one-dimensional monetary indicators, be it cost measures or foregone GDP growth.

Against this background, Prins and Rayner (2007) argue in favor of "placing investment in energy R&Don a wartime footing". Earlier, former US Vice-President and Nobel Peace laureate Al Gore made a similar call for a "global Marshall Plan". Various others have referred to the Manhattan Project and New Green Deal in this context. Even the US Bush administration expressed interest in directly stimulating energy-related R&D—rather than implementing stringent environmental (climate) regulation. Sufficient R&D on de-carbonized energy technologies and a transition to sustainable energy technologies are indeed not guaranteed by environmental regulation alone. One important reason is the lock-in features of fossil fuel energy and related technologies like vehicles with combustion engines.

Case studies of historical transitions show that a number of conditions need to be met for a transition to occur (Geels 2005). One of these is public investment in infrastructure and basic (fundamental) research. The history of nuclear fission shows this clearly; it received strong support through direct subsidies and military R&D (in the USA). Several other technologies have benefited greatly from public R&D, particularly investments in military R&D. Notable in this respect are information and communication technologies (ICT), supporting technologies like solid state electronics, semi-conductors, transistors, integrated circuits, data transmission networks, and of course basic software codes. All these have received massive funding from the (American) military complex, usually with the motivation of the Cold War. In many countries, agriculture also has received a great deal of public support, both to maintain the status quo (protection) and to foster certain transitions (Green revolution). For example, the post-war transition in Dutch agriculture was extensively funded by the government through investment subsidies, financial compensation for taking out land, public investment in land consolidation, and the creation and maintenance of drainage systems. This was motivated by a strong urge to achieve food security and self-sufficiency. Similarly, if one recognizes a stable climate as a basic condition for human life and activity, one needs to seriously invest in it. But perhaps reducing dependence on imported oil is a more effective motivation for fundamental changes in the energy sector. This is in any case illustrated by the transition to nuclear energy in France during the 1970s (after the oil crisis), and more recently by government support for bio-ethanol agriculture in the US.

12.5.8 PERSPECTIVE 8: BEHAVIOR, LEARNING AND SUBSTITUTION

Closely related to the previous transition perspective is a behavioral perspective. Economists are generally optimistic about prices as signals of scarcity that stimulate appropriate changes in households' and firms' behavior. Their model assumptions may well underestimate individuals' actual responses to stringent climate policy. More generally, many substitution opportunities at the level of inputs, sectors, and demand are insufficiently recognized by existing models because of aggregation and limits of empirical data. Notably, stringent climate policymoves prices outside ranges historically observed, so that, for instance, the empirical price elasticities of demand may underestimate potential responses. Aggregation is relevant as shown by metaanalyses of different model studies, which indicate particularly that inclusion of more fuel types and energy technologies leads to lower cost estimates (Söderholm 2007, Table 4.2). Generally, the more substitution opportunities exist, the easier it is for systems to adapt in a way so as to reach a similar performance level without much additional cost. Moreover, models often do not reflect the fact that in the long run people can change fundamental choices that affect their energy use. For example, road pricing (or toll roads) is often resisted on the basis that it would make life very hard for many car users. However, in practice people can respond in very many ways to a higher (variable) cost of driving a car: changing the time they drive (outside peak hours), carpooling, using other means of transport (walking, biking, public transport), traveling less, being more efficient in combining trips, and in the longer run changing jobs or houses to reduce commuting distances.

Behavior is not adequately dealt with in current climate policy analyses, since they mostly assume that agents behave rationally in the sense of perfectly maximizing utility or profits (Laitner et al. 2000; Gowdy 2008; Brekke and Johansson-Stenman 2008). However, people may act as citizens or as consumers, which are more characterized by habits, imitation, social pressure (in terms of both status and conformity), cooperation, and altruism, whereas firms may be better described as showing routinelike behavior (van den Bergh et al. 2000). Moreover, one should recognize the diversity of behaviors within both consumer and producer populations. In fact, some consumers show a great deal of altruism, citizenship, and solidarity with the future. Current studies are inaccurate as they insufficiently reflect the diversity and bounded rationality of behavior. Based on a review of economic and psychological studies of environmental behavior by households with regard to energy, water, and waste, van den Bergh (2008a) finds that existing econometric-statistical empirical studies entail mostly an incomplete assessment of the motivations and factors behind behaviors like waste collection, energy conservation, and a prudent use of water. In particular, integrated studies of economic and psychological factors are rare. However, at the same time the statistical findings of such studies—notably estimated price and income elasticities—often form the basis for more complex (partial or general) equilibriumtype economic analyses in the context of climate policy studies. All in all, it is very likely that substitution opportunities are not well represented in current climate policy studies.

A particular aspect of the behavior of firms and individuals is learning and innovation. Sagar and van der Zwaan (2006) examine learning-by-doing in relation to renewable energy and note various learning mechanisms: at the individual worker level (education, learning-by-operating so as to develop tacit skills), within a firm (learning-by-manufacturing), within the industry (learning by copying), across different industries, and within supply-demand interactions (learning-by-implementing, such as integrating PV systems into buildings, on roofs, which involves institutional structures such as for financing and equipment maintenance). Feedback from users to producers and from products to processes, along with systemic improvements (adjustment of all elements, such as institutions, markets, integrated building components, production chain) lead to falling overall costs of the renewable energy technology. Generally, the literature shows that adding endogeneity of growth, i.e. R&D or learning instead of ex-ogenous technological change, reduces policy cost estimates (Söderholm 2007).

Finally, some types of bounded rationality may lead to higher estimates for certain policy cost categories than the rational agent assumption. The energy gap literature illustrates this. Firms do not always invest in profit-able energy conservation opportunities for various reasons. One is that agents do not have full information; another is that they do not minimize overall costs but instead focus on what they regard as main activities or investments, which does not include energy conservation (DeCanio 1998); and habitual behavior has also been suggested as an explanation. Informa-tion provision and other strategies to stimulate more rational responses as part of climate policy may increase energy conservation (rebound effects not considered) and thus reduce the cost of effective policy.Agood trans-lation of insights from behavioral to environmental, energy, and climate economics is currently lacking and would be needed to shed more light on these issues (Gowdy 2008; Brekke and Johansson-Stenman 2008).

12.5.9 PERSPECTIVE 9: ANCILLARY BENEFITS

As discussed in Section 2, CBA studies of climate policy have omit-ted many benefits or avoided cost categories. The euphemistic term em-ployed for some of these is ancillary benefits or co-benefits of policy. One that has received ample attention is that the reduction of GHGs generated by fossil fuel combustion will sometimes go along with reduc-tions in other emissions, notably acidifying substances (nitrogen oxides and sulfur dioxide). For example, HEAL (2008) estimates that if the Eu-ropean Union raised its GHG emission target from the current 20% to 30% (in line with IPCC recommendations), then additional co-benefits in the range of e6.5 to 25 billion per year would result from health sav-ings arising from an associated reduction in emissions of fine particles, nitrogen oxide, and sulfur dioxide.

All avoided cost categories in CBA studies of climate policy can be regarded as ancillary benefits. An important one is avoidance of human conflict due to climate change. Such conflict would be stimulated, for example, if climate change causes water to become scarcer and agriculture to loose productivity, which may result in increasing land pressure and migration. Although it is difficult to definitely prove, theDarfur crisis has been attributed to less rainfall due to climate change. Population pressure in Sahel countries like Sudan may already have gone beyond the carrying capacity in view of low agricultural productivity, making these countries extremely vulnerable to even slow changes in climate conditions.

Another category of ancillary benefits of climate policy is omitted large-scale biodiversity loss. This is enhanced by shifting climate zones from which certain species cannot escape. Synergy between multiple causes of biodiversity loss—including also overexploitation, hunting, fragmentation due to land use and road infrastructure, invasion of exotic species, and environmental pollution—adds to the direct or pure impact of climate change on biodiversity loss.

The strong connection between scarce fossil fuel resources and greenhouse gas emissions from combusting fossil fuels also creates a relevant co-benefit. Notably, solving emissions problems by creating new sources of energy (renewable) will mean reducing problems of energy resource scarcity, avoiding potential fierce oil peak shocks, enhancing energy security, and avoiding conflicts over scarce energy resources. For example, a study assessing the social cost of the OPEC oil cartel to the US identified four cost categories, namely wealth transfer to OPEC, cost of strategic petroleum reserve, total GNP loss due to price shocks and shortages, and military costs. This resulted in an estimated cost ranging from about US\$150 to 400 billion per year (1990\$) during the period 1974–1985 (Green and Leiby 1993).

Ancillary benefits further arise when adaptation options are being created as a result of mitigation activities. One example is that planting forests to capture CO_2 will in turn allow for protection of biodiversity, water regulation, and reduced vulnerability to flooding or storms.

Finally, reducing GHG emission through taxes generates tax revenues which can be used to reduce distorting taxes on capital and especially labor. Even though the debate on the double dividend of shifting taxes from labor to environment suggests that one should not hope for too large effects of this kind, some positive effects are likely. Notably, the employment benefits due to fewer tax distortions in labor markets are robust, even though in welfare terms they are considerably smaller than the associated environmental benefits (de Mooij 1999; Patuellia et al. 2005).

An argument against considering certain ancillary benefits in climate policy evaluation is that they might have been achieved more directly and cheaply by specific, appropriate policies (e.g., acid rain policy). However, whenever the side effects are an inevitable consequence of the climate policy under consideration, one can regard them as efficient and relevant to the evaluation of this policy.

12.5.10 PERSPECTIVE 10: UPWARD BIAS IN EX ANTE ESTIMATES OF REGULATION COST

Various studies indicate that there is often a gap and sometimes even a large gap between ex ante and ex post estimates of the costs of environmental regulation, including both private and public-administrative costs (Harrington et al. 1999). MacLeod et al. (2009) find this for a wide range of environmental policies in European countries, including policies aimed at water and air pollution, health, food safety, fuel standards, directives on combustion plants, and animal welfare. There are two important reasons why ex ante cost assessments may deliver overestimates. First, information on actual costs is often provided by firms having an interest or stake. As a result, those being regulated may provide overly high estimates of individual abatement costs. This can be due to strategic behavior to resist implementation of stringent regulations, or simply to individual uncertainty about (future) abatement costs. Standard environmental economics somehow recognizes these problems, regarding price regulation as having the advantage that it decentralizes the problem of environmental regulation, and not requiring governments to have full information about pollution abatement technologies and associated costs (Baumol and Oates 1988). A second reason for ex ante overestimates is that they may neglect or underrate the potential for reduction of abatement costs through polluters' innovation, learning, and adaptation. As this was already discussed in Sections 4 and 5.8, we will not enter into further details here.

12.5.11 PERSPECTIVE 11: INTERNATIONAL COOPERATION AND AGREEMENTS

An additional important factor influencing cost estimates of climate policy is the presence (or absence) of international agreements, or more generally international cooperation between countries on climate policy and related technological diffusion. If international agreements are absent or weakly constrain individual countries, vast differences in policy may exist between countries. As a result, the costs of stringent climate policy for industries or consumers may be high since it will mean a loss in the international competitive position of industries as well as leakage of emissions from countries with stringent to those with less stringent policies. Instead, a stringent climate policy agreed upon by all countries in the world would mean a level playing field that reduces the policy cost, as competitive disadvantages and emission spillover is avoided (Neuhoff 2006).

The relationship between policy cost and international cooperation is like a vicious circle.As long as governments think that the cost of safe climate policy is high, they will refrain from committing themselves to a stringent international climate agreement. However, as long as such an agreement is lacking, the cost of unilaterally stringent climate policy will be excessively high because of the loss of competitive position. One way out is to design clever strategies in negotiations for international agreements (Barrett 2007). Note in this respect that the Kyoto agreement does not count as a stringent agreement and as a result is quite ineffective (McKibbin and Wilcoxen 2002): the Kyoto limits are far removed from what is needed to stabilize the CO_2 concentration in the atmosphere (at any close to a safe level); they entailed no restrictions whatsoever for Germany (unification) and Russia (economic collapse); and they do not bind all developing countries or the largest economy on the planet (USA).

Brekke and Johansson-Stenman (2008) argue that taking into account social preferences (a form of bounded rationality) implies that successful international agreements are estimated as being more likely than when assuming purely selfish motives. This may involve altruistic rewards and punishment, reciprocity, the greater altruistic capacity of teams than of individuals (think of climate negotiating teams), shame, and citizen (voting) rather than consumer behavior. Unfortunately, bounded rationality may also involve opposite effects. Brekke and Johansson-Stenman mention as an example 'cognitive dissonance', i.e. inconsistency between beliefs and behaviors causing an uncomfortable psychological tension which can sometimes lead to a change in beliefs rather than behavior, so the two match up. With regard to climate change, some people who (indirectly) cause many GHG emissions may show such 'cognitive dissonance' by denying or playing down the facts and risks of climate change.

12.5.12 PERSPECTIVE 12: LACK OF INSURANCE AGAINST CLIMATE CHANGE

Currently, private insurance with premiums that reflect the risk of extreme events like those possibly caused by climate change, such as flooding and hurricanes, is largely lacking in most countries (Botzen and van den Bergh 2008). This has three consequences for judging the cost of climate policy. First, it means that there is no efficient sharing of climate-related risks which would reduce the overall costs of the consequences of both climate change and climate policy. Second, the absence of insurance means that appropriate incentives for adequate adaptation to climate risks and changes is lacking. Third, it also means disoptimal incentives for stimulating producers, consumers, (re)insurance companies, and even governments to efficiently reduce greenhouse gas emissions. At present, insurers are already actively involved in promoting reductions in greenhouse gas emissions (Botzen et al. 2009). Such efforts are likely to become stronger if more climate change risks were covered through private insurance. Both insured and insurers have incentives to limit climate risk in case increases in the frequency and severity of natural hazards are reflected in a higher cost of offering insurance and higher premiums. Moreover, with insurance, adaptation at the individual and social level will be more adequate so that climate mitigation policy may need to be less stringent and thus less expensive. In other words, with adequate insurance arrangements in the face of climate-related risks, safe climate mitigation policies will turn out to be more efficient, i.e. less expensive. This is especially true since climate insurance would imply many indirect economic effects because insurance affects the direct and indirect costs of economic activities and therefore works as a price signal of risk. If climate policy is undertaken in the presence of adequate insurance arrangements for risks related to climate change, or if such a policy includes incentives for insurance companies to undertake these arrangements, then the cost of climate policy will be lower than without such arrangements. This is likely not a large effect, but for completeness one should take it into consideration.

Table 1 summarizes the perspectives.

12.6 CONCLUSIONS

This paper has argued that both cost–benefit analysis and cost assessment or accounting of climate policy using quantitative models are overly ambitious, despite the fact that we can evidently learn much from them. The multi-perspective approach to evaluating the cost of a safe, precautionary climate policy as presented here can be regarded as a way out of the never-ending debate on the usefulness and feasibility of cost–benefit analyses of climate policy. Indeed, if climate policy is seen as a precautionary strategy to avoid unpredictable and irreversible natural as well as economic catastrophes rather than as a way to optimize social welfare (or GDP growth) in the face of GHG emission–climate–economic damage feedback, then a focus on qualitative risk analysis and cost assessment of climate policy makes more sense than a quantitative cost–benefit analysis. This is true both for methodological reasons—CBA possibly represents an overly risk-loving decision-maker—and for practical reasons—quantification of extreme events with small probabilities simply is not feasible.

The paper has tried to credibly defend, using various arguments, that a safe or precautionary approach to climate policy is indeed rational. The twelve perspectives trying to assess the cost of a safe climate policy together provide a strong case for the view that such a policy is affordable and cheaper than most previous studies have suggested. This is subject to the usual conditions and provisos: nothing is certain when talking about such a complex issue as climate change and climate policy. Moreover, since 'expensive' is a relative concept, various perspectives have precisely aimed at putting the cost in a particular context, such as comparing it with the GDP of OECD countries or with expenditures on other large public projects, arguing that (national) expenditures on energy are not very high, and interpreting climate policy from the angle of human happiness rather than economic (GDP) growth.

The happiness or subjective well-being perspective on the cost of climate policy emerges as possibly the most important new view. It is pertinent to introduce it into the debate on climate policy to arrive at a correct picture of what we really gain and sacrifice if we undertake a stringent, safe climate policy worldwide. The discussion of this perspective showed that the implications may be quite different for rich and poor countries. Climate policy may have a cost in the sense of slowing down the rate of GDP growth, thus reaching a given level of GDP just a few years later in the distant future (Section 5.3). Happiness research (Section 5.4) indicates that this is not worrisome at all. Four considerations are important here. First, GDP is not a reliable measure of welfare or happiness, notably for rich countries. Second, climate policy to avoid extreme events means preventing serious reductions in happiness. Such reductions are not captured or are insufficiently captured by GDP analyses as these events involve many non-market effects related to extreme and highly uncertain impacts on resource availability (clean water), human health, vulnerability of poor people in regions with low productivity, migration, and violent conflicts. Third, climate change for people in poor countries may mean that their basic needs will be threatened. Fourth, climate policy concerns a period of many decades to several hundreds of years in the future, during which the GDP of rich countries will certainly have grown far beyond any welfare or happiness maximizing level. Therefore, in terms of happiness or real welfare, climate policy looks much less expensive than in terms of lost GDP, while climate change causing catastrophes may be evaluated as much more expensive in terms of happiness than in terms of GDP.

Finally, on the basis of various quantitative indicators it was argued that energy is currently not very expensive (Section 5.6), so there is considerable leeway for increasing its price through climate policy. Indeed, an effective and safe climate policy cannot avoid raising energy prices considerably, certainly if one wants to reduce the rebound effects of energy conservation and efficiency improvements and stimulate a transition to renewable energy sources.

Of course, while the costs of a safe climate policy may be manageable at global and national levels, as argued here, such a policy will pose serious challenges for particular economic sectors. But this is entirely logical and acceptable, since higher energy costs will regulate and restructure the economy and affect energy-intensive products, processes, firms, and industries relatively severely. Higher energy prices and costs will thus set into motion a process of creative destruction, which is an inevitable component in the transition to a low-carbon economy. Postponing such a transition will only make it more expensive, while safe levels of atmospheric GHG concentration will get out of reach. In other words, the optimal timing of a safe climate policy is right now.

REFERENCES

1.   Ackerman F, Finlayson IJ (2007) The economics of inaction on climate change: a sensitivity analysis. Climate Policy 6(5):509–526
2.   Ackerman F, DeCanio SJ, Howarth RB, Sheeran K (2009) Limitations of integrated assessment models of climate change. Clim Change 95:297–315. http://www.springerlink.com/content/ c85v5581x7n74571/fulltext.pdf
3.   Alcamo J, Shaw R, Hordijk L (eds) (1990) The RAINS model of acidification. In: Science and strategies in Europe. Kluwer, Dordrecht, The Netherlands
4.   Anderson K, Bows A (2008) Reframing the climate change challenge in light of post-2000 emission trends. Philos Trans R Soc A 366(1882):3863–3882
5.   Arrow KJ (1973) Rawls's principle of just saving. Swed J Econ 75:323–335
6.   Arrow KJ (2007) Global climate change: a challenge to policy. The Economists' voice 4(3):Article 2. Available at: http://www.bepress.com/ev/vol4/iss3/art2
7.   Arrow KJ, Fisher AC (1974) Environmental preservation, uncertainty, and irreversibility. Q J Econ 88(2):312–319
8.   Ayres RU, Walters J (1991) Greenhouse effects: damages, costs and abatement. Environ Resour Econ 1:237–270
9.   Azar C (1998) Are optimal CO2 emissions really optimal? Four critical issues for economists in the greenhouse. Environ Resour Econ 11(3–4):301–315
10. Azar C, Lindgren K (2003) Catastrophic events and stochastic cost–benefit analysis of climate change. Clim Change 56(3):245–255
11. Azar C, Schneider SH (2002) Are the economic costs of stabilising the atmosphere prohibitive? Ecol Econ 42:73–80

12. Azar C, Schneider SH (2003) Are the economic costs of stabilising the atmosphere prohibitive? A response to Gerlagh and Papyrakis. Ecol Econ 46:329–332
13. Bala G, Caldeira K, Mirin A, WickettM, Delire C (2005) Multicentury changes in global climate and carbon cycle: results from a coupled climate and carbon cycle model. J Clim 18:4531–4544
14. Barker T (1996) A review of managing the global commons: the economics of global change, by WD Nordhaus. Energy Environ 7(1):85–88
15. Barker T, Köhler J, Villena M (2002) The costs of greenhouse gas abatement: a meta-analysis of post-SRES mitigation scenarios. Environ Econ Policy Stud 5:135–166
16. Barker T, QureshiMS, Köhler J (2006) The costs of greenhouse gas mitigation with induced technical change: a meta-analysis of estimates in the literature. Report prepared for the HM Treasury Stern review on the economics of climate change. Available at www.sternreview.org.uk, Department of Land Economy, University of Cambridge, UK
17. Barrett S (2007) Proposal for a new climate change treaty system. The Economists' voice 4(3):Article 6. Available at: http://www.bepress.com/ev/vol4/iss3/art6
18. Baumol WJ, Oates WE (1988) The theory of environmental policy, 2nd edn. Cambridge University Press, Cambridge, UK
19. Blanchflower DG, Oswald AJ (2004) Well-being over time in Britain and the USA. J Public Econ 88(7–8):1359–1386
20. Bollen JC, Manders AJG, Veenendaal PJJ (2004) How much does a 30% reduction cost? Macroeconomic effects of post-Kyoto climate policy. CPB document 64, RIVM rapport 500035001
21. Botzen W, van den Bergh JCJM (2008) Insurance against climate change and flooding in The Netherlands: present, future and comparison with other countries. Risk Anal 28(2):413–426
22. Botzen W, van den Bergh JCJM (2009) Bounded rationality, climate risks and insurance: is there a market for natural disasters? Land Econ 85(2):266–279
23. Botzen WJW, Gowdy JM, van den Bergh JCJM (2008) Cumulative CO2 emissions: shifting international responsibilities for climate debt. Climate Policy 8:569–576
24. Botzen WJW, van den Bergh JCJM, Bouwer LM (2009) Climate change and increased risk for the insurance sector: a global perspective and an assessment for The Netherlands. Nat Hazards. http://www.springerlink.com/content/u237166571587112/fulltext.pdf
25. Brekke KA, Howarth RB (2002) Status, growth and the environment: goods as symbols in applied welfare economics. Edward Elgar, Cheltenham
26. Brekke KA, Johansson-Stenman O (2008) The behavioural economics of climate change. Oxf Rev Econ Policy 24(2):280–297
27. Brereton F, Clinch JP, Ferreira S (2008) Happiness, geography and the environment. Ecol Econ 65(2):386–396
28. Broome J (1992) Counting the cost of global warming. White House Press, Cambridge
29. Bryden H, Longwort H, Cunningham S (2005) Slowing of the Atlantic meridional overturning circulation at 25° N. Nature 438:655–657

30. Brzoska M (2008) Trends in global military and civilian (R&D) and their changing interface. Institute for Peace Research and Security Policy, University of Hamburg. Available at: http://www.ifsh.de/pdf/aktuelles/india_brzoska.pdf

31. Chichilnisky G (1996) An axiomatic approach to sustainable development. Soc Choice Welf 13(2):219–248

32. Clark AE, Frijters P, Shields MA (2008) Relative income, happiness, and utility. J Econ Lit 46(1):95–144

33. Cline WR (1992) The economics of global warming. Institute for International Economics, Washington

34. ClineWR (2007) Comments on the Stern Review. In: Yale symposium on the Stern Review, chapter 6. http://www.ycsg.yale.edu/climate/forms/FullText.pdf

35. Cohen MD (1995) Risk aversion in concepts in expected- and non-expected-utility models. Geneva Pap Risk Insur, Theory 20:73–91

36. Cohen MA, Vandenbergh MP (2008) Consumption, happiness, and climate change. RFF discussion paper 08–39, Oct 2008. Resources for the Future, Washington, DC

37. Daily GC, Ehrlich PR, Mooney HA, Ehrlich AH (1991) Greenhouse economics: learn before you leap. Ecol Econ 4:1–10

38. Daly HE (1977) Steady-state economics. Freeman, San Francisco

39. Daly HE, Cobb J (1989) For the common good: redirecting the economy toward community, the environment, and a sustainable future. Beacon, Boston

40. Dasgupta P (2001) Human well-being and the natural environment. Oxford University Press, Oxford

41. Dasgupta P (2007) Commentary: the Stern Review's economics of climate change. Natl Inst Econ Rev 199:4–7

42. Dasgupta P, Maskin E (2005) Uncertainty and hyperbolic discounting. Am Econ Rev 95(4): 1290–1299

43. Davidson MD (2006) A social discount rate for climate damage to future generations based on regulatory law. Clim Change 76(1–2):55–72

44. de Groot HLF, Nahuis R, Tang PJG (2004) Is the American model miss world? Choosing between the Anglo-Saxon model and a European-style alternative. CPB discussion paper 40, Centraal Planbureau, Den Haag

45. de Mooij RA (1999) The double dividend of an environmental tax reform. In: van den Bergh JCJM (ed) Handbook of environmental and resource economics. Edward Elgar, Cheltenham

46. DeCanio SJ (1998) The efficiency paradox: bureaucratic and organizational barriers to profitable energy-saving investments. Energy Policy 26(5):441–454

47. DeCanio SJ (2003) Economic models of climate change: a critique. Palgrave-Macmillan, New York

48. Diecidue E, Wakker PP (2001) On the intuition of rank-dependent utility. J Risk Uncertain 23(3):281–298

49. Dietz S, Anderson D, Stern N, Taylor C, Zenghelis D (2007) Right for the right reasons: a final rejoinder on the Stern Review. World Econ 8(2):229–258

50. Easterlin RA (1974) Does economic growth improve the human lot? Some empirical evidence. In: David PA, Reder MW (eds) Nations and households in economic growth: essays in honour of Moses Abramowitz. Academic, New York

51. Easterlin RA (2001) Income and happiness: towards a unified theory. Econ J 111:465–484

52. Easterling DR, Meehl GA, Parmesan C, Changnon SA, Karl TR, Mearns LO (2000) Climate extremes: observations, modeling, and impacts. Science 289(5487):2068–2074

53. EIA (2008) Annual energy outlook 2008. Energy Information Administration, Washington, DC. Available at http://www.eia.doe.gov/oiaf/aeo/

54. Fankhauser S, Tol RSJ (1996) Climate change costs–recent advancements in the economic assessment. Energy Policy 24:665–673

55. Fankhauser S, Tol RSJ (1997) The social costs of climate change: the IPCC second assessment report and beyond. Mitig Adapt Strategies Glob Chang 1:385–403 Climatic Change (2010) 101:339–385 381

56. Fargione J, Hill J, Tilman D, Polasky S, Hawthorne P (2008) Land clearing and the biofuel carbon debt. Science 319:1235–1238

57. Ferrer-i-Carbonell A (2005) Income and well-being: an empirical analysis of the comparison income effect. J Public Econ 89(5–6):997–1019

58. Ferrer-i-Carbonell A, Frijters P (2004) How important is methodology for the estimates of the determinants of happiness? Econ J 114:641–659

59. Ferrer-i-Carbonell A, Gowdy JM (2008) Environmental degradation and happiness. Ecol Econ 60(3):509–516

60. Fisher C, Morgenstern RD (2006) Carbon abatement costs: why the wide range of estimates? Energy J 27(2):73–86

61. Fouquet R, Pearson PJG (2006) Seven centuries of energy services: the price and use of light in the United Kingdom (1300–2000). Energy J 27(1):139–177

62. Frank RH (1985) Choosing the right pond: human behavior and the quest for status. Oxford University Press, New York

63. Frank RH (2004) Positional externalities cause large and preventable welfare losses. Am Econ Rev 95(2):137–141

64. Frederick S, Loewenstein G (1999) Hedonic adaptation. In: Kahneman D, Diener E, Schwartz N (eds.) Well-being: the foundations of hedonic psychology. Russell Sage Foundation, New York, pp 302–329

65. Frederick S, Loewenstein G, O'Donoghue T (2002) Time discounting and time preference: a critical review. J Econ Lit 40:351–401

66. Frey BS, Stutzer A (2002) What can economists learn from happiness research? J Econ Lit 40: 402–435

67. Frijters P, van Praag BMS (1998) The effects of climate on welfare and wellbeing in Russia. Clim Change 39:61–81

68. Frondel M, Peters J, Vance C (2008) Identifying the rebound: evidence from a German household panel. Energy J 29(4):145–163

69. Fudenberg D, Levine DM (2006) A dual-self model of impulse control. Am Econ Rev 96:1449–1476

70. Funtowicz S, Ravetz JR (1993) Science for the postnormal age. Futures 25:739–755

71. Geels FW (2005) Technological transitions and system innovations: a co-evolutionary and sociotechnical analysis. Edward Elgar, Cheltenham

72. Gollier C, Jullien B, Treich N (2000) Scientific progress and irreversibility: an economic interpretation of the precautionary principle. J Public Econ 75:229–253

73. Gowdy JM (2008) Behavioral economics and climate change policy. J Econ Behav Organiz 68: 632–644
74. Green DL, Leiby PN (1993) The social costs to the US of monopolization of the world oil market 1972–1991. Oak Ridge National Laboratory, Report 6744, Oak Ridge, TN
75. Greening LA, GreeneDL, Difiglio C (2000) Energy efficiency and consumption—the rebound effect: a survey. Energy Policy 28:389–401
76. Guo JK, Hepburn C, Tol RSJ, Anthoff D (2006) Discounting and the social cost of carbon: a closer look at uncertainty. Environ Sci Policy 9(5):203–216
77. Ha-Duong M (1998) Quasi-option value and climate policy choices. Energy Econ 20(5–6):599–620
78. Harrington W, Morgenstern RD, Nelson P (1999) On the accuracy of regulatory cost estimates. Resources for the Future, Washington
79. Hartley K (2006) The costs of war. Centre for Defence Economics, The University of York, UK. http://www.york.ac.uk/depts/econ/documents/research/iraq_conflict_vox_june2006.pdf
80. Hawellek J, Kemfert C, Kremers H (2007) A quantitative comparison of economic cost assessments implementing the Kyoto protocol. Department of Economics and Statistics, Carl von Ossietzky Universität Oldenburg, Oldenburg
81. HEAL (2008) The co-benefits to health of a strong EU climate change policy. Health and Environment Alliance (HEAL), Climate Action Network Europe (CAN-E) and WWF. Available at http://www.climnet.org/Co-benefits%20to%20health%20report%20-september%202008.pdf
82. Heal G, Kriström B (2002) Uncertainty and climate change. Environ Resour Econ 22(1–2): 3–39
83. Helliwell J (2003) How's life? Combining individual and national variations to explain subjective well-being. Econ Model 20:331–360
84. Hirsch F (1976) Social limits to growth. Harvard University Press, Cambridge
85. Holling CS (1986) The resilience of terrestrial ecosystems: local surprise and global change. In: Clark WC, Munn RE (eds) Sustainable development of the biosphere. Cambridge University Press, Cambridge
86. Hueting R (1974) Nieuwe Schaarste and Economische Groei. Elsevier, Amsterdam (English edition 1980, New Scarcity and Economic Growth, North-Holland, Amsterdam)
87. IPCC (2007) Fourth assessment report. Intergovernmental Panel on Climate Change, Geneva
88. Ireland NJ (2001) Optimal income tax in the presence of status effects. J Public Econ 81:193–212
89. Jaccard M, Nyboer J, Bataille C, Sadownik B (2003) Modeling the cost of climate policy: distinguishing between alternative cost definitions and long-run cost dynamics. Energy J 24(1):49–73
90. Johansson P-O (1987) The economic theory and measurement of environmental benefits. Cambridge University Press, Cambridge
91. Jorgenson DW, Goettle RJ, Wilcoxen PJ, Ho MS (2008) The economic costs of a market-based climate policy. DW Jorgenson Associates for Pew Center on Global Climate Change, Arlington

92. Kahneman D, Tversky A (1979) Prospect theory: an analysis of decision under risk. Econometrica 47(2):263–291
93. Kahneman D, Krueger A, Schkade D, Schwarz N, Stone A (2004) Toward national well-being accounts. Am Econ Rev Pap Proc 94:429–434
94. Kasting J (1998) The carbon cycle, climate, and the long-term effects of fossil fuel burning. Consequences 4:15–27
95. Kelly DL, Kolstad CD (1999) Integrated assessment models for climate change control. In: Folmer H, Tietenberg T (eds) The international yearbook of environmental and resource economics 1999/2000. Edward Elgar, Cheltenham
96. Kimball MS (1990) Precautionary saving in the small and in the large. Econometrica 58:53–73
97. Kolstad C (1996) Learning and stock effects in environmental regulations: the case of greenhouse gas emissions. J Environ Econ Manage 31:1–18
98. Kuik O, Brander L, Tol RSJ (2008) Marginal abatement costs of carbon-dioxide emissions: a metaanalysis. ESRI working paper no 248, Economic and Social Research Institute, Dublin, Ireland
99. Kump L (2002) Reducing uncertainty about carbon dioxide as a climate driver. Nature 419:188–190
100. Laitner JA, DeCanio SJ, Peters I (2000) Incorporating behavioural, social, and organizational phenomena in the assessment of climate change mitigation options. In: Jochem E, Sathaye J, Bouille D (eds) Society, behaviour, and climate change mitigation. Kluwer, Dordrecht, pp 1–64
101. Lawn P, Clarke M (2008) Sustainable welfare in the Asia-Pacific: studies using the genuine progress indicator. Edward Elgar, Cheltenham
102. Layard R (2005) Happiness: lessons from a new science. Penguin, London
103. Llavador H, Roemer JE, Silvestre J (2008) Dynamic analysis of human welfare in a warming planet. Cowles Foundation discussion paper no 1673, Yale University
104. Loulou R, Kanudia A (1999) Minimax regret strategies for greenhouse gas abatement: methodology and application. Oper Res Lett 25:219–230
105. Lubowski RN (2008) The role of REDD in stabilizing greenhouse gas concentrations: lessons from economic models. Center for International Forestry Research (CIFOR), Bogor, Indonesia. http://www.cifor.cgiar.org/globalredd
106. Luechinger S (2007) Valuing air quality using the life satisfaction approach. https://editorialexpress. com/cgi-bin/conference/download.cgi?db_name=res2008&paper_id=680
107. Lynas M (2007) Six degrees: our future on a hotter planet. Fourth Estate/HarperCollins, London
108. MacLeod M, Ekins P, Vanner R, Moran D (eds) (2009) Understanding the costs of environmental regulation in Europe. Edward Elgar, Cheltenham
109. Maddison DJ (2003) The amenity value of the climate: the household production function approach. Resour Energy Econ 25:155–175
110. March JG (1991) Exploration and exploitation in organizational learning. Organ Sci 2:71–87
111. Maréchal K (2007) The economics of climate change and the change of climate in economics. Energy Policy 35:5181–5194

112. Margolis RM, Kammen DM (1999) Underinvestment: the energy technology and R&D policy challenge. Science 285(5428):690–692

113. Matthews HD, Caldeira K (2008) Stabilizing climate requires near-zero emissions. Geophys Res Lett 35:1–5

114. McKibbin WJ, Wilcoxen PJ (2002) The role of economics in climate change policy. J Econ Perspect 16(2):107–129

115. McKinsey & Company (2009). Pathways to a low-carbon economy. New York. https://solutions. mckinsey.com/climatedesk/CMS/Default.aspx

116. Meinshausen M (2006) What does a 2∘C target mean for greenhouse gas concentrations? A brief analysis based on multi-gas emission pathways and several climate sensitivity uncertainty estimates. In: Schellnhuber HJ, Cramer W, Nakicenovic N, Wigley T, Yohe G (eds) Avoiding dangerous climate change. Cambridge University Press, Cambridge

117. Messner S (1997) Endogenized technological learning in an energy systems model. J Evol Econ 7:291–313

118. Mishan EJ (1967) The cost of' economic growth. Staples, London

119. Montenegro A, Brovkin V, Eby M, Archer D, Weaver AJ (2007) Long term fate of anthropogenic carbon. Geophys Res Lett 34:L19707. doi:10.1029/2007GL030905

120. Nemet GF (2008) Interpreting interim deviations from cost projections for publicly supported energy technologies. Working paper no 2008-012, La Follette School of Public Affairs, University of Wisconsin, Madison. http://www.lafollette.wisc.edu/publications/workingpapers

121. Neuhoff K (2006) Where can international cooperation support domestic climate policy? EPRG working paper 06/27, Faculty of Economics, University of Cambridge

122. Neumayer E (1999) Global warming: discounting is not the issue, but substitutability is. Energy Policy 27:33–43

123. Ng Y-K (2003) From preference to happiness: towards a more complete welfare economics. Soc Choice Welf 20:307–350

124. Nordhaus WD (1977) Economic growth and climate: the case of carbon dioxide. Am Econ Rev 67(1):341–346

125. Nordhaus WD (1991) To slow or not to slow? The economics of the greenhouse effect. Econ J 101:920–937

126. Nordhaus WD (2007) Comments on the Stern review. In: Yale symposium on the Stern review, chapter 5. http://www.ycsg.yale.edu/climate/forms/FullText.pdf

127. Nordhaus WD, Boyer J (2000) Warming the world: economic models of global warming. The MIT Press, Boston

128. Nordhaus WD, Tobin J (1972) Is growth obsolete? Economic Growth. 50th anniversary colloquium V. In: Moss M (ed) Columbia University Press for the National Bureau of Economic Research, New York. Reprinted in The Measurement of Economic and Social Performance, Studies in Income and Wealth, vol 38. National Bureau of Economic Research, 1973

129. OECD (2006) Going for growth. OECD, Paris

130. OECD (2008) OECD in figures 2007. OECD, Paris. www.oecd.org/infigures

131. Padilla E (2004) Climate change, economic analysis and sustainable development. Environ Values 13:523–544

132. Patuellia R, Nijkamp P, Pels E (2005) Environmental tax reform and the double dividend: a metaanalytical performance assessment. Ecol Econ 55(4):564–583
133. PBL (2008) Milieubalans 2008. Netherlands Environmental Assessment Agency, Bilthoven
134. Polimeni JM, Mayumi K, Giampietro M, Alcott B (2008) The Jevons paradox and the myth of resource efficiency improvements. Earthscan, London
135. Prins G, Rayner S (2007) Time to ditch Kyoto. Nature 449:973–975
136. Quiggin J (2008) Stern and his critics on discounting and climate change: an editorial essay. Clim Change 89(3–4):195–205
137. Rawls J (1972) A theory of justice. Harvard University Press, Cambridge
138. Rehdanz K, Maddison DJ (2004) The amenity value of climate to German households. FEEM working paper no 57.04, March 2004, FEEM, Venice
139. Rehdanz K, Maddison DJ (2005) Climate and happiness. Ecol Econ 52:111–125
140. Reilly J, Stone PH, Forest CE, Webster MD, Jacoby HD, Prinn RG (2001) Uncertainty and climate change assessments. Science 293(5529):430–433
141. Repetto R, Austin D (1997) The costs of climate protection: a guide for the perplexed. World Resources Institute, Washington
142. Rind D (1999) Complexity and climate. Science 284(5411):105–107
143. Roberts S (2008) Energy, equity and the future of the fuel poor. Energy Policy 36(12):4471–4474
144. Roland A (2001) The military-industrial complex. American Historical Association, Washington
145. Royal Society (2005) Ocean acidification due to increasing atmospheric carbon dioxide. Policy document 12/05. www.royalsoc.ac.uk/displaypagedoc.asp?id=13539
146. Sagar AD, van der Zwaan B (2006) Technological innovation in the energy sector:R&D, deployment and learning-by-doing. Energy Policy 34:2601–2608
147. Sandén BA, Azar C (2005) Near-term technology policies for long-term climate targets—economy wide versus technology specific approaches. Energy Policy 33:1557–1576
148. SchimmelpfennigD(1995) The option value of renewable energy: the case of climate change. Energy Econ 17(4):311–317
149. Schock RN, FulkersonW, Brown ML, San Martin RL, Greene DL, Edmonds J (1999) How much is energy research and development worth as insurance? Annu Rev Energy Environ 24:487–512
150. Scitovsky T (1976) The joyless economy. Oxford University Press, New York
151. Searchinger T, Heimlich R, Houghton RA, Dong F, Elobeid A, Fabiosa J, Tokgoz S, Hayes D, Yu T-H (2008) Croplands for biofuels increases greenhouse gases through emissions from land-use change. Science 319:1238–1240
152. Sen A (1976) Real national income. Rev Econ Stud 43(1):19–39
153. Söderholm P (2007) Modelling the economic costs of climate policy. Research report 2007-17, Department of Business Administration and Social Sciences, Luleå University of Technology
154. Solow RM (1974) Intergenerational equity and exhaustible resources. Rev Econ Stud 41:29–45
155. Sorrell S (2007) The rebound effect: an assessment of the evidence for economy-wide energy savings from improved energy efficiency. UK Energy Research Centre,

London. http://www.ukerc.ac.uk/ Downloads/PDF/07/0710ReboundEffect/0710Re boundEffectReport.pdf

156. Sorrell S (2009) Jevons' paradox revisited: the evidence for backfire from improved energy efficiency. Energy Policy 37(4):1456–1469

157. Spash CL (2002) Greenhouse economics. Routledge, London

158. Spash CL (2007) Climate change: need for new economic thought. Economic and Political Weekly 10th February, 2007, pp 483–490

159. Spash CL (2008) Deliberative monetary valuation (DMV) and the evidence for a new value theory. Land Econ 84(3):469–488

160. Stern N (2007) Findings of the Stern Review on the economics of climate change. Yale symposium on the Stern Review, chapter 1. http://www.ycsg.yale.edu/climate/ forms/FullText.pdf

161. Stern N, Peters S, Bakhshi V, Bowen A, Cameron C, Catovsky S, Crane D, Cruick- shank S, Dietz S, Edmonson N, Garbett S-L, Hamid L, Hoffman G, Ingram D, Jones B, Patmore N, Radcliffe H, Sathiyarajah R, Stock M, Taylor C, Vernon T, Wanjie H, Zenghelis D (2006) Stern Review: the economics of climate change. HM Treasury, London. http://www.hm-treasury.gov.uk/stern_review_climate_change.htm. (Later published as Stern N (2007) The economics of climate change: the Stern Review. Cambridge University Press, Cambridge)

162. Stiglitz JE, Bilmes L (2008) The three trillion dollar war: the true cost of the Iraq conflict. Norton, New York

163. Tol RSJ (2005) The marginal damage costs of carbon dioxide emissions: an assess- ment of the uncertainties. Energy Policy 33(16):2064–2074

164. Tol RSJ (2006) The Stern Review of the economics of climate change: a comment. Energy Environ 17(6):977–981

165. Tol RSJ (2008a) The social cost of carbon: trends, outliers, and catastrophes. Eco- nomics, the openaccess, open-assessment. E-Journal 2(25):1–24

166. Tol RSJ (2008b) The economic impact of climate change. Working paper no 255, September 2008. Economic and Social Research Institute, Dublin

167. Tol RSJ, Yohe GW (2007) A stern reply to the reply of the review of the Stern Re- view.World Econ 8(2):153–159

168. Ulph A, Ulph D (1997) Global warming, irreversibility and learning. Econ J 107:636–650

169. van Beers C, van den Bergh JCJM (2009) Environmental harm of hidden subsidies: global warming and acidification. AMBIO (in press)

170. van den Bergh JCJM (2004) Optimal climate policy is a utopia: from quantitative to qualitative cost– benefit analysis. Ecol Econ 48:385–393

171. van den Bergh JCJM (2008a) Environmental regulation of households? An empiri- cal review of economic and psychological factors. Ecol Econ 66:559–574

172. van den Bergh JCJM (2008b) Optimal diversity: increasing returns versus recombi- nant innovation. J Econ Behav Organ 68:565–580

173. van den Bergh JCJM (2009) The GDP paradox. J Econ Psychol 30(2):117–135

174. van den Bergh JCJM, Ferrer-i-Carbonell A, Munda G (2000) Alternative mod- els of individual behaviour and implications for environmental policy. Ecol Econ 32(1):43–61

175. van den Heuvel STA, van den Bergh JCJM (2009) Multilevel assessment of diversity, innovation and selection in the solar photovoltaic industry. Struct Chang Econ Dyn 20(1):50–60

176. van der Zwaan B, Rabl A (2003) Prospects for PV: a learning curve experience. Sol Energy 74(1): 19–31

177. van der Zwaan B, Rabl A (2004) The learning potential of photovoltaics: implications for energy policy. Energy Policy 32:1545–1554

178. van Praag B, Ferrer-i-Carbonell A (2004) Happiness quantified: a satisfaction calculus approach. Oxford University Press, Oxford

179. Weitzman M (2001) Gamma discounting. Am Econ Rev 91(1):260–271

180. Weitzman ML (2007) A review of the Stern Review on the economics of climate change. J Econ Lit 45(3):703–724

181. Welsch H (2006) Environment and happiness: valuation of air pollution using life satisfaction data. Ecol Econ 58(4):801–813

182. Woodward RT, Bishop RC (2000) Efficiency, sustainability and global warming. Ecol Econ 14: 101–111

There is one table that is not available in this version of the article. To view this additional information, please use the citation on the first page of this chapter.

Author Notes

CHAPTER 1

Acknowledgments

I would like to thank colleagues at Landcare Research, especially Phil Cowan, for discussions underlying the development of the proposed methodology, and Robbie Andrew, Anne Austin, Annette Cowie, Édouard Périé and Katsumasa Tanaka and anonymous reviewers for many useful and insightful comments on the manuscript.

CHAPTER 2

Funding

Financial support was provided by the Howard Hughes Medical Institute (Interfaces grant to JHB, JGO, and WZ), National Institutes of Health (grant T32EB009414 to JHB and JRB), National Science Foundation (grant OISE-0653296 to ADD and grant DEB-0541625 to MJH), and Rockefeller Foundation (grant to MJH). The content is solely the responsibility of the authors and does not necessarily reflect the views of the organizations listed above. The funders had no role in study design, data collection and analysis, decision to publish, or preparation of the manuscript.

Acknowledgments

This paper is a product of the Human Macroecology Group, an informal collaboration of scientists associated with the University of New Mexico and the Santa Fe Institute. It benefits from discussions with many colleagues on the important but controversial topic of sustainability. We thank Woody Woodruff, Matthew Moerschbaecher, and John Day for providing helpful comments on the manuscript.

CHAPTER 3

Financial & Competing Interests Disclosure

The authors have no relevant affiliations or financial involvement with any organization or entity with a financial interest in or financial conflict with the subject matter or materials discussed in the manuscript. This includes employment, consultancies, honoraria, stock ownership or options, expert testimony, grants or patents received or pending, or royalties.
No writing assistance was utilized in the production of this manuscript.

CHAPTER 4

Acknowledgements

The author would like to thank all the collaborators and co-authors of the all manuscripts mentioned in this paper, and Stephanie Seddon-Brown for English proof reading of this manuscript.

CHAPTER 5

Acknowledgements

The authors acknowledge the PRIN10-11 projects "Mechanisms of activation of CO_2 for the design of new materials for energy and resource efficiency" and "Innovative processes for the conversion of algal biomass for the production of jet fuel and green diesel" for the financial support.

CHAPTER 6

Acknowledgments

This work was made possible by funding from the Federal Aviation Administration (FAA), Air Force Research Laboratory (AFRL) and the Defense Logistics Agency-Energy (DLA Energy), under Project 47 of the Partnership for Air Transportation Noise and Emissions Reduction (PARTNER). The authors would like to thank Dr James I Hileman and Dr Mohan Gupta at the FAA for their guidance on technical matters. Any views or opinions expressed in this work are those of the authors and not the FAA, AFRL or DLA-Energy. We also thank the reviewers for their comments.

CHAPTER 7

Acknowledgments
Chih-Chun Kung would like to thank the financial support from the National Natural Science Foundation of China (#41161087; #41061049 #71173095; #71263018), National Social Science Foundation of China (#12&ZD213), China Postdoctoral Foundation (2013M531552) and University Social Science Project of Jiangxi (JJ1208). We also sincerely appreciate the great assistance and valuable comments from Bruce A. McCarl at Texas A&M University and Chi-Chung Chen at National Chung-Hsing University.

CHAPTER 8

Funding
This research was funded by the U.S. Department of Agriculture (USDA), Agricultural Research Service (ARS) including funds from the USDA-ARS GRACEnet effort, and partly by the USDA Natural Resources Conservation Service. The funders had no role in study design, data collection and analysis, decision to publish, or preparation of the manuscript.

Acknowledgments
Authors would also like to acknowledge the contribution of Edward J. Wolfrum and his staff at the National Renewable Energy Laboratory (Golden, CO) in conducting the corn stover analysis. Mention of trade names or commercial products in this publication does not imply recommendation or endorsement by the U.S. Department of Agriculture. USDA is an equal opportunity provider and employer.

Author Contributions
Conceived and designed the experiments: KV GV RF. Performed the experiments: KV GV RF RM MS VJ. Analyzed the data: MS. Contributed

reagents/materials/analysis tools: MS KV RM RF. Wrote the paper: MS KV GV RF RM VJ.

CHAPTER 9

Funding

This work is supported by the National Science Foundation Grant CBET-1137677. The funders had no role in study design, data collection and analysis, decision to publish, or preparation of the manuscript.

Acknowledgments

We would like to acknowledge the valuable suggestions of Drs. Chris Kucharik and Anita Thompson. This material is based upon work supported by the National Science Foundation Grant CBET-1137677. Any opinions, findings, and conclusions or recommendations expressed in this material are those of the author(s) and do not necessarily reflects the views of the National Science Foundation.

Author Contributions

Conceived and designed the experiments: RA CA SK. Performed the experiments: CA SK. Analyzed the data: SK DH. Contributed reagents/materials/analysis tools: SK. Wrote the paper: SK RA CA.

CHAPTER 11

Competing Interests

The authors declare that they have no competing interests.

Author Contributions

SA initiated the study and performed the literature review. LDL created the diagrams and performed the policy analysis. Both authors substantially contributed to the analysis and the drafting of the manuscript. All authors read and approved the final manuscript.

Author Information

Serina Ahlgren holds a PhD in agronomy and is currently working as a researcher at the Swedish University of Agricultural Sciences, mainly with life cycle assessment studies of bioenergy systems. Lorenzo Di Lucia is a post-doctoral fellow at Lund University, where he is working on the governance of biofuel systems in the EU and developing countries.

CHAPTER 12

Acknowledgements

Wouter Botzen, Laurens Bouwer, Frank Geels, John Gowdy, Stijn van den Heuvel, Richard Tol and two anonymous reviewers provided useful suggestions.

Index